Selected Titles in This Series

618 **Karl H. Hofmann and Wolfgang A. F. Ruppert,** Lie groups and subsemigroups with surjective exponential function, 1997

617 **Robin Hartshorne,** Families of curves in $\mathbb{P}^3$ and Zeuthen's problem, 1997

616 **Serguei G. Bobkov and Christian Houdré,** Some connections between isoperimetric and Sobolev-type inequalities, 1997

615 **Michael A. Dritschel and Hugo J. Woerdeman,** Model theory and linear extreme points in the numerical radius unit ball, 1997

614 **Richard Warren,** The structure of k-CS-transitive cycle-free partial orders, 1997

613 **D. L. Flannery,** The finite irreducible linear 2-groups of degree 4, 1997

612 **Joan Porti,** Torsion de Reidemeister pour les variétés hyperboliques, 1997

611 **D. Ginzburg, I. Piatetski-Shapiro, and S. Rallis,** L functions for the orthogonal group, 1997

610 **Mark Hovey, John H. Palmieri, and Neil P. Strickland,** Axiomatic stable homotopy theory, 1997

609 **Liviu I. Nicolaescu,** Generalized symplectic geometries and the index of families of elliptic problems, 1997

608 **Christina Q. He and Michel L. Lapidus,** Generalized Minkowski content, spectrum of fractal drums, fractal strings, and the Riemann zeta-functions, 1997

607 **Adele Zucchi,** Operators of class C_0 with spectra in multiply connected regions, 1997

606 **Moshé Flato, Jacques C. H. Simon, and Erik Taflin,** Asymptotic completeness, global existence and the infrared problem for the Maxwell-Dirac equations, 1997

605 **Liangqing Li,** Classification of simple C^*-algebras: Inductive limits of matrix algebras over trees, 1997

604 **Hajnal Andréka, Steven Givant, and István Németi,** Decision problems for equational theories of relation algebras, 1997

603 **Bruce N. Allison, Saeid Azam, Stephen Berman, Yun Gao, and Arturo Pianzola,** Extended affine Lie algebras and their root systems, 1997

602 **Igor Fulman,** Crossed products of von Neumann algebras by equivalence relations and their subalgebras, 1997

601 **Jack E. Graver and Mark E. Watkins,** Locally finite, planar, edge-transitive graphs, 1997

600 **Ambar Sengupta,** Gauge theory on compact surfaces, 1997

599 **Tai-Ping Liu and Yanni Zeng,** Large time behavior of solutions for general quasilinear hyperbolic-parabolic systems of conservation laws, 1997

598 **Valentina Barucci, David E. Dobbs, and Marco Fontana,** Maximality properties in numerical semigroups and applications to one-dimensional analytically irreducible local domains, 1997

597 **Ragnar-Olaf Buchweitz and John J. Millson,** CR-geometry and deformations of isolated singularities, 1997

596 **Paul S. Bourdon and Joel H. Shapiro,** Cyclic phenomena for composition operators, 1997

595 **Eldar Straume,** Compact connected Lie transformation groups on spheres with low cohomogeneity, II, 1997

594 **Solomon Friedberg and Hervé Jacquet,** The fundamental lemma for the Shalika subgroup of $GL(4)$, 1996

593 **Ajit Iqbal Singh,** Completely positive hypergroup actions, 1996

592 **P. Kirk and E. Klassen,** Analytic deformations of the spectrum of a family of Dirac operators on an odd-dimensional manifold with boundary, 1996

591 **Edward Cline, Brian Parshall, and Leonard Scott,** Stratifying endomorphism algebras, 1996

590 **Chris Jantzen,** Degenerate principal series for symplectic and odd-orthogonal groups, 1996

(*Continued in the back of this publication*)

MEMOIRS
of the
American Mathematical Society

Number 618

Lie Groups and Subsemigroups
with Surjective Exponential
Function

Karl H. Hofmann
Wolfgang A. F. Ruppert

November 1997 • Volume 130 • Number 618 (second of 4 numbers) • ISSN 0065-9266

American Mathematical Society
Providence, Rhode Island

1991 *Mathematics Subject Classification.*
Primary 22–02, 22E15, 22E67, 22A15;
Secondary 17B30, 17B20.

Library of Congress Cataloging-in-Publication Data
Hofmann, Karl Heinrich.
   Lie groups and subsemigroups with surjective exponential function / Karl H. Hofmann, Wolfgang A. F. Ruppert.
      p. cm. — (Memoirs of the American Mathematical Society, ISSN 0065-9266 ; no. 618)
   "November 1997, volume 130, number 618 (second of 4 numbers)."
   Includes bibliographical references and index.
   ISBN 0-8218-0641-6
   1. Lie groups. 2. Loops (Group theory) 3. Topological semigroups. I. Ruppert, Wolfgang, 1948– . II. Title. III. Series.
QA3.A57 no. 618
[QA387]
510 s—dc21
[512′.55]   97-30686
                                                                                                                              CIP

Memoirs of the American Mathematical Society

This journal is devoted entirely to research in pure and applied mathematics.

Subscription information. The 1998 subscription begins with volume 131 and consists of six mailings, each containing one or more numbers. Subscription prices for 1998 are $435 list, $348 institutional member. A late charge of 10% of the subscription price will be imposed on orders received from nonmembers after January 1 of the subscription year. Subscribers outside the United States and India must pay a postage surcharge of $30; subscribers in India must pay a postage surcharge of $43. Expedited delivery to destinations in North America $35; elsewhere $110. Each number may be ordered separately; *please specify number* when ordering an individual number. For prices and titles of recently released numbers, see the New Publications sections of the *Notices of the American Mathematical Society*.

Back number information. For back issues see the *AMS Catalog of Publications*.

Subscriptions and orders should be addressed to the American Mathematical Society, P. O. Box 5904, Boston, MA 02206-5904. *All orders must be accompanied by payment.* Other correspondence should be addressed to Box 6248, Providence, RI 02940-6248.

Copying and reprinting. Individual readers of this publication, and nonprofit libraries acting for them, are permitted to make fair use of the material, such as to copy a chapter for use in teaching or research. Permission is granted to quote brief passages from this publication in reviews, provided the customary acknowledgment of the source is given.

Republication, systematic copying, or multiple reproduction of any material in this publication (including abstracts) is permitted only under license from the American Mathematical Society. Requests for such permission should be addressed to the Assistant to the Publisher, American Mathematical Society, P. O. Box 6248, Providence, Rhode Island 02940-6248. Requests can also be made by e-mail to reprint-permission@ams.org.

Memoirs of the American Mathematical Society is published bimonthly (each volume consisting usually of more than one number) by the American Mathematical Society at 201 Charles Street, Providence, RI 02904-2294. Periodicals postage paid at Providence, RI. Postmaster: Send address changes to Memoirs, American Mathematical Society, P. O. Box 6248, Providence, RI 02940-6248.

© 1997 by the American Mathematical Society. All rights reserved.
This publication is indexed in *Science Citation Index*®, *SciSearch*®, *Research Alert*®, *CompuMath Citation Index*®, *Current Contents*®/*Physical, Chemical & Earth Sciences*.
Printed in the United States of America.

∞ The paper used in this book is acid-free and falls within the guidelines
established to ensure permanence and durability.
Visit the AMS home page at URL: http://www.ams.org/

10 9 8 7 6 5 4 3 2 1   02 01 00 99 98 97

CONTENTS

Chapter 1. Introduction

1. The Problem and its Background	1
2. A Survey of Methods and Techniques	5
3. The Major Results	8
4. Suggestions for the Reader	10

Chapter 2. The Basic Theory of Exponential Semigroups in Lie Groups

1. Basic Definitions and Examples	12
2. The Exponential Function and Divisibility	20
3. The Exponential Image of Lie Wedges and Regularity	24
4. Maximal Rank Subgroups	29
5. Interior Points in the Tangent Wedge	31

Chapter 3. Weyl Groups and Finiteness Properties of Cartan Subalgebras

1. Cartan Dense Ideals	38
2. Algebraic Hulls	41
3. Generalized Weyl Groups	44
4. Intersections of Cartan Subalgebras	48
5. Porcupine Varieties	61
6. Porcupine Varieties are Proper	65
7. The Codimension of Porcupine Varieties	67

Chapter 4. Lie Semialgebras

1. Semialgebras Revisited	75
2. Dispersion of Weakly Exponential Semigroups	79
3. More about Subtangent Vectors	80
4. Reductions by Factoring Normal Subgroups	83
5. Porcupine Varieties and Lie Semialgebras	86
6. Groups with a Unique Maximal Compact Subgroup	90
7. More about Lean Sets	92

Chapter 5. More Examples

1. Examples Connected with the Special Linear Group ... 95
2. Some Results on Injective Images of Lie Subsemigroups ... 98
3. Examples Connected with the Motion Algebra ... 102
4. Examples Connected with the Oscillator Group ... 104
5. Examples Connected with Compact Lie Algebras ... 107

Chapter 6. Test Algebras and Groups

1. Extensions of the Motion Algebra and of the Oscillator Algebra ... 109
2. Test Objects ... 111
3. Locally Direct Products of Lie Groups ... 115
4. The Proof of the Testing Theorem ... 119

Chapter 7. Groups Supporting Reduced Weakly Exponential Semigroups

1. 'Occam's Razor' and Reduced Weakly Exponential Subsemigroups ... 126
2. The Spectrum of $\operatorname{ad} w$ is Contained in $\mathbb{R} \cup i\mathbb{R}$ ... 128
3. Root Spaces with Respect to Real-Valued Roots ... 132
4. Weakly Exponential Subsemigroups and Eggert Algebras ... 135

Chapter 8. Roots and Root Spaces

1. Roots and Root Spaces ... 141
2. The Answer ... 145

Chapter 9. Appendix: The Hyperspace of a Locally Compact Space

1. Continuous Lattices, Lawson Topology, and Hyperspaces ... 154
2. Continuous Functions ... 158
3. The Use of Nets ... 165
4. Applications to Topological Algebras ... 168

References ... 169

Index ... 173

ABSTRACT

A closed subsemigroup S in a Lie group G determines a closed convex wedge $\mathfrak{L}(S)$ in the Lie algebra $\mathfrak{L}(G)$ of G. If S is a subgroup, then $\mathfrak{L}(S)$ is the Lie algebra of S; in the general case $\mathfrak{L}(S)$ is called the Lie wedge of S. The subsemigroup S of a Lie group is called reduced, if it is closed and does not contain a nonsingleton normal subgroup. The exponential function $\exp_G \colon \mathfrak{L}(G) \to G$ induces an exponential function $\exp_S \colon \mathfrak{L}(S) \to S$. The subsemigroup S is called exponential, respectively, weakly exponential, if $\exp_S \mathfrak{L}(S) = S$, respectively, $\overline{\exp_S \mathfrak{L}(S)} = S$. This definition applies, in particular, with $S = G$ and allows the formulation of the following problems in the structure theory of Lie groups:

Problem 1. Characterize all exponential Lie groups.

Problem 2. Characterize all weakly exponential Lie groups.

Problem 1 is open while Problem 2 is solved. We address the analogous problems for reduced subsemigroups of Lie groups:

Problem 3. Characterize all exponential reduced subsemigroups of Lie groups.

Problem 4. Characterize all weakly exponential reduced subsemigroups of Lie groups.

Problems 3 and 4 are completely settled in this memoir. In the process it is shown that all weakly exponential reduced subsemigroups are exponential and

---

1991 *Mathematics Subject Classification.* Primary 22-02, 22E15, 22E67, 22A15.
Secondary 17B30, 17B20, 20M99, 22E25, 22E46.

Key words and phrases. Lie group, Lie algebra, Cartan subalgebra, Weyl group, root spaces, regular points, exponential function, surjectivity of the exponential function, exponential Lie group, exponential Lie semigroup, weakly exponential Lie group, weakly exponential Lie semigroup, divisible semigroup, Lie wedge, Lie subsemigroup, Lie semialgebra.

We thank JIMMIE D. LAWSON and KARL-HERMANN NEEB for numerous enlightening discussions on this topic which were invaluable to us. The results presented in this memoir rest on the work of quite a few authors, notably on that of NORBERT DÖRR, ANSELM EGGERT, JOACHIM HILGERT, and DIRK MITTENHUBER. LAWSON is a coauthor of earlier papers and books on the subject. Among this group of people, the problem whose solution we present in these pages was notorious as "the Divisibility Problem for Subsemigroups of Lie Groups" since the mid-eighties.

We are sincerely grateful to the *Deutsche Forschungsgemeinschaft* for funding the project "Lie groups," to the *Vereinigung von Freunden der Technischen Hochschule Darmstadt* for giving a grant and to the *Alexander von Humboldt Foundation* for awarding a fellowship to WOLFGANG RUPPERT, to the *Fachbereich Mathematik der Technischen Hochschule Darmstadt* for providing hospitality and the infrastructure for our joint work, and to the *Universität für Bodenkultur* in Vienna for travel money. The support from all of these organisations allowed RUPPERT to visit THD regularly during the protracted work on this project.

that the Lie wedge $\mathfrak{L}(S)$ of an exponential reduced subsemigroup S of a Lie group G is a Lie semialgebra in $\mathfrak{L}(G)$. Lie semialgebras have been classified.

Received by the editor July 26, 1994, and in revised form April 9, 1996.

CHAPTER 1

INTRODUCTION

1. The Problem and its Background

The present memoir is devoted to the question: when is the exponential function surjective? This question has been discussed for various classes of Lie groups, it is studied here for *Lie semigroups*.

The theory of Lie semigroups, sometimes also called "Geometric Semigroup Theory," has emerged during the last decade, in close connection with such fields of mathematics as geometric control theory, holomorphic representation theory, causality and chronogeometry on pseudoriemannian manifolds. Its main features and some of its applications have been treated systematically and are documented in the following books:

Hilgert, J., Hofmann, K. H., and J. D. Lawson, "Lie groups, convex cones and semigroups," [27];

Hilgert, J., and K.-H. Neeb, "Lie Semigroups and their Applications," [28];

Hofmann, K. H., J. D. Lawson and J. S. Pym, Eds., "The Analytical and Topological Theory of Semigroups," [37].

These sources also list several articles devoted to certain fundamental special aspects of the theory. The following three may serve as a sample:

Hofmann, K. H., and J. D. Lawson, *Foundations of Lie semigroups*, [35];

Hofmann, K. H. and W. A. F. Ruppert, *On the interior of subsemigroups of Lie groups*, [43];

Neeb, K.-H., *On the foundations of Lie semigroups*, [69].

Lie semigroup theory attaches to every closed subsemigroup S in a Lie group G a tangent object $W = \mathfrak{L}(S)$ in the Lie algebra $\mathfrak{g}$ of G, defined by

$$\mathfrak{L}(S) = \{X \in \mathfrak{g} \mid \exp \mathbb{R}^+ \cdot X \subseteq S\}.$$

This set W is easily verified to be a so-called *Lie wedge:* an additively and topologically closed convex subset of $\mathfrak{g}$ such that $e^{\mathrm{ad}(W \cap -W)} W = W$.

If S is a closed *subgroup* then $\mathfrak{L}(S)$ is exactly the Lie algebra of S. We say that a subsemigroup S of a Lie group G is a *Lie semigroup* if it is closed and $S = \overline{\langle \exp \mathcal{L}(S) \rangle}$, i.e., the subsemigroup $\langle \exp \mathcal{L}(S) \rangle$, which is algebraically generated by the exponential image of $\mathcal{L}(S)$, is dense in S. It is an immediate consequence of the definition that every Lie semigroup is connected. A subgroup S of G is a Lie semigroup if and only if it is a closed connected Lie subgroup, or, equivalently, a closed connected subgroup of G.

Every Lie group possesses an identity neighborhood U which is diffeomorphic under the exponential function to an open euclidean ball around zero in the associated Lie algebra. We may choose U in such a fashion that each element $u \in U$ lies on a local one-parameter semigroup $t \mapsto \exp t \cdot X : [0,1] \to U$. Thus every Lie group G is *locally ruled by local one-parameter subsemigroups*: there are arbitrarily small open neighborhoods U of 1 in which every point can be joined with 1 by a local one-parameter subsemigroup *staying inside U*. This motivates the following definition:

Definition. A topological semigroup with identity is called *locally exponential* if 1 has arbitrarily small open neighborhoods U, such that for every point $u \in U$ there exists a continuous one-parameter subsemigroup $f : \mathbb{R}^+ \to S$ satisfying $f(0) = 1$, $f(1) = u$ and $f([0,1]) \subseteq U$.

Every closed subgroup of a Lie group is a Lie subgroup, so closed subgroups of a Lie group are automatically locally exponential. However, Lie *subsemigroups* are rarely locally exponential. Indeed, although the set $\exp \mathcal{L}(S)$ is assumed to generate S as a closed subsemigroup it is usually not a 1-neighborhood in S (cf. [27], Chapter V, Section 4, e.g., V.4.12, V.4.13).

It can be shown that the Lie wedge $W = \mathcal{L}(S)$ of a locally exponential subsemigroup S satisfies the following condition:

(∗) *There is an open convex neighborhood B of 0 in $\mathfrak{g}$ on which the Campbell-Hausdorff-Dynkin multiplication $*$ is defined and satisfies $(W \cap B) * (W \cap B) \subseteq W$.*

Note that if there exists one such convex neighborhood B then the condition on B in (∗) is satisfied for any smaller open convex 0-neighborhood as well, so B can be choosen arbitrarily small. Thus if W satisfies (∗) then $(W \cap B, *)$ is a *locally exponential local semigroup*.

A closed wedge W in a Lie algebra $\mathfrak{g}$ satisfying condition (∗) is called a *Lie semialgebra* or, if no confusion is possible, a *semialgebra*.

There is a subtle point here, which might be overlooked at first sight: The Lie wedge of a locally exponential semigroup is a Lie semialgebra, but it is not at all clear that, conversely, a Lie subsemigroup S is locally exponential if its Lie wedge $W = \mathcal{L}(S)$ is a Lie semialgebra. A direct proof that S is locally exponential could be obtained from the fact that $(W \cap B, *)$ is a locally exponential local semigroup *provided* that exp injects B diffeomorphically into G and $\exp(W \cap B)$ is a 1-neighborhood in S—but it is far from obvious that this is the case. In [69] K.-H. NEEB showed that for sufficiently small convex neighborhoods B the exp-image $\exp(W \cap B)$ is indeed a 1-neigborhood in S if W is a semialgebra. Hence *a Lie semigroup is locally exponential if and only if its Lie wedge is a semialgebra*.

Right from the beginning of the Lie theory of semigroups, Lie semialgebras played a prominent role (see e.g. [35], [26]). Considerable effort went into their characterization, culminating in the classification theorems of EGGERT [18]. Since the locally exponential Lie semigroups are those whose Lie wedge is a semialgebra we therefore also have a classification of the locally exponential Lie semigroups.

EGGERT showed that every Lie semialgebra is an intersection of semialgebras belonging to one of two main types: invariant wedges and half-space semialgebras.

A wedge W in a Lie algebra $\mathfrak{g}$ is called *invariant*, if $e^{\operatorname{ad} x}W = W$ for all $x \in \mathfrak{g}$, i.e., if W is invariant under *all* inner automorphisms of $\mathfrak{g}$ (and not only under those implemented by elements $x \in W \cap -W$ from the edge of a Lie wedge). A vector subspace of $\mathfrak{g}$ is an invariant wedge if and only if it is an ideal. Every invariant wedge is a semialgebra (cf. [27], p. 89, II.2.15). The characterization and classification of invariant wedges is a theory in its own right ([27], Chapter III [71], [22]), constituting a core piece of the Lie theory of semigroups.

A half-space of a Lie algebra is a Lie wedge if and only if its bounding hyperplane is a subalgebra, and every half-space Lie wedge is a Lie semialgebra. A half-space semialgebra is invariant if and only if the bounding hyperplane is an ideal, and this is the case if and only if it contains the commutator algebra $\mathfrak{g}'$. Thus half-spaces can be invariant only in a trivial way. Hyperplane subalgebras of Lie algebras are classified in [30], so the half-space semialgebras can be considered as known.

After these comments on locally exponential semigroups let us now take a global viewpoint!

We first note that a locally exponential Lie semigroup need not be globally exponential. This is well known for Lie groups. For instance in the special linear group $\mathrm{Sl}(2,\mathbb{R})$ no matrix with strictly negative entries can have a square root in $\mathrm{Sl}(2,\mathbb{R})$ and thus cannot belong to the exponential image. Note that these elements form an open subset $\mathcal{O}$ of $\mathrm{Sl}(2,\mathbb{R})$. The following example shows that there are also invariant Lie subsemigroups which have no invertible elements $\neq 1$ and which are not globally exponential.

Write the elements of $\mathfrak{sl}(2,\mathbb{R})$ as 2×2 matrices

$$X = \begin{pmatrix} a & b \\ c & -a \end{pmatrix}.$$

Then the quadratic form obtained by polarization from the function $X \mapsto -\det X = a^2 + cd$ is invariant under the inner automorphisms of $\mathfrak{sl}(2,\mathbb{R})$. Since this is a Lorentzian form with signature $++-$, its zero set bounds a solid 'double cone,' one half of which is the invariant wedge $W = \{X \in \mathfrak{sl}(2,\mathbb{R}) \mid \det(X) \leq 0,\ a \geq 0\}$. In the simply connected covering group G of $\mathrm{Sl}(2,\mathbb{R})$ the exponential image $\exp W$ generates a closed invariant semigroup S and $\exp W$ is an identity neighborhood in S. Since W is invariant (hence is a semialgebra) S is locally exponential. However, the inverse image of $\mathcal{O}$ under the covering map $G \to \mathrm{Sl}(2,\mathbb{R})$

intersects S in a nonempty open subset $\mathcal{O}^*$ of G (I.1.7), no point of which has a square root in G, let alone in S.

The following definition is convenient:

Definition. A closed subsemigroup S of a Lie group is called *exponential* if $S = \exp \mathfrak{L}(S)$. It is called *weakly exponential* if $S = \overline{\exp \mathfrak{L}(S)}$. □

One has to be careful with this terminology: some authors call a Lie group exponential if the exponential function $\exp\colon \mathfrak{g} \to G$ is a diffeomorphism. *In the terminology we adopt in these notes an exponential Lie group or semigroup is one for which the exponential function is surjective.* Classical results say that the general linear groups $\mathrm{Gl}(n, \mathbb{C})$ as well as all compact or nilpotent connected Lie groups are exponential. By a theorem of HOFMANN and LAWSON every divisible Lie subsemigroup is exponential [36]. Thus the terms 'divisible Lie subsemigroup' and 'exponential Lie subsemigroup' could be used synonymously. (For more details on the connection with divisibility we refer to Section 2 of Chapter 1.)

To this day, we still lack reasonable necessary and sufficient conditions for a connected Lie group to be exponential, in spite of the existence of a considerable body of literature on this subject. But we have excellent information about the weakly exponential case. We mention that the surjectivity of the exponential function was investigated in [10], [77] (the case of simply connected solvable Lie groups), [80, 83], [67] (the case of connected solvable Lie groups) [13, 14] (the case of classical matrix groups), [64, 65], [13], [81] (the case of centerless groups), [66] [82] (the case of algebraic groups, and that of complex splittable groups). Weakly exponential groups were studied in [42] (basic information; in particular, all complex and all solvable connected Lie groups are weakly exponential, the latter essentially known after [10]), [33] (more on weakly exponential groups; a group is weakly exponential iff its Cartan subgroups are connected), [72] and [17] (the classification of weakly exponential Lie groups). A general approach is initiated in [32] (regular points and the exponential function), [33], [80] (near Cartan subalgebras and near Cartan subgroups are proposed as a tool).

While the problem of classifying exponential Lie groups is still open we go one step further, formulating its analog for semigroups:

The Classification Problem. *Classify exponential Lie semigroups.*

From a strictly formal point of view the classification of exponential Lie semigroups is an even more ambitious task than the classification of exponential Lie groups. But there are important intrinsic structural features setting apart the semigroup case from the group case as something different in methods and results.

In this memoir we offer a solution of the Classification Problem under a natural mild restriction which allows us to concentrate on those aspects which properly pertain to the semigroup case.

2. A Survey of Methods and Techniques

In the following we discuss briefly the main lines of arguments leading to our final results: reduction arguments, local and infinitesimal arguments bringing semialgebra theory into the play, intersections with maximal rank groups, and arguments based on general Lie group and Lie algebra theory.

1. Reductions. If a closed subsemigroup S of a Lie group G is exponential, then obviously so is the closed semigroup S/N of G/N, where N is the largest closed normal subgroup of G contained in S. The closed semigroup S/N in the Lie group G/N does not contain any nontrivial normal subgroups, and it is a reasonable strategy to solve the Classification Problem first for closed exponential semigroups S not containing nontrivial normal subgroups. On the level of Lie wedges this means that $\mathfrak{L}(S)$ does not contain any nonzero ideals of $\mathfrak{g}$.

The second reduction is much more delicate. It is clear that the arcwise connected set $\exp \mathfrak{L}(S)$ algebraically generates a pathwise connected subgroup A. By a theorem of YAMABE's A is an analytic subgroup of G, which in general is not closed. With its intrinsic Lie group topology, A becomes a Lie group A_{Lie}. We shall show in I.5.8 that A_{Lie} contains S as a closed exponential subsemigroup, and that S generates A_{Lie} algebraically. Moreover $\mathfrak{a} = \mathfrak{L}(S) - \mathfrak{L}(S)$, i.e., $\mathfrak{L}(S)$ has inner points in $\mathfrak{a}$. In particular, S has inner points in A_{Lie}. Thus it is also reasonable to assume that $\mathfrak{g}$ is algebraically generated by $\mathfrak{L}(S)$. If S is itself a group, then $\mathfrak{L}(S)$ is a subalgebra, so $\mathfrak{g} = \mathfrak{L}(S)$ and thus $S = G$.

If S is itself a group then $N = S$, so a closed *subgroup* S can be reduced only in the trivial case $S = G = \{1\}$.

The existence of inner points in Lie semigroups is a theme with many variations some of which are known from geometric control theory. One of the pertinent references is the article by HOFMANN and RUPPERT [44] mentioned in the beginning.

Reduced Subsemigroups. In view of the two reduction steps just discussed we shall say that a closed subsemigroup S of a connected Lie group G is *reduced* if it algebraically generates G and does not contain any nondegenerate normal subgroups of G. Note that a closed *subgroup* S can be reduced only if $G = S = \{1\}$.

Now we formulate a weaker version of the classification problem, which certainly will have to be addressed if one aims for a general solution:

The Classification Problem, reduced. *Classify all reduced exponential Lie semigroups.*

Such a classification we shall present at the end of these notes in a rather satisfactory and complete form. It will turn out that the solution of the reduced Classification Problem inevitably requires the solution of a more general problem, which looks even harder:

The Classification Problem, reduced and generalized. *Classify all reduced weakly exponential Lie semigroups.*

Very remarkably, having achieved such a general classification we shall see that these two problems actually coincide: *every reduced weakly exponential closed subsemigroup is exponential*. Thus the difference between 'weakly exponential' and 'exponential' disappears if we assume that S contains no nondegenerate subgroup whose normalizer contains S. (This is *not* true in the non-reduced case.)

2. Local and infinitesimal arguments.

It is a general strategy in Lie theory to solve structural problems on Lie groups by passing to associated problems on the level of Lie algebras. In order to apply this strategy one has to find local consequences of global properties. So it is natural to ask whether every exponential Lie semigroup is locally exponential, or, equivalently, whether the Lie wedge W of an exponential Lie subsemigroup S must be a semialgebra. Geometrically speaking, if every point in a Lie semigroup S lies on a one-parameter subsemigroup, can we find an open neighborhood U of the identity in S where every element lies on a local one-parameter subsemigroup *not leaving U*?

If we had an affirmative answer to this question the ample information we have on semialgebras would greatly simplify our classification task. Back in 1983 HOFMANN and LAWSON already observed that the answer to the above question is indeed 'yes' if S has no invertible elements [36] (cf. also [27], p. 462). But the difficulty of the general case stems exactly from the presence of nontrivial invertible elements in S.[1]

In the general case the situation is much more subtle, and in fact the final affirmative answer depends on the solution of the Classification Problem. Thus the following result (Theorem IV.2.3 and Theorem IV.2.14) will emerge only at the very end of our chain of arguments, as a constituent part of the classification.

Theorem A. *The Lie wedge $\mathfrak{L}(S)$ of a closed weakly exponential semigroup S is a Lie semialgebra. Every weakly exponential (hence a fortiori every exponential) closed subsemigroup is locally exponential. If the semigroup S, in addition, does not contain nondegenerate subgroups which are normal in $\langle S \cup S^{-1} \rangle$, then it is even exponential.*

If S is a (not necessarily closed) subsemigroup with

$$S \subseteq \overline{\bigcup \{f(\mathbb{R}^+) \mid f \in \mathrm{Hom}(\mathbb{R}^+, S)\}}$$

then $\overline{S}$ is a closed weakly exponential subsemigroup, so by Theorem A its Lie wedge is a Lie semialgebra. It can be shown that the closure of a divisible subsemigroup is weakly exponential (Corollary I.2.7(i)).

We emphasize that not every ideal free semialgebra occurs as the Lie wedge of an exponential reduced closed semigroup. Following EGGERT we can write a Lie semialgebra W as the intersection $W = W_{sec} \cap W_{inv}$ of a semialgebra W_{sec} which is the intersection of half-space semialgebras and an invariant wedge W_{inv}. Our results will show that an invariant reduced Lie semigroup S in a Lie group is not exponential (and not even weakly exponential) unless it is contained in

---

[1] Complications induced by invertible elements are a familiar phenomenon in the Lie theory of semigroups. Another example is the construction of local semigroups with a given tangent object in [27] (compare the proofs of Theorem V.4.8 and of Corollary V.8.7).

a maximally almost periodic Lie subgroup. As an illustration of this fact we remark that the two invariant Lie subsemigroups contained in $\widetilde{\mathrm{Sl}}(2,\mathbb{R})$ are not exponential. However, every invariant Lie subsemigroup is locally exponential.

We noted above that the general version of Theorem A is not an intermediate step in the proof of the Classification Theorem, but is rather a consequence of the latter. Special cases of Theorem A, however, are instrumental for certain steps in the proof of the Classification Theorem.

3. Maximal rank subgroups. In dealing with exponential subsemigroups S of a Lie group G we encounter a characteristic difficulty: If M is a closed connected subgroup of G then the closed subsemigroup $S \cap M$ need not be connected, let alone exponential. Thus we cannot pass to an exponential subsemigroup of lower dimension by simply intersecting S with an arbitrary connected Lie subgroup of G, a familiar technique allowing inductive arguments.

If, however, we intersect a weakly exponential closed subsemigroup S with a *maximal rank subgroup* M, i.e., an analytic subgroup whose Lie algebra contains a Cartan subalgebra of $\mathfrak{g}$ (such subgroups are automatically closed), and if M meets the interior of S then $S \cap M$ is weakly exponential, too (I.4.3). This is seen by a careful analysis of the one-parameter subsemigroups passing through regular elements. Our strategy is to study such intersections with maximal rank subgroups of certain special types, where the problem is more amenable, and to deduce piece by piece the essential structural properties of G and S.

We can *not* conclude from the onset that exponential closed subsemigroups intersect maximal rank subgroups in exponential subsemigroups. Even if we are interested only in exponential subsemigroups we are therefore forced to consider weakly exponential subsemigroups as well.

4. General theory of Lie groups and Lie algebras. Lie group and Lie algebra methods are pervasive in the entire proof. It is noteworthy that in the process of preparing the necessary tools we had to gather new results which are of independent interest. These include

(i) a general theory of Weyl groups for arbitrary real or complex Lie groups and Finiteness Theorems for Cartan subalgebras,
(ii) an algebraic geometric theory of certain subvarieties of Lie algebras (the so-called 'Porcupine Varieties'),
(iii) the spectral theory for roots in a not necessarily semisimple real Lie algebra with respect to a given Cartan subalgebra,
(iv) information on the surjectivity of the exponential function for some special types of Lie groups,
(v) hyperspace techniques applied to Lie groups and Lie algebras.

The material pertaining to (i) and (ii) is contained in Chapter II and is published in [38], [39], and [46]; Hyperspace techniques are discussed in the appendix of this memoir. A major portion of the arguments of this memoir consists of a detailed analysis of selected lower dimensional test situations, notably universal coverings of the special linear group in two dimensions, the group of motions in the euclidean plane, and the oscillator group.

3. The Major Results

The first part of the Classification Theorem involves the classification of the Lie algebras of those Lie groups G which contain a reduced weakly exponential closed subsemigroup. Its formulation requires some preliminary definitions:

We say that a Lie algebra $\mathfrak{d}$ is *diagonally metabelian* if $[\mathfrak{d},\mathfrak{d}]$ is abelian and if there is a Cartan algebra $\mathfrak{h}$ such that each of the operators $\operatorname{ad} m$, $m \in \mathfrak{h}$, is diagonalizable over the reals. The structure of diagonally metabelian Lie algebras is completely known ([30]).

Theorem B. *If a connected Lie group G contains a weakly exponential reduced subsemigroup then there are ideals $\mathfrak{s}_j$, $j = 1, \ldots, k$, all isomorphic to $\mathfrak{sl}(2,\mathbb{R})$, a diagonally metabelian and centerfree ideal $\mathfrak{d}$, and a compact ideal $\mathfrak{k}$ such that $\mathfrak{g} = \mathfrak{s}_1 \oplus \cdots \oplus \mathfrak{s}_k \oplus \mathfrak{d} \oplus \mathfrak{k}$.*

Conversely, if G is a simply connected Lie group whose Lie algebra is of the type described in Theorem B then G contains a weakly exponential reduced subsemigroup. We note that the radical of $\mathfrak{g}$ is $\mathfrak{d} \oplus \mathfrak{z}(\mathfrak{k})$ and that $\mathfrak{g}$ has a unique Levi complement $\mathfrak{s}_1 \oplus \cdots \oplus \mathfrak{s}_k \oplus \mathfrak{k}'$. All subalgebras isomorphic with $\mathfrak{sl}(2,\mathbb{R})$, and in particular all $\mathfrak{sl}(2,\mathbb{R})$-triples ([5], p. 159) are contained in the ideal $\mathfrak{s}_1 \oplus \cdots \oplus \mathfrak{s}_k \cong \bigl(\mathfrak{sl}(2,\mathbb{R})\bigr)^k$. Once more, this emphasizes the very special role of $\mathfrak{sl}(2,\mathbb{R})$ in Lie theory.

Once the structure of $\mathfrak{g}$ is known we can formulate the main result on the Lie wedge of an exponential reduced subsemigroup S of G.

Theorem C. *Let G be a connected Lie group containing a reduced weakly exponential subsemigroup S with Lie wedge W. Then $S = \exp W$ and, in the notation of Theorem B, the following conclusions hold:*
  (i) *$W = (\mathfrak{s}_1 \cap W) \oplus \cdots \oplus (\mathfrak{s}_k \cap W) \oplus W_0$, where the intersections $\mathfrak{s}_j \cap W$ are intersection algebras, and $W_0 = W \cap (\mathfrak{d} \oplus \mathfrak{k})$.*
  (ii) *W_0 is described as follows:*
      Set $W_{inv} = \overline{W_0 + \mathfrak{d}'}$. Then the wedge W_{inv} is the smallest invariant wedge containing W_0. There is an intersection algebra W_{sec} containing $\mathfrak{k}'$ such that $W_0 = W_{inv} \cap W_{sec}$.
  (iii) *$W \cap -W$ is a metabelian subalgebra of $\mathfrak{g}$ with $W \cap -W \cap \mathfrak{k} = \{0\}$. More specifically, $W \cap -W = \mathfrak{a} \oplus \mathfrak{m}$, where $\mathfrak{a}$ is an abelian subalgebra of $\mathfrak{s}_1 \oplus \mathfrak{s}_2 \oplus \cdots \oplus \mathfrak{s}_k$, and $\mathfrak{m}$ is a subalgebra of the sum $\mathfrak{d} \oplus \mathfrak{z}(\mathfrak{k})$ of $\mathfrak{d}$ with the center $\mathfrak{z}(\mathfrak{k})$ of $\mathfrak{k}$.*
  (iv) *The group of invertible elements in S is exponential, that is, $S \cap S^{-1} = \exp(W \cap -W)$.*

Interdependence Chart

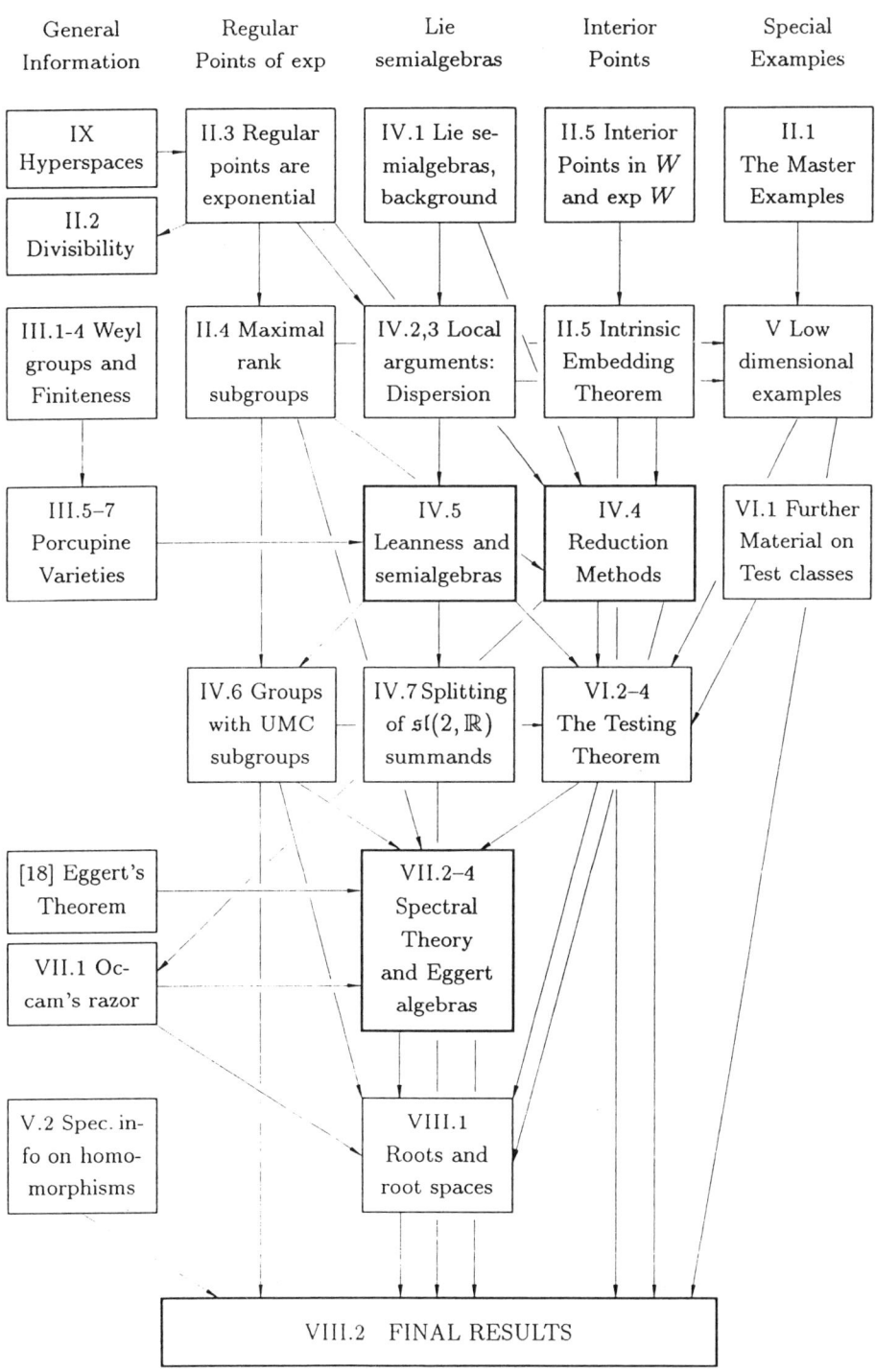

Assertion (iii) of Theorem C shows that the group of units $S \cap S^{-1}$ of a reduced weakly exponential Lie semigroup must belong to a well understood class of metabelian Lie groups, and in particular does not contain any element $\neq 1$ which can be represented as the limit of compact elements. This fact is indeed crucial: if we knew it beforehand then local arguments such as those used in the paper by HOFMANN and LAWSON [36] would have implied immediately that W must be a semialgebra.

Let $p: \widetilde{G} \to G$ be the universal covering homomorphism of G and $\widetilde{S} = \exp_{\widetilde{G}} W$.

Theorem D. *Under the hypotheses of* Theorem C, *the set $\widetilde{S}$ is a closed exponential semigroup and $S = p(\widetilde{S})$.*

In other words, exponential closed subsemigroups of G can be lifted to the universal covering group $\widetilde{G}$, a fact which a priori is not clear at all. In many discussions concerning exponential semigroups this permits us to assume that G is simply connected.

4. Suggestions for the Reader

We now give some hints concerning the reading of this memoir.

Hint 1. Prerequisites. The reader is assumed to be reasonably familiar with Lie algebras and Lie groups. As a general source, BOURBAKI [3, 4, 5] suffices. Certain basic facts on Lie semigroups, such as are contained in [27] will have to be assumed, notably the foundations of the theory of Lie semialgebras as it is discussed there. Regarding the complete theory of Lie semialgebras we shall use the results of EGGERT which he exposed in his monograph [18]. However, the reader will understand the substance of what we use from his work without having to go back to the original source. We shall use hyperspaces, i.e., spaces of closed subsets of locally compact spaces. For the convenience of the reader we discuss these in an appendix including information which is not readily available elsewhere.

Hint 2. The logical structure of our arguments is represented in the accompanying chart.

Hint 3. In Section 2 of Chapter II we discuss, for closed subsemigroups of Lie groups, the relations between the algebraic property of being divisible and the analytic property of being exponential. Some of the results and some of the proofs are new, but they will not be used in the sequel. A reader primarily interested in the analytical aspects of the problem may decide to skip this section without endangering the understanding of the remainder.

Hint 4. We strongly recommend the study of the examples given in II.1. It will turn out that all reduced divisible Lie semigroups are essentially composed of these. It is also instructive to study the "Negative Examples" of this section and the examples in Chapter V. These are instrumental for Chapter VI which in turn is an important part of the main argument. In principle, all examples are presented without invoking the body of the base theory provided in Chapter II and Chapter IV and therefore may be perused independently. Orientation on the presence or absence of such interrelations may be found in the chart.

Hint 5. In Section II.3 we invoke material from the theory of hyperspaces from the Appendix (Chapter IX).

Hint 6. Chapter III concerns Lie algebras only. Its goal is to establish one single result at the end of the chapter on the dimension of a special subvariety of a Lie algebra. This result is a main ingredient of our arguments; a reader taking this fact for granted may skip this chapter; however, it contains some information of independent interest concerning, e.g., Weyl groups.

CHAPTER 2

THE BASIC THEORY OF EXPONENTIAL SEMIGROUPS IN LIE GROUPS

1. Basic Definitions and Examples

1.1. DEFINITION. Let $\mathfrak{g}$ denote a finite dimensional real vector space. Consider a subset $\mathfrak{a} \subseteq \mathfrak{g}$. We say that an element $z \in \mathfrak{g}$ is *a subtangent vector of $\mathfrak{a}$ at the origin* if there is a sequence $x_n \in \mathfrak{a}$ converging to 0 in $\mathfrak{g}$ and a sequence of positive real numbers r_n such that $z = \lim r_n \cdot x_n$. We shall write $L(\mathfrak{a})$ for the set of all subtangent vectors of $\mathfrak{a}$ at the origin. □

1.2. DEFINITION. Let $\mathfrak{g}$ be a real Lie algebra and suppose that $\exp: \mathfrak{g} \to G$ is the exponential function of a Lie group G. If $A \subseteq G$, then we set $\mathfrak{L}(A) = L(\exp^{-1}(A))$ and call $\mathfrak{L}(A)$ the *tangent object* of A at the origin. (The sets $L(\mathfrak{a})$ or $\mathfrak{L}(A)$ may be empty.) If $f: G \to H$ is a morphism of Lie groups then we shall write $\mathfrak{L}(f): \mathfrak{L}(G) \to \mathfrak{L}(H)$ for the morphism of Lie algebras induced by f. (Some authors write $d(f)$ instead.) □

We note that $\mathfrak{g} = \mathfrak{L}(G)$.

1.3. DEFINITION. (i) For any subset W of a vector space $\mathfrak{g}$ we shall write $H(W) = W \cap -W$.
 (ii) A subset W of a finite dimensional real vector space is called a *wedge* if it is topologically closed and is algebraically closed under addition and nonnegative scalar multiplication. The set $H(W)$ is then called the *edge of the wedge*.
 (iii) A wedge W in a finite dimensional real Lie algebra $\mathfrak{g}$ is called a *Lie wedge* if $e^{\operatorname{ad} H(W)} W \subseteq W$.
 (iv) A wedge W in a finite dimensional Lie algebra $\mathfrak{g}$ is called *invariant* if $e^{\operatorname{ad} \mathfrak{g}} W \subseteq W$. □

A vector subspace of a Lie algebra is a Lie wedge if and only if it is a Lie subalgebra. A wedge is invariant in a Lie algebra if and only if it is invariant under all inner automorphisms. Obviously every invariant wedge is a Lie wedge.

For geometric formulations of the above definitions and additional details we refer to section II.1 of the book by HILGERT, HOFMANN and LAWSON [27], p. 76ff.

The following lemma records the basic facts about tangent objects of subsemigroups of Lie groups. Using the terminology of [27] we write $\langle A \rangle$ for the semigroup algebraically generated by a subset A.

1.4. LEMMA. *Let S be a subsemigroup of a Lie group G. Then*
(i) $\mathfrak{L}(S) = \{x \in \mathfrak{g} \mid \exp \mathbb{R}^+ \cdot x \subseteq \overline{S}\}$.
(ii) $\mathfrak{L}(S)$ *is a Lie wedge.*
(iii) $\langle \exp \mathfrak{L}(S) \rangle \subseteq \overline{S}$. □

We note that $\mathfrak{L}(S) = \mathfrak{L}(\overline{S})$. If S is a closed subsemigroup then we shall call $\mathfrak{L}(S)$ the *Lie wedge of S*. At this point a mild warning is in order. If S is a closed subgroup of G then it is a Lie subgroup and $\mathfrak{L}(S)$ is its Lie algebra, as expected. If, however, S is an analytic subgroup, then $\mathfrak{L}(S)$ is the Lie algebra of the closure of S in G (which is a Lie group). But the Lie algebra of the analytic group S is definitely smaller than that of $\overline{S}$ if S is not closed in G. We shall not be concerned here with this complication, because we shall only consider Lie wedges of closed subsemigroups. However, a more subtle theory of Lie wedges can be developed for semigroups, as is shown in [27], p. 373ff.

1.5. DEFINITION. Suppose that S is a subsemigroup of a Lie group G.
(i) S is called a *Lie semigroup* if $S = \overline{\langle \exp \mathfrak{L}(S) \rangle}$.
(ii) S is called *weakly exponential* if $S = \overline{\exp \mathfrak{L}(S)}$.
(iii) S is called *exponential* if $S = \exp \mathfrak{L}(S)$. □

All connected compact and all connected nilpotent Lie groups are exponential. All connected solvable Lie groups and all underlying real Lie groups of complex Lie groups are weakly exponential (cf. e.g. [42, 32]); the 3-dimensional Lie group $\mathrm{Sl}(2, \mathbb{R})$ is not weakly exponential.

We now introduce a few examples which are not merely illustrations of the above concepts but in fact constitute a complete list of 'Archetypes' for weakly exponential and exponential subsemigroups. As we shall see in the final chapter every weakly exponential subsemigroup can be constructed with the aid of (fairly straightforward generalizations of) these examples. At many points of the subsequent discussions it is exactly this set of model examples which motivates our concepts and ideas. Thus recognizing their guiding role for the theory we shall call them 'the Master Examples.'

1.6. EXAMPLES. **The Master Examples.**
(i) *The Abelian Master Examples.* It is straightforward that every abelian Lie semigroup is exponential. As a prototype we consider the real half-line $\mathbb{R}^+$. Slightly less trivial examples are the images of the two-dimensional cones

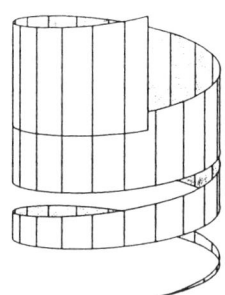 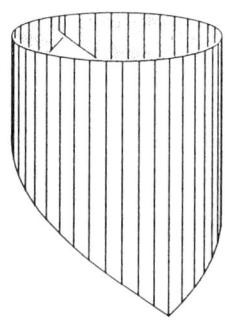

FIGURE 1. Lie subsemigroups of the direct product $G = (\mathbb{R}/\mathbb{Z}) \times \mathbb{R}$. The Lie wedge of the exponential Lie subsemigroup at the left is the wedge $\{(x,y) \in \mathfrak{g} = \mathbb{R}^2 \mid 0 \leq 0.2x \leq y \leq 0.6x\}$, the Lie wedge for the semigroup at the right is $\{(x,y) \in \mathfrak{g} = \mathbb{R}^2 \mid y \geq |x|\}$. Basically the same intuitive view lies behind the Master Example discussed in (v) below.

$W_1 = \{(x,y) \in \mathbb{R}^2 \mid 0 \leq 0.2x \leq y \leq 0.6x\}$ and $W_2 = \{(x,y) \in \mathfrak{g} = \mathbb{R}^2 \mid y \geq |x|\}$ under the exponential map $\exp: \mathbb{R}^2 \to (\mathbb{R}/\mathbb{Z}) \times \mathbb{R}$ (Fig. 1).

(ii) *The Metabelian Master Examples.* (Fig. 2) The next simple case concerns Lie semigroups in the affine ('$ax + b$'-) group $G = (\mathbb{R}, +) \rtimes (\mathbb{R}^+, \cdot)$ whose Lie algebra $\mathfrak{g}$ is the nonabelian Lie algebra in two dimensions. This Lie algebra is *metabelian* ($\mathfrak{g}'' = \{0\}$) and all operators $\operatorname{ad} x|[\mathfrak{g}, \mathfrak{g}]$ are diagonalizable. Every half-space in $\mathfrak{g}$ is bounded by a one-dimensional subalgebra (hence is a Lie wedge) and its exponential image is a closed subsemigroup. All these subsemigroups are exponential and so is the intersection of every two of them. If the bounding subalgebra happens to be the unique one-dimensional ideal then the corresponding half-spaces are invariant wedges. If such a semigroup S is not a half-line then it is *absolutely closed*: If φ is a continuous homomorphism mapping S injectively into a Lie group G then $\varphi(S)$ is closed.

This is a consequence of the fact that $\widetilde{G}$ itself is absolutely closed. (Exercise)

(iii) *The half-space in* $\widetilde{\operatorname{Mot}}$. (Fig. 3) Let $\mathfrak{g}$ be the motion algebra, $\mathfrak{g} = \{(\zeta, \alpha) \mid \zeta \in \mathbb{C}, \alpha \in \mathbb{R}\}$ with Lie brackets $[(\zeta, \alpha), (\zeta', \alpha')] = (2i(\alpha\zeta' - \alpha'\zeta), 0)$, and let G be the associated simply connected Lie group, $G = \{(z,t) \mid z \in \mathbb{C}, t \in \mathbb{R}\}$ with multiplication $(z,t)(z',t') = (ze^{-it'} + e^{it}z', t + t')$. Then the half-space semigroup $G^+ = \{(z,t) \in G \mid t \geq 0\}$ is not the exponential image of its Lie wedge $W = \{(\zeta, \alpha) \mid \alpha \geq 0\}$, since $\exp \mathfrak{g}$ does not contain any of the points $(c, k\pi)$ with $k \in \mathbb{N}, c \neq 0$. Nevertheless we have $G^+ = \overline{\exp W}$. Note that W is a half-space bounded by the unique two-dimensional ideal in $\mathfrak{g}$, hence is invariant. The semigroup G^+ is not absolutely closed.

(iv) *The semigroup* $\operatorname{Sl}(2)^+$. (Fig. 5) Let $\mathfrak{g} = \mathfrak{sl}(2, \mathbb{R})$, realized as the Lie algebra of all real matrices with trace 0, and define

$$\mathfrak{sl}(2, \mathbb{R})^+ := \left\{ \begin{pmatrix} a & b \\ c & -a \end{pmatrix} \mid a \in \mathbb{R}, b, c \geq 0 \right\}$$

This is the set of all matrices m in $\mathfrak{sl}(2, \mathbb{R})$ such that $\exp r \cdot m$ has nonnegative entries for all $r \geq 0$ (in Fig. 4 it is denoted by W_+ for short). Then for *any* Lie

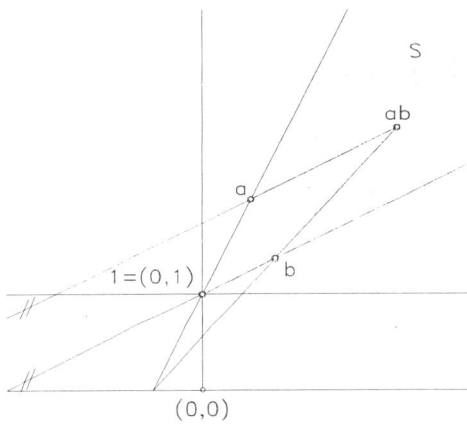

FIGURE 2. Lie subsemigroups of the $ax+b$ group $G = \mathbb{R} \rtimes (\mathbb{R}_+, \cdot)$. The one-parameter subgroups of G are exactly the lines passing through the identity $(0,1)$. It is easy to see that the product ab of any two elements $a, b \in G$ always lies in the convex cone generated by the line segments connecting the identity with a and b. Thus every half-space is an exponential subsemigroup of G and so are all intersections of half-spaces.

group G with Lie algebra $\mathfrak{g}$ the exponential image $\exp \mathfrak{sl}(2,\mathbb{R})^+$ is a closed subsemigroup of G; hence it is an exponential Lie semigroup $G^+ := \exp \mathfrak{sl}(2,\mathbb{R})^+$. For any choice of G we have $G^+ \cong \mathrm{Sl}(2)^+$. We also note that $\mathrm{Sl}(2)^+$ is *absolutely closed*, this will be proved in V.2.1. More detailed information about subsemigroups of Lie groups with Lie algebra $\mathfrak{sl}(2,\mathbb{R})$ will be given in Chapter IV.

(v) *The Master Example in* $\mathrm{SO}(3) \times \mathbb{R}$. Let $G = \mathrm{SO}(3) \times \mathbb{R}$ and write $\mathfrak{g}$ as the direct sum $\mathfrak{g} = \mathfrak{so}(3) \oplus \mathbb{R}$ (where $\mathbb{R}$ has trivial brackets). Our construction is inspired by the exponential subsemigroup $\exp W_1$ of $(\mathbb{R}/\mathbb{Z}) \times \mathbb{R}$ in (i): we define a one-parameter semigroup of invariant closed 1-neighborhoods $U(t)$ in $\mathrm{SO}(3)$ and put $S := \bigcup \{U(t) \times \{t\} \mid t \in \mathbb{R}^+\}$. It is straightforward to check that S is closed and a semigroup.

To define $U(t)$ we use the operator norm $\|\bullet\|$ of $\mathfrak{gl}(3,\mathbb{R})$. Note that the operator norm of an element $k \in \mathfrak{so}(3)$ coincides with the spectral radius of k, that is, for $k \neq 0$ we have $\|k\| = |\lambda|$, where λ is one of the two non-zero eigenvalues of k. For any $t > 0$ we now define $B(t) = \{k \in K \mid \|k\| \leq t\}$, these sets are 0-neighborhoods and invariant under the adjoint action of G. (See Figure 6.) The natural action of G on the unit sphere $\mathbb{S}^2 = \{p \in \mathbb{R}^3 \mid \|p\| = 1\}$ is effective and for every $p \in \mathbb{S}^2$ and $0 \leq t \leq \pi$ the orbit $(\exp B(t))p$ is the 'scullcap' $C(p,t) = \{q \in \mathbb{S}^2 \mid \|p - q\| \leq 2|\sin \frac{t}{2}|\}$. Conversely, $\exp B(t) = \{g \in G \mid g.p \in C(p,t) \text{ for all } p \in \mathbb{S}^2\}$. For $t \geq \pi$ we have $C(p,t) = \mathbb{S}^2$. For any pair $t_1, t_2 \in \mathbb{R}^+$

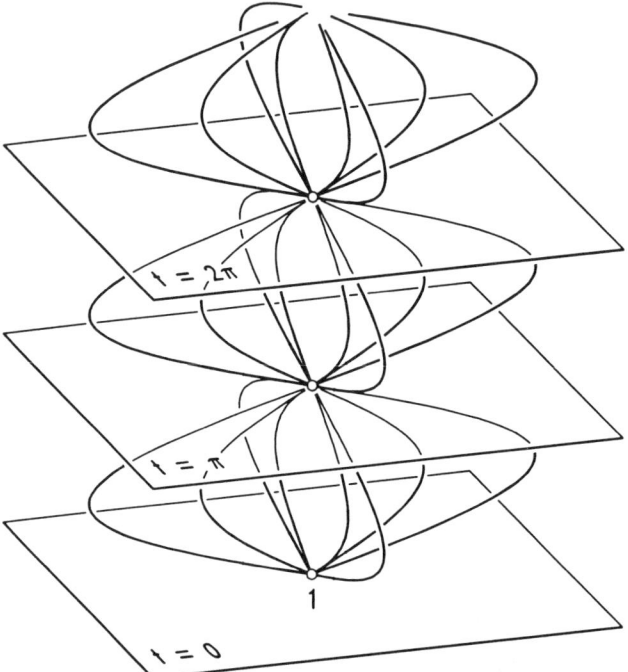

FIGURE 3. The half-space $S = \widetilde{\operatorname{Mot}}^+$. ($G = \widetilde{\operatorname{Mot}}$ is the point set $\mathbb{C} \times \mathbb{R}$ with multiplication $(z, t)(z', t') = (ze^{-it'} + e^{it}z', t + t')$, and $S = \mathbb{C} \times \mathbb{R}^+$.) Every one-parameter subgroup of G either is contained in the normal subgroup $\mathbb{C} \times \{0\}$ or else maps onto one of the curves $\{(z\sin t, t) \mid t \in \mathbb{R}\}$, $z \in \mathbb{C}$. Thus the exponential image in G misses the sets $\{(c, k\pi) \mid c \in \mathbb{C}^\times\}$, $k \in \mathbb{N}$. The half-space semigroup S is weakly exponential, but not exponential.

and any $p \in \mathbb{S}^2$ we find by successive application of the definition

$$(\exp B(t_1))(\exp B(t_2))p = (\exp B(t_1))C(p, t_2)$$
$$= \bigcup_{q \in C(p, t_2)} C(q, t_1) = C(p, t_1 + t_2) = (\exp B(t_1 + t_2))p,$$

which implies that $\exp B(t_1 + t_2) = (\exp(B(t_1)))(\exp(B(t_2)))$. In other words, the map $t \mapsto U(t) := \exp B(t)$ defines a one-parameter subsemigroup of the semigroup of all closed invariant subsets of $\operatorname{SO}(3)$. (Note that $U(t) = \operatorname{SO}(3)$ for every $t \geq \pi$.)

Now the wedge $W = \{x + t \mid x \in \mathfrak{k}, t \in \mathbb{R}^+, \text{ with } \|x\| \leq t\}$ is a pointed invariant cone, and its exponential image $S = \exp W = S = \bigcup \{U(t) \times \{t\} \mid t \in \mathbb{R}^+\}$. is closed and a subsemigroup. Thus S is an invariant exponential subsemigroup of G. □

Note that with the exception of the half-space in $\widetilde{\operatorname{Mot}}$ all of these examples are exponential. As we shall see in the final chapter, a weakly exponential subsemigroup with inner points in a Lie group is exponential whenever it does not contain a non-singleton normal subgroup.

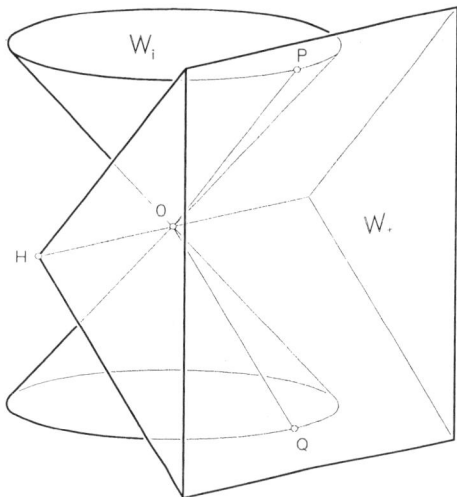

FIGURE 4. The Standard Lorentzian Cone and semialgebras in $\mathfrak{g} = \mathfrak{sl}(2,\mathbb{R})$. We write $\mathfrak{g}$ as the span of three elements H, P, Q with $[H, P] = P$, $[H, Q] = -Q$ and $[P, Q] = H$. The Standard Lorentzian Cone is the null set $\{X \in \mathfrak{g} \mid k(X, X) = 0\}$ of the Killing form, it is invariant under the adjoint action. Its 'upper' part generates the invariant wedge W_i. The Lie wedge $W_+ = \mathbb{R} \cdot H + \mathbb{R}^+ \cdot P + \mathbb{R}^+ \cdot Q$ of the semigroup $\mathrm{Sl}(2,\mathbb{R})^+$ of matrices with nonnegative entries is a semialgebra.

Let us also record two 'negative' examples, Lie subsemigroups which are locally but not globally exponential, in fact not even weakly exponential. They contain 'holes,' open subsets into which no one-parameter subsemigroup enters.

We shall add more details to these two examples in Chapter IV.

1.7. EXAMPLES. **Negative Examples.**

(i) *The standard invariant subsemigroup of* $\widetilde{\mathrm{Sl}}(2,\mathbb{R})$. (Fig. 5) Let G be a simply connected Lie group with Lie algebra $\mathfrak{g} = \mathfrak{sl}(2,\mathbb{R})$. We define the 'Standard Invariant Wedge' in $\mathfrak{g}$ by

$$W_i := \left\{ \begin{pmatrix} a & b \\ c & -a \end{pmatrix} \middle| b \geq 0,\, a^2 + bc \leq 0 \right\}.$$

The center of G is the subgroup $Z = \exp \mathbb{Z} \cdot z$, where $z = \begin{pmatrix} 0 & \pi \\ -\pi & 0 \end{pmatrix}$. The exponential image $\exp W_i$ is not a semigroup, it generates the semigroup

$$S_i = \exp W_i \cup \exp(\mathbb{N} \cdot z) \exp \left(\left\{ \begin{pmatrix} a & b \\ c & -a \end{pmatrix} \middle| a^2 + bc \geq 0 \right\} \right),$$

which is closed in G. The exponential image of $\mathfrak{g}$ (hence, a fortiori the exponential image of W_i) misses the nonvoid open set

$$\exp \mathbb{N} \cdot z \exp \left(\left\{ \begin{pmatrix} a & b \\ c & -a \end{pmatrix} \middle| a^2 + bc > 0 \right\} \right).$$

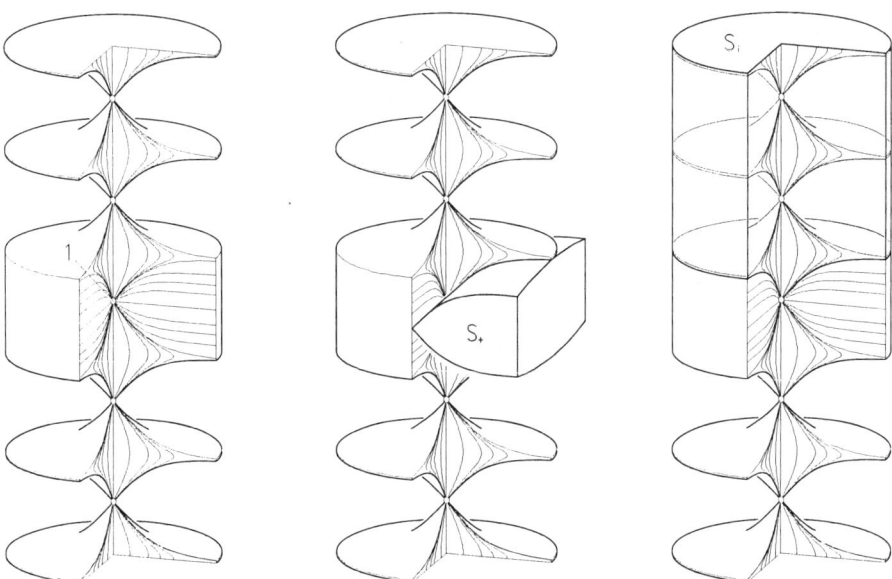

FIGURE 5. Exponential and non-exponential subsemigroups of $G = \widetilde{\mathrm{Sl}}(2,\mathbb{R})$. In the parametrization of [27] the group G is defined on the space $\mathbb{R}^3$ in such a way that the one-parameter subgroups are planar curves. The exponential image $\exp\mathfrak{sl}(2,\mathbb{R})$ (above left) is invariant under rotations about the z-axis, it definitely does not cover all of G. The connected subsemigroup S corresponding to the set $\mathrm{Sl}(2,\mathbb{R})^+$ of all matrices in $\mathrm{Sl}(2,\mathbb{R})$ with nonnegative entries (above middle) lies in $\exp\mathfrak{sl}(2,\mathbb{R})$, it is a non-invariant exponential Lie subsemigroup of G. (Note that its Lie wedge is a Lie semialgebra.) The exponential image of the upper half of the invariant wedge $\{X \in \mathfrak{sl}(2,\mathbb{R}) \mid k(X,X) < 0\}$ generates an invariant subsemigroup S_i which is not even weakly exponential (above right).

Thus S_i is not weakly exponential and so is every subsemigroup containing it.

(ii) *The standard invariant subsemigroup of* Osc. For the oscillator algebra $\mathfrak{osc}$ and the simply connected oscillator group Osc we use the following parametrisation from DÖRR's paper [12] (see also [27] for a slightly different version):

We define the oscillator algebra $\mathfrak{osc}$ on the space $\mathbb{R} \oplus \mathbb{C} \oplus \mathbb{R}$, with Lie brackets

$$(1) \qquad [(x,c,r),(x',c',r')] = \left(\mathrm{Im}(\bar{c}c'), 2i\cdot \det\begin{pmatrix} r & r' \\ c & c' \end{pmatrix}, 0\right).$$

The associated simply connected Lie group Osc is then defined on the space $\mathbb{R} \times \mathbb{C} \times \mathbb{R}$ with multiplication

$$(2) \quad (x,c,r)\cdot(x',c',r') = \left(x + x' + \frac{1}{2}\mathrm{Im}(e^{i(r+r')}\bar{c}c'), e^{ir}c' + e^{-ir'}c, r+r'\right).$$

In this parametrization the exponential function is expressed as follows:

$$(3) \qquad \exp(x,c,r) = \begin{cases} \left(\frac{|c|^2+4rx}{4r} - \frac{|c|^2\sin r \cos r}{4r^2}, \frac{c\sin r}{r}, r\right) & \text{for } r \neq 0 \\ (x,c,0) & \text{for } r = 0. \end{cases}$$

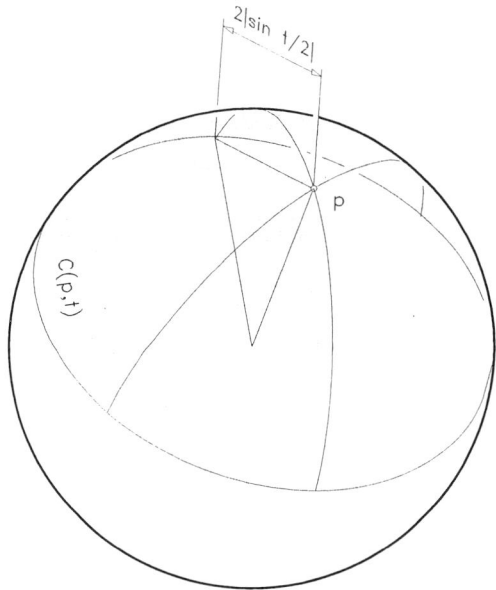

FIGURE 6. The orbit $C(p,t) = \exp(B(t))p$.

The *Standard Lorentzian Cone in* osc is defined as the set (Fig. 7)

$$W_{\text{Lor}} := \{(x,c,r) \in \mathfrak{osc} \mid |c|^2 + 4rx \leq 0,\ r \geq 0,\ x \leq 0\}.$$

The pair (osc, W_{Lor}) is sometimes also called the *standard pair*. It is seen readily that W_{Lor} is an invariant wedge. In view of (3) and (4) we have $(u,v,w) = \exp(x,c,r)$ with $(x,c,r) \in W_{\text{Lor}}$ if and only if

(4) $\quad u \leq -\dfrac{|v|^2}{4}\cot w,\ w \in \mathbb{R}^+ \setminus \pi\mathbb{Z} \quad \text{or} \quad u \leq 0 = v \leq w \in \pi\mathbb{Z} \setminus \{0\} \quad \text{or} \quad w = 0.$

The closed subsemigroup generated by the set $\exp W_{\text{Lor}}$ is

$$S_{\text{Lor}} = \exp W_{\text{Lor}} \cup \{(u,v,w) \mid w \geq \frac{\pi}{2}\}.$$

(A short proof of this fact and additional information is given in V.4.4 and V.4.5.) But S_{Lor} contains the open sets

$$\mathbb{R} \times \mathbb{C} \times\,]2k\pi, (2k+1)\pi[,\quad k = 1, 2, \ldots,$$

which are not contained in $\exp W_{\text{Lor}}$ (cf. Fig. 8 above). Thus S_{Lor} is not weakly exponential. This was first observed by DÖRR [12]. (Cf. Figures 8, 9.) □

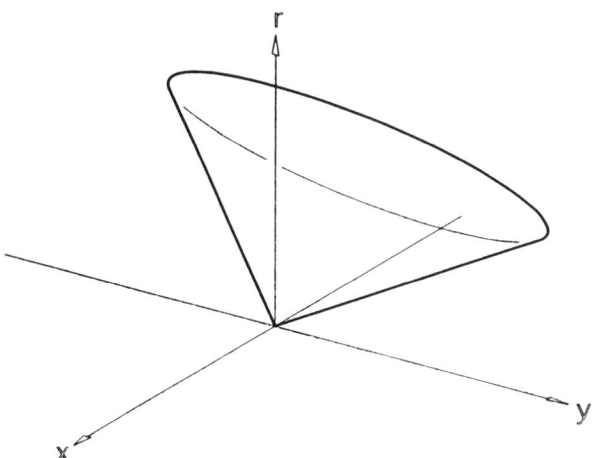

FIGURE 7. The Standard Lorentzian Cone W_{Lor}. This picture shows the set $\{(x,y,r) \in \mathbb{R}^3 \mid (x, y\cdot c, r) \in W_{\text{Lor}}\}$, for arbitrary $c \in \mathbb{C}$.

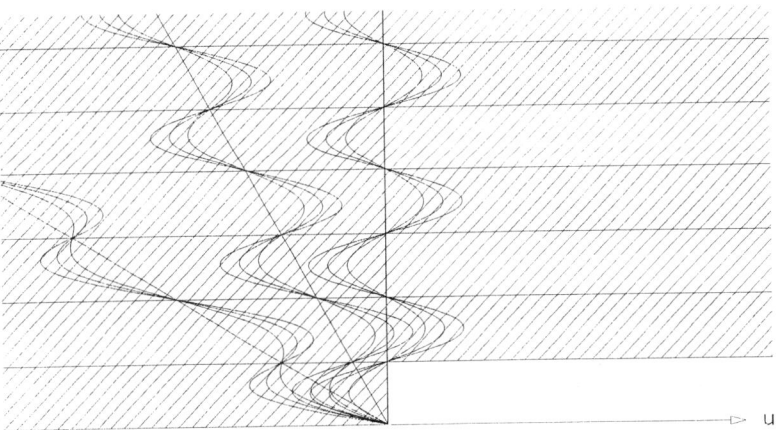

FIGURE 8. The one-parameter subsemigroups of $S_{\text{Lor}} = \overline{\langle \exp W_{\text{Lor}} \rangle}$. This picture shows the images of $\exp W_{\text{Lor}}$ (shaded region) and S_{Lor} (hatched region) under the projection $p \colon \text{Osc} \to \mathbb{R} \times \mathbb{R}$, $(u,v,w) \mapsto (u,w)$. By formula (3) the one-parameter subsemigroups $\mathbb{R}^+ \to \text{Osc}$ are, up to parametrization, the maps

$$t \mapsto (at + \tfrac{|b|^2}{8}\sin 2t, b\sin t, t) \text{ and } t \mapsto (at, bt, 0), \quad \text{with } a \in \mathbb{R},\, b \in \mathbb{C}.$$

Such a one-parameter subsemigroup lies in S_{Lor} if and only if $a \leq 0$. Thus $p(\exp W_{\text{Lor}})$ is the union of the left upper quadrant with the strips
$$\mathbb{R}^+ \times \,](2k+1)\tfrac{\pi}{2}, (2k+2)\tfrac{\pi}{2}[, \; k = 0, 1, 2, \ldots.$$
However the projection $p(S_{\text{Lor}})$ contains $p(\exp W_{\text{Lor}})$ as well as the whole half-plane $\{(u,w) \mid w \geq \tfrac{\pi}{2}\}$. Thus S_{Lor} cannot be weakly exponential.

2. The Exponential Function and Divisibility

The present section is an excursion into classical topological algebra, exploring the relations between divisibility and the exponential function in subsemigroups of Lie groups as a field of independent interest. The results collected here will

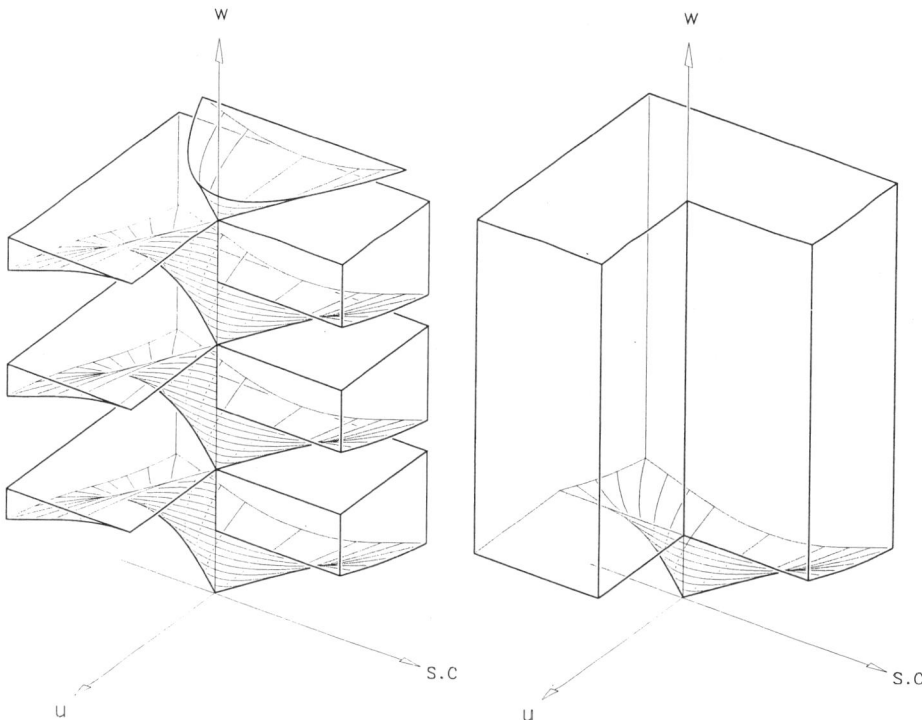

FIGURE 9. Pictures of $\exp W_{\text{Lor}}$ (left) and S_{Lor} (right), intersected with the set $\mathbb{R} \times \mathbb{R}c \times \mathbb{R}$, for arbitrary $0 \neq c \in \mathbb{C}$. The semigroup S_{Lor} contains $\exp W_{\text{Lor}}$ as well as the half space $\{(u, v, w) \in \text{Osc} \mid w \geq \frac{\pi}{2}\}$ and thus is not exponential.

play no role in later arguments, thus the purely analysis-oriented reader might safely skip this section. It is included here as an additional source of motivation and applications. Occasionally it will allow us to express analytical properties of subsemigroups, such as being exponential, in terms of topological algebra. The main result of this section (Theorem 2.4), which appears to be new, provides alternative proofs for some well established facts in divisibility theory.

2.1. DEFINITION. Consider a semigroup S and an element $s \in S$.
(i) The element s is called *divisible* if $(\forall n \in \mathbb{N})(\exists x \in S) \quad x^n = s$.
(ii) The semigroup S is called *divisible* if all of its elements are divisible. □

We let $\mathbb{Q}_+$ denote the additive semigroup of positive rational numbers.

2.2. LEMMA. *If S is a divisible semigroup then for each $g \in S$ there is a homomorphism $\xi: \mathbb{Q}_+ \to S$ with $\xi(1) = g$. If S is a group, then ξ extends uniquely to a group morphism $\mathbb{Q} \to S$.*

Proof. We pick a sequence of successive roots x_n of g, choosen recursively so that $x_1 = g$, $x_2^2 = x_1, \ldots, x_n^n = x_{n-1}$, $(n > 1)$. Note that $\frac{m}{n!} = \frac{m'}{n'!}$ always implies

$x_n^m = x_{n'}^{m'}$, so we may define a function $\xi : \mathbb{Q}_+ \to S$ unambiguously by $\xi(\frac{m}{n!}) = x_n^m$. It is straightforward to verify that ξ is a morphism and that $\xi(1) = d$. If S is a group then we extend ξ by defining $\xi(0) = 1$ and $\xi(-q) = \xi(q)^{-1}$ for all $q \in \mathbb{Q}_+$. This extension is readily seen to be a morphism, and it is the only homomorphic extension. □

Thus divisibility of a semigroup can be defined alternatively in the following way: *A semigroup S is divisible if for each $s \in S$ there is a morphism of semigroups $\xi : \mathbb{Q}_+ \to S$ such that $\xi(1) = s$.*

In Lie groups the algebraic condition of divisibility is connected with the analytic structure since clearly every element in the image of the exponential function is divisible. In [59] MCCRUDDEN has proved the converse:

2.3. PROPOSITION. *Every divisible element g in a connected Lie group G is contained in $\exp \mathfrak{g}$.* □

As a consequence we now prove the following result which is reminiscent of A. Weil's Lemma on the image of $\mathbb{Z}$ under a morphism into a locally compact topological group.

2.4. THEOREM. *Let $f : \mathbb{Q}_+ \to G$ be a morphism of semigroups into a connected Lie group. Then $\overline{f(\mathbb{Q}_+)}$ is singleton, or is a torus, or is isomorphic to the direct product of a torus (possibly of dimension 0) and a half-line $\mathbb{R}^+$. In particular, $\overline{f(\mathbb{Q}_+)}$ is connected and contains the identity.*

Proof. Suppose that $f : \mathbb{Q}_+ \to G$ is a morphism, with image D. We let $D_n = f(\mathbb{N}\frac{1}{n!})$. Then $D_n \subseteq D_{n+1}$, for all n, and $D = \bigcup D_n$. For a subset A of G denote by $C(A)$ the centralizer of A in G. Then $C(D) = \bigcap_{n \in \mathbb{N}} C(D_n)$. The descending chain $C(D_1)_0 \supseteq C(D_2)_0 \supseteq \cdots$ becomes stationary after finitely many steps. Thus there is an index k with $C(D)_0 = \bigcap_{n \leq k} C(D_n)_0 = C(D_k)_0$.

By MCCRUDDEN's result 2.3 we can find for each n a one-parameter subgroup E_n containing $f(\frac{1}{n!})$. By construction, $D_n \subseteq E_n \subseteq C(\{f(\frac{1}{n!})\})_0 = C(D_n)_0$. It follows that for all $n \geq k$ we have $D_n \subseteq C(D_n)_0 = C(D_k)_0 = C(D)_0$. Hence $D = \bigcup_{n \geq k} D_n \subseteq C(D)_0$. Thus D is central in $C(D)_0$. The center of $C(D)_0$ is contained in a connected abelian Lie subgroup A (s. [29], p.189), we take A of minimal dimension. Note that $D \subseteq C(D)_0 \subseteq A$ and $DD^{-1} = f(\mathbb{Q}) \subseteq A$. In order to simplify our notation we assume that $G = A$ and that f is the restriction of a group morphism $\mathbb{Q} \to G$, which we also call f. Let K denote the maximal torus of G and $p : G \to G/K$ the quotient map. Then G/K is a vector group (hence uniquely divisible) and thus $p \circ f : \mathbb{Q} \to G/K$ extends uniquely to a continuous group morphism $F : \mathbb{R} \to G/K$. Since $G = A$ is of minimal dimension by our choice above we have $G/K = F(\mathbb{R})$. It follows that $\dim G/K \leq 1$. Moreover,

$$p\Big(\bigcap_{0 < \varepsilon} \overline{f(]0, \varepsilon[\cap \mathbb{Q})}\Big) \subseteq \bigcap_{0 < \varepsilon} \overline{(p \circ f)(]0, \varepsilon[\cap \mathbb{Q})} = \{1\}$$

in G/K since $F = p \circ f$ is continuous. Thus
$$\bigcap_{0<\varepsilon} \overline{f(]0,\varepsilon[\cap \mathbb{Q})} \subseteq K.$$
It follows that $N := \bigcap_{0<\varepsilon} \overline{f(]0,\varepsilon[\cap \mathbb{Q})}$ is a compact divisible group, thus is connected and therefore a torus. Also $f(]0,\varepsilon[\cap \mathbb{Q})$ is contained in the compact set $p^{-1}F([0,1])$. Now either $\overline{f(\mathbb{Q}_+)}$ is a compact group (if $\dim G/K = 0$) or else $\overline{f(\mathbb{Q}_+)} \cong N \times \mathbb{R}^+$. (For these matters we refer, e.g., to [25], p. 218, Theorem 3.5.1; see also [41], proof of 3.5 on p. 100 and the following results.) In particular, this shows that $\overline{f(\mathbb{Q}_+)}$ is connected. □

We now draw some consequences of 2.4.

2.5. COROLLARY. *If $f: \mathbb{Q} \to G$ is an algebraic morphism mapping $\mathbb{Q}$ into a connected Lie group then $\overline{f(\mathbb{Q})}$ is connected.*

Proof. Since $\overline{f(\mathbb{Q})} = \overline{f(\mathbb{Q}_+)}(\overline{f(\mathbb{Q}_+)})^{-1}$ this follows immediately from Proposition 2.4. □

2.6. COROLLARY. *A discrete divisible subsemigroup of a connected Lie group is singleton.*

Proof. Let D be a discrete divisible subsemigroup of G and $d \in D$. Then there is a surjective morphism $f: \mathbb{Q}_+ \to D$ with $f(1) = d$. Then $E := f(\mathbb{Q}_+) \cup f(\mathbb{Q}_+)^{-1} \cup \{1\} = f(\mathbb{Q}_+ \cup -\mathbb{Q}_+ \cup \{0\}) = f(\mathbb{Q})$ is a discrete divisible subgroup and is, therefore, closed. However, by Theorem 2.4, $\overline{f(\mathbb{Q}_+)}$ is connected and thus is singleton. This implies $d = 1$ and proves the claim. □

For our purposes the following consequence of Proposition 2.4 is instructive:

2.7. COROLLARY. *If D is a divisible subsemigroup of a connected Lie group G, then $D \subseteq \exp \mathfrak{L}(\overline{D})$. In particular we have the following consequences:*
  (i) *The closure of a divisible subsemigroup of a connected Lie group is weakly exponential and therefore connected.*
  (ii) *A closed subsemigroup D of a connected Lie group is divisible if and only if it is exponential.*

Proof. Let D be a divisible subsemigroup of G. Let $d \in D$. By Proposition 2.4 there is an exponential subsemigroup T contained in $\overline{D}$ and containing d. It is obvious that the closure of the exponential image of a Lie wedge is connected and that an exponential subsemigroup is divisible. □

As we noted in Example 1.6(i) any abelian weakly exponential closed subsemigroup of a Lie group is exponential. Thus the above Corollary shows that the closure of an abelian divisible subsemigroup is exponential, hence divisible. This recovers a result by HOFMANN and LAWSON in [36]. In the same paper the equivalence in (ii) was established without using MCCRUDDEN's result.

In general the closure of a divisible subsemigroup need not be divisible. This can be seen from a variant of Example 1.6(iii):

2.8. EXAMPLE. Let $G = \mathbb{C} \rtimes \mathbb{R}$ be the simply connected covering group of the motion group of the euclidean plane with multiplication $(z,t)(z',t') = (z + e^{it}z', t+t')$. Write $\mathbb{R}$ as a direct sum of $2\pi\mathbb{Q}$ and a rational vector subspace $R \cong \mathbb{R}$. Set $D = \mathbb{C} \rtimes R$. Then D is uniquely divisible, dense and normal in G such that G/D is algebraically isomorphic to $\mathbb{Q}$. (In particular, D has countable index in G.) But G is not divisible.

Let $R^+ = \{r \in R \mid 0 \leq r\}$. Then $D^+ := \mathbb{C} \rtimes R^+$ is a *uniquely divisible* dense subsemigroup of the *nondivisible* half-space semigroup G^+ (cf. 1.6(iii)). □

2.9. REMARK. Some care is indicated with the use of divisibility of elements versus that of subsemigroups. An abelian group (additively written) may have nonzero divisible elements while the maximal divisible subgroup is singleton. Thus divisibility of an element g in a semigroup does not a priori allow us to embed it into a subsemigroup $\xi(\mathbb{Q}_+)$ as constructed in Lemma 2.2. Even MC-CRUDDEN's theorem is to be used with care: Suppose that G is a connected Lie group, D a subsemigroup and d an element which is divisible in D. Then it is not clear that we find a one-parameter semigroup connecting 1 with d and lying in $\overline{D}$. □

3. The Exponential Image of Lie Wedges and Regularity

In this section we show that in a weakly exponential subsemigroup S of a Lie group G the exponential image of the Lie wedge $\mathcal{L}(S)$ always covers all *regular points* contained in S. In the subsequent section this result will then be extended to intersections with maximal rank subgroups.

Our strategy is to construct one-parameter subsemigroups passing through regular elements as Γ-limits of one-parameter subsemigroups in $\exp W$. We start by recalling the basic definitions.

3.1. DEFINITION. Let G be a Lie group, $\mathfrak{g}$ its Lie algebra. An element $x \in \mathfrak{g}$ [respectively, $g \in G$] is called *regular* in $\mathfrak{g}$ [respectively, G] if the rank of the linear map $\operatorname{ad} x \colon \mathfrak{g} \to \mathfrak{g}$ [respectively, the linear map $\operatorname{Ad}(g) - 1 \colon \mathfrak{g} \to \mathfrak{g}$] is maximal among such maps, hence equals the rank of $\mathfrak{g}$. (cf. BOURBAKI [5] chap. VII, §2 and §4). □

3.2. NOTATION. (i) The set of regular elements in $\mathfrak{g}$ [respectively, G] is denoted by $\operatorname{reg}(\mathfrak{g})$ [respectively, $\operatorname{Reg}(G)$]. The complement of $\operatorname{Reg}(G)$ in G will be written $\operatorname{Irr}(G)$.

(ii) We write $\operatorname{reg} \exp$ for the set of all points in $\mathfrak{g}$ where the exponential function is regular; its complement in $\mathfrak{g}$ is denoted by $\operatorname{irr} \exp$. Similarly, we write $\operatorname{reg} \mathfrak{g}$ for the set of all algebraically regular elements in $\mathfrak{g}$, and $\operatorname{irr} \mathfrak{g}$ for the set of algebraically nonregular elements.

(iii) We denote with $\operatorname{REG}(\mathfrak{g})$ the set $\operatorname{reg} \mathfrak{g} \cap \operatorname{reg} \exp$ of all regular elements of $\mathfrak{g}$ at which $\exp$ is nonsingular. The complement of this set in $\mathfrak{g}$ is written $\operatorname{IRR} \mathfrak{g}$. We notice at once that $\operatorname{IRR} \mathfrak{g} = \operatorname{irr} \mathfrak{g} \cup \operatorname{irr} \exp$.

(iv) For any $g \in G$ we let $\mathfrak{h}(g)$ denote the nilspace of $\operatorname{Ad}(g) - 1$ in $\mathfrak{g}$ and $\mathfrak{a}(g)$ the Fitting one-component of this operator, i.e., the largest vector subspace of $\mathfrak{g}$ on which it operates as an automorphism. Furthermore, we write $H(g)$ for the analytic subgroup of G generated by $\exp \mathfrak{h}(g)$. □

The following standard facts are used frequently in the sequel, without further notice.

3.3. PROPOSITION. *Let $\mathfrak{g}$ be the Lie algebra of a connected Lie group G.*
(i) *An element $g \in G$ is regular if and only if the nilspace $\mathfrak{h}(g)$ of $\operatorname{Ad}(g) - 1$ is a Cartan subalgebra of $\mathfrak{g}$. Thus for $g \in \operatorname{Reg}(G)$ we have $H(g) = \exp \mathfrak{h}(g)$ and $H(g)$ is closed.*
(ii) *The set $\operatorname{Reg}(G)$ is open and dense in G.*
(iii) *$\operatorname{REG} \mathfrak{g} = \exp^{-1} \operatorname{Reg}(G)$ and for any $x \in \operatorname{REG} \mathfrak{g}$ the subalgebra $\mathfrak{h}(\exp x)$ is exactly the Cartan subalgebra generated by x.*

Proof. Assertions (i) and (ii) are established by Proposition 4.3.8 (p. 38) and Proposition 4.1.1 (p. 34) of BOURBAKI [5].
Assertion (iii) is Lemma 2 of HOFMANN [32]. □

For exp-regular elements x the solutions of the equation $\exp x = \exp y$ are closely connected with exp-compact elements in $\mathfrak{g}$, which are defined next.

3.4. DEFINITION. Let $\exp \mathfrak{g} \to G$ be the exponential function of a Lie group G. An element $x \in \mathfrak{g}$ is called exp-*compact* if $\overline{\exp \mathbb{R} \cdot x}$ is compact. The set of all exp-compact elements is denoted by $\operatorname{comp}_G(\mathfrak{g})$. □

Note that every exp-compact element in $\mathfrak{g}$ is compact (that is, $\overline{e^{\operatorname{ad} \mathbb{R} \cdot x}}$ is compact in $\operatorname{Aut} \mathfrak{g}$), but not vice versa.

3.5. LEMMA. *Suppose that $\exp x = \exp y$ for two elements x, y in the Lie algebra $\mathfrak{g}$ of a Lie group G. If the exponential function is regular at x or at y then $[x, y] = 0$, $\exp(x - y) = 1$ and $\exp[0, 1] \cdot (x - y)$ is a circle subgroup of G. In particular, $x - y \in \operatorname{comp}_G(\mathfrak{g})$.*

Proof. See e.g. [27], Lemma V.6.7 on p. 461, or [36]. □

3.6.. **The Γ-topology.** Let X be a locally compact topological space. The set of all closed subsets of X carries a compact Hausdorff topology, called the *hyperspace topology* (see Chapter IX). The resulting *hyperspace* will be denoted by $\Gamma(X)$. For the convenience of the reader we list here some properties of the hyperspace needed in the sequel.

3.7. PROPOSITION. *Let X be a locally compact space.*
(i) *If X is second countable and metrizable then $\Gamma(X)$ is metrizable.*
(ii) *The Γ-limit of connected closed subsets of a compact subspace of X is compact and connected.*
(iii) *If $f: X \to Y$ is an open and continuous map between locally compact spaces then the induced map $\Gamma(f): \Gamma(Y) \to \Gamma(X)$, $A \mapsto f^{-1}A$, is continuous. If f is surjective then $\Gamma(f)$ is injective, hence is a homeomorphic embedding.*
(iv) *If $f: X \to Y$ is a proper continuous map between locally compact spaces, then the function $A \mapsto f(A): \Gamma(X) \to \Gamma(Y)$ is continuous.*
(v) *Suppose that $\mathcal{U}$ is any open cover of X. Then the function*

$$A \mapsto (U \cap A)_{U \in \mathcal{U}} : \Gamma(X) \to \prod_{U \in \mathcal{U}} \Gamma(U)$$

is a homeomorphism onto the image. In particular, a net (Y_n) in $\Gamma(X)$ converges to a closed subspace Y if and only if $\lim_\Gamma U \cap Y_n = U \cap Y$ in the space $\Gamma(U)$, for every $U \in \mathcal{U}$.

Proof. For a proof we refer to Chapter IX. Specifically, for (i) see IX.1.9, for (ii) see IX.3.8, for (iii) see IX.2.6, for (iv) see IX.2.11, and for (v) see IX.2.9. $\square$

3.8. COROLLARY.
(i) *Suppose that X is a locally compact topological group. Then the Γ-limit of closed subsemigroups [respectively, subgroups] is a closed subsemigroup [respectively, subgroup].*
(ii) *If H is a closed subgroup of a locally compact topological group G and κ the quotient map $G \to G/H$ then $\Gamma(\kappa)$ is a homeomorphic embedding of $\Gamma(G/H)$ into $\Gamma(G)$. If, in addition, H is compact then κ is closed and the map $\Gamma(G) \to \Gamma(G/H)$, $A \mapsto \kappa(A)$ is continuous.*

Proof. For (i) see IX.4.3, and for (ii) see IX.4.4. $\square$

3.9. PROPOSITION. *Let G be a connected nilpotent Lie group, M its maximal torus, with Lie algebras $\mathfrak{g}$ and $\mathfrak{m}$, respectively. Let $\mathfrak{e}_n$ denote a sequence of closed half-lines in $\mathfrak{g}$ and write $E_n = \overline{\exp \mathfrak{e}_n}$. Suppose that $E = \lim_\Gamma E_n$ exists in $\Gamma(G)$. Then there are three cases:*
(a) *No Γ-cluster point of the sequence $\mathfrak{e}_n$ is contained in $\mathfrak{m}$. Then $\mathfrak{e}_n$ converges in $\Gamma(\mathfrak{g})$ and $E = \exp(\lim_\Gamma \mathfrak{e}_n)$.*
(b) *There is a subsequence $\mathfrak{e}_{n(m)}$ with $\mathfrak{e}_{n(m)} \subseteq \mathfrak{m}$ eventually. Then E is a torus in M.*
(c) *There is a subsequence $\mathfrak{e}_{n(m)}$ converging to $\mathfrak{e} \subseteq \mathfrak{m}$ such that $\mathfrak{e}_{n(m)} \nsubseteq \mathfrak{m}$. Then there is a half-line $\mathfrak{d}$ and a torus $T \subseteq M$ such that*

$$(t, X) \mapsto t \exp X : T \times \mathfrak{d} \to E$$

is an isomorphism of topological semigroups.

In all three cases E is exponential and $\exp \mathcal{L}(E) = E$.

Proof. It is obvious that the case distinction is complete.

For convenience we endow $\mathfrak{g}$ with a Euclidean inner product.

Case (a): The Campbell-Hausdorff multiplication $*$ on $\mathfrak{g}$ makes $(\mathfrak{g}, *)$ into the universal covering group of G with $\exp \colon \mathfrak{g} \to G$ as the covering map. We shall show that for each cluster point $\mathfrak{e}$ of $\mathfrak{e}_n$ we have $\mathfrak{e} = \mathcal{L}(E)$. Since $\mathcal{L}(E)$ is uniquely determined, this means that $\mathfrak{e}_n$ has only one cluster point and therefore must converge, since $\Gamma(\mathfrak{g})$ is compact.

By passing to a convergent subsequence and renaming it we enforce $\mathfrak{e} = \lim_\Gamma \mathfrak{e}_n$. Let $K = \ker \exp$. Since $\mathfrak{e} \not\subseteq \mathfrak{m} = \mathrm{span}(K)$ we conclude from Weil's Lemma that $\mathfrak{e} + K$ is closed. Likewise, for all sufficiently large n we also have $\mathfrak{e}_n \not\subseteq \mathfrak{m}$, so $\mathfrak{e}_n + K$ is closed for these n, too. We next show that $\mathfrak{e} + K = \lim_\Gamma(\mathfrak{e}_n + K)$. Observe first that for any open ball B around zero there is a finite subset F of K such that $(\mathfrak{e}+K) \cap B = (\mathfrak{e}+F) \cap B$ and $(\mathfrak{e}_n+K) \cap B = (\mathfrak{e}_n+F) \cap B$ for all sufficiently large n.

Indeed, let ρ be the radius of B and let $v_n \in \mathfrak{e}_n$ with $\|v_n\| = 1$, for every $n \in \mathbb{N}$. Then $B \cap \mathfrak{e}_n + k = \emptyset$ whenever $\|k\| > \rho/(1 - \|p(v_n)\|^2)$, where $p(v_n)$ denotes the orthogonal projection of v_n onto $\mathrm{span}(K) = \mathfrak{m}$. Also, $\lim v_n = v \in \mathfrak{e} \not\subseteq \mathfrak{m}$, so $(1 - \|p(v_n)\|^2)$ is bounded away from 0 for all sufficiently large indexes n, which establishes our claim.

Now $\mathfrak{e} + F = \lim_\Gamma(\mathfrak{e}_n + F)$. Thus with respect to the topology of $\Gamma(B)$ we have

$$(\mathfrak{e} + K) \cap B = (\mathfrak{e} + F) \cap B = \lim_\Gamma(\mathfrak{e}_n + F) \cap B = \lim_\Gamma((\mathfrak{e}_n + K) \cap B)$$
$$= \left(\lim_\Gamma(\mathfrak{e}_n + K)\right) \cap B.$$

Choosing balls B with radius $1, 2, 3, \ldots$ and applying 3.7(v) we see that $\mathfrak{e}_n + K = \exp^{-1} \exp \mathfrak{e}_n$ converges to $\mathfrak{e} + K$. Now the assertion $E = \exp \mathfrak{e}$ follows from 3.8(i).

Before considering the remaining cases we observe that, as a limit of a sequence of closed semigroups, E is a closed semigroup, by 3.8(ii). Hence $E \cap M$ is a central compact subgroup.

Case (b). Since $\mathfrak{e}_{n(m)} \subseteq \mathfrak{m}$ eventually, we have $E_{n(m)} \subseteq M$ eventually. Thus E is a closed subsemigroup of the compact group M and is therefore a group. Also, being the Γ-limit of connected closed subsets of the compact subspace M, the group E is connected, hence a torus.

Case (c). After renaming our subsequence we suppose that $\mathfrak{e}_n \not\subseteq \mathfrak{m}$ and $\mathfrak{e} = \lim_\Gamma \mathfrak{e}_n \subseteq \mathfrak{m}$. We set $T = (E \cap M)_0$. Then T is a torus and $\mathcal{L}(T) = L(E) \cap \mathfrak{m}$. Let $p \colon G \to G/T$ denote the quotient map and $\mathcal{L}(p) \colon \mathfrak{g} \to \mathfrak{g}/\mathfrak{t}$ the corresponding morphism on the Lie algebra level. Since $\mathfrak{e}_n \not\subseteq \mathfrak{m}$ we have $\mathcal{L}(p)(\mathfrak{e}_n) \neq \{0\}$. By passing to a further subsequence and renaming we obtain $\mathfrak{e}^* = \lim_\Gamma \mathcal{L}(p)(\mathfrak{e}_n)$. Note that, as the Γ-limit of half-lines, $\mathfrak{e}^*$ is a half-line.

Since T is compact we see from 3.8(i) that $p(E) = p(\lim_\Gamma E_n) = \lim_\Gamma p(E_n) = \lim_\Gamma p(\overline{\exp \mathfrak{e}_n}) = \lim_\Gamma \overline{p(\exp \mathfrak{e}_n)} = \lim_\Gamma \overline{\exp \mathfrak{e}'_n}$. Hence $\exp \mathfrak{e}^* \subseteq p(E)$, and therefore $\mathfrak{e}^* \subseteq \mathcal{L}(p(E)) = \mathcal{L}(p)(\mathcal{L}(E))$. Thus $\mathcal{L}(p)^{-1}(\mathfrak{e}^*) \subseteq \mathcal{L}(E)$ since $\mathfrak{t} \subseteq \mathcal{L}(E)$.

We claim that $\mathfrak{e}^*$ is not contained in $p(\mathfrak{m})$. Indeed, if it were, then

$$\mathcal{L}(p)^{-1}(\mathfrak{e}*) \subseteq \mathcal{L}(p)^{-1} \mathcal{L}(p)(\mathfrak{m}) \cap \mathcal{L}(E) = \mathfrak{m} \cap \mathcal{L}(E) = \mathfrak{t}$$

and therefore $\mathfrak{e}^* = \{0\}$, a contradiction.

Now Case (a) applies to G/T and $\mathfrak{L}(p)(\mathfrak{e}_n)$, yielding $p(E) = \exp \mathfrak{e}^*$. Let $\mathfrak{d}$ be any half-line in $\mathfrak{L}(p)^{-1}(\mathfrak{e}^*)$ with $\mathfrak{L}(p)(\mathfrak{d}) = \mathfrak{e}^*$. Then the map $(t, X) \mapsto t \exp X : T \times \mathfrak{d} \to E$ is a surjective continuous morphism of semigroups. This function is also injective: Let $t_1 \exp X_1 = t_2 \exp X_2$ with $X_2 - X_1 \in \mathfrak{d}$. Then $t_2^{-1}t_1 = \exp(X_2 - X_1) \in T \cap \exp \mathfrak{d} = \{\mathbf{1}\}$, and therefore $t_1 = t_2$, $X_1 = X_2$. The differential of the analytic function $(t, X) \mapsto t \exp X : T \times (\mathfrak{d} - \mathfrak{d}) \to EE^{-1}$ is bijective everywhere. Hence the function is a diffeomorphism and its restriction to $T \times \mathfrak{d}$ is a diffeomorphism onto E. □

Our next Lemma shows that if the Γ-limit of one-parameter subsemigroups contains a regular element then this regular element lies on a one-parameter subsemigroup of the limit semigroup. An important consequence is that the intersection $\mathrm{Reg}(G) \cap \exp \mathfrak{g}$ is closed in $\mathrm{Reg}(G)$. In short: the limit of exponential images is an exponential image *if it is regular*. (Note that because of the compactness of $\Gamma(G)$, the existence of the Γ-limit in the formulation of the Lemma can always be enforced by passing to a suitable subsequence.)

3.10. LEMMA. *Let G be a Lie group and $g \in G$ a regular element satisfying $g = \lim \exp Z_n$. Suppose that $E = \lim_\Gamma \overline{\exp \mathbb{R}^+ \cdot Z_n}$ exists. Then E is a closed subsemigroup of G and $g = \exp X$ with $X \in \mathfrak{L}(E)$.*

Proof. The key idea of the proof is to apply inner automorphisms to move the elements Z_n into the exp-image of a fixed Cartan subalgebra and then to apply Lemma 3.9.

For all sufficiently large n we have $\exp Z_n \in \mathrm{Reg}(G)$ since $\mathrm{Reg}(G)$ is open. We assume $\exp Z_n \in \mathrm{Reg}(G)$ for all n. Then every element Z_n lies in $\exp^{-1} \mathrm{Reg}(G) = \mathrm{REG}(\mathfrak{g})$, hence is regular in $\mathfrak{g}$ and therefore generates a Cartan subalgebra $\mathfrak{h} = \mathfrak{h}(\exp Z_n)$ containing it, so $\exp Z_n \in \exp \mathfrak{h}(\exp Z_n) = H(\exp Z_n)$. By [32], Theorem 9, the map $\mathrm{Reg}(G) \to \Gamma(G)$, $x \mapsto H(x)$, is continuous. Hence $g = \lim \exp Z_n \in \lim_\Gamma H(\exp Z_n) = H(g)$.

Also, Lemma 8 of [32] says that the map

$$\mathfrak{a}(g) \times \mathfrak{h}(g) \to G, \quad (A, B) \mapsto \exp A (g \exp B) \exp -A,$$

maps a neighborhood of $(0,0)$ diffeomorphically onto a neighborhood of g. Thus there is a sequence $(A_n, B_n) \in \mathfrak{a}(g) \times \mathfrak{h}(g)$ converging to $(0,0)$ and an index $N \in \mathbb{N}$ such that

$$\exp Z_n = \exp A_n (g \exp B_n) \exp -A_n \in \mathrm{Reg}(G) \quad \text{for } n \geq N.$$

Furthermore, since $g \in H(g)$,

$$g \exp B_n \in gH(g) = H(g).$$

Thus $\exp e^{-\operatorname{ad} A_n} Z_n = (\exp -A_n) \exp Z_n \exp A_n = \exp Y_n$ with a suitable $Y_n \in \mathfrak{h}(g)$ for $n \geq N$. But $\exp Y_n \in \mathrm{Reg}(G)$ since $\exp Z_n \in \mathrm{Reg}(G)$ for $n \geq N$.

Since $\text{Reg}(G) \cap \exp \mathfrak{g} = \exp \text{REG}(\mathfrak{g})$ we conclude that $Y_n \in \text{REG}(\mathfrak{g})$, therefore $\mathfrak{h}(g)$ is the Cartan algebra generated by Y_n. Now we set $\alpha_n = e^{-\text{ad } A_n}$ and $X_n = \alpha_n(Z_n)$. Then $\lim \alpha_n = 1$ in $\text{Aut } \mathfrak{g}$. Furthermore, since X_n is exp-regular and $\exp X_n = \exp Y_n$, Lemma 3.5 shows that $[X_n, Y_n] = 0$ by 3.5. Hence $X_n \in \mathfrak{h}(g) = \mathfrak{h}(\exp Y_n)$ and $g = \lim \exp Z_n = \lim \exp X_n$ where g and $\exp X_n$ are in the connected nilpotent Lie group $H(g)$. We define $\mathfrak{e}_n = \mathbb{R}^+ \cdot X_n$ and $E_n = \overline{\exp \mathfrak{e}_n}$ and obtain a sequence E_n in the compact space $\Gamma(H(g))$.

By assumption $E = \lim_\Gamma E_n$ in $\Gamma(G)$. We set $\beta_n(x) = (\exp A_n) x (\exp -A_n)$ for $x \in G$. Then β_n converges uniformly on compact sets to id_G. Hence $E = \lim_\Gamma E_n = \lim_\Gamma \beta_n E_n = \lim_\Gamma (\exp A_n) E_n \exp -A_n$. We know that $g = \lim \exp Z_n \in \lim_\Gamma (\exp A_n) E_n \exp -A_n = E$. Now by Proposition 3.9 we finally find an element $X \in \mathfrak{L}(E)$ with $g = \exp X$. □

3.11. THEOREM. *Let S be a weakly exponential subsemigroup of a Lie group G, and write $W := \mathfrak{L}(S)$. Then $S \cap \text{Reg}(G) \subseteq \exp W$.*

In slogan form: *Regular points in a weakly exponential subsemigroup are exponential.*

Proof. Let $g \in S \cap \text{Reg}(G)$. By hypothesis we find a sequence $Z_n \in W$ such that $g = \lim \exp Z_n$. By passing to a suitable subsequence and renaming it we enforce the existence of $E = \lim_\Gamma \overline{\exp \mathbb{R}^+ \cdot Z_n}$ in the compact space $\Gamma(G)$. By Lemma 3.10 there is an $X \in \mathfrak{L}(E)$ with $g = \exp X$. But $\exp \mathbb{R}^+ \cdot Z_n \subseteq \exp W \subseteq S$. Hence $E \subseteq S$ and so $X \in \mathfrak{L}(E) \subseteq \mathfrak{L}(S) = W$. □

The above result cannot be so improved that the conclusion of the theorem reads $S = \exp W$. The Master Example in $\widetilde{\text{Mot}}$ 1.6(ii) shows that a weakly exponential subsemigroup need not be exponential.

4. Maximal Rank Subgroups

Recall that an analytic subgroup M of a connected Lie group G is said to be of *maximal rank* if its Lie algebra contains a Cartan algebra of $\mathfrak{g}$. (Note that this is not equivalent with the equation $\text{rank } M = \text{rank } G$.) Maximal rank subgroups are always closed. Trivially, the exponential image of a Cartan subalgebra and the group G itself are maximal rank subgroups.

4.1. LEMMA. *Let G denote a connected Lie group and M a closed connected subgroup of maximal rank. Then*
  (i) $M \cap \text{Reg}(G) \subseteq \text{Reg}(M)$,
  (ii) $M \cap \text{Reg}(G)$ *is open dense in* M.

Proof. (i) Let $g \in M \cap \text{Reg } G$. Then $\mathfrak{h}(g) = \mathfrak{g}^1(\text{Ad}(g)) = \mathcal{N}(\text{Ad } g - 1)$ is a Cartan algebra of $\mathfrak{g}$. In $\mathfrak{m}$ any nilspace such as $\mathfrak{h}_M(g) = \mathfrak{m} \cap \mathcal{N}(\text{Ad } g - 1) \subseteq \mathfrak{h}(g)$ has a dimension which is greater than or equal to the rank of $\mathfrak{m}$. This rank equals that of $\mathfrak{g}$, i.e., the dimension of $\mathfrak{h}(g)$. It follows that $\mathfrak{h}_M(g) = \mathfrak{h}(g)$. Thus $g \in \text{Reg}(M)$.

(ii) The complement of $\operatorname{Reg}(G)$ is the zero set of an analytic function on G which does not vanish identically on M. Hence $M \cap \operatorname{Reg}(G)$ is open dense in M. □

4.2. THEOREM. *Let G be a connected Lie group and S a weakly exponential subsemigroup with Lie wedge $W = \mathfrak{L}(S)$. Suppose that M is a closed connected maximal rank subgroup. Then the following assertions hold:*
(i′) $(\operatorname{int} S) \cap M \cap \operatorname{Reg}(G) \subseteq \exp(W \cap \mathfrak{m})$.
(i) $(\operatorname{int} S) \cap M \subseteq \overline{\exp(W \cap \mathfrak{m})} \subseteq S \cap M$.
(ii) *If $\mathbf{1} \in \overline{\operatorname{int} S \cap M}$ —in particular, if $\mathfrak{m}$ meets the interior of W— then*
$$S \cap M = \overline{\exp(W \cap \mathfrak{m})}.$$

Proof. If $\operatorname{int} S = \emptyset$ then assertion (i′) is trivial, so suppose that $\operatorname{int} S \neq \emptyset$. Then $\operatorname{int} S$ is a dense ideal of S. The set $M \cap \operatorname{Reg}(G)$ is dense open in M by Lemma 4.1. Hence $(\operatorname{int} S) \cap M \cap \operatorname{Reg}(G)$ is dense in $(\operatorname{int} S) \cap M$. By Theorem 3.11 we have $S \cap \operatorname{Reg}(G) \subseteq \exp W$, therefore

(1) $\qquad\qquad (\operatorname{int} S) \cap M \cap \operatorname{Reg}(G) \subseteq \exp W.$

Now let $m \in (\operatorname{int} S) \cap M \cap \operatorname{Reg}(G)$. Then by (1) there is an $X \in W$ with $m = \exp X$. By 3.3(iii), $X \in \exp^{-1} \operatorname{Reg}(G) = \operatorname{REG} \mathfrak{g}$ and $\mathfrak{h}(m)$ is the Cartan algebra generated by X.

But $\mathfrak{h}(m) = \mathfrak{h}_M(m)$ (as we have seen in the proof of Lemma 4.1) is contained in $\mathfrak{m}$. Hence $X \in \mathfrak{h}(m) \subseteq \mathfrak{m}$. It follows that $X \in W \cap \mathfrak{m}$ and thus $m \in \exp(W \cap \mathfrak{m})$. This proves (i′).

The right hand inclusion of (i) is clear. Since $M \cap \operatorname{Reg}(G)$ is dense in M bei 4.1, assertion (i) follows from (i′).

If $\mathbf{1} \in \overline{(\operatorname{int} S) \cap M}$ then the ideal $(\operatorname{int} S) \cap M$ of $S \cap M$ is dense in $S \cap M$. Thus (ii) follows from (i) in this case. The remainder of the assertion is clear. □

The following theorem can be expressed by the slogan: *The intersection of a weakly exponential subsemigroup with a maximal rank subgroup is weakly exponential.*

4.3. THEOREM. ('The Maximal Rank Theorem') *Let G be a Lie group containing a weakly exponential subsemigroup S with Lie wedge W. Furthermore, suppose that $\mathfrak{m}$ is a maximal rank subalgebra of $\mathfrak{g}$ and that M is the associated analytic subgroup. If $\mathfrak{m} \cap \operatorname{int} W \neq \emptyset$ then*

$$S \cap M \subseteq \exp(W \cap \mathfrak{m}) \cup \operatorname{Irr}(M),$$

and therefore $S \cap M$ is weakly exponential.

Proof. By Theorem 4.2 we know that $S \cap M = \overline{\exp(W \cap \mathfrak{m})}$. Now Theorem 3.11, applied to $S \cap M$ and $W \cap \mathfrak{m}$, proves the assertion. □

Note that $S \cap M$ has dense interior in M, since $\operatorname{int}_{\mathfrak{m}} W \cap \mathfrak{m} \neq \emptyset$.

4.4. COROLLARY. *Let G be a Lie group containing a weakly exponential subsemigroup S with Lie wedge W. We assume that $\mathfrak{h}$ is an abelian Cartan subalgebra of $\mathfrak{g} = \mathfrak{L}(G)$ which meets the interior of W, and write $H = \exp \mathfrak{h}$. Then $S \cap H = \exp(W \cap \mathfrak{h})$.*

Proof. Since H is a maximal rank subgroup of G we conclude from the Maximal Rank Theorem 4.3 that $S \cap H \subseteq \exp(W \cap \mathfrak{h}) \cup \mathrm{Irr}(H)$. But H is abelian, so $\mathrm{Irr}(H) = \emptyset$, hence $S \cap H \subseteq \exp(W \cap \mathfrak{h})$. The reverse inclusion $\exp(W \cap \mathfrak{h}) \subseteq S \cap H$ is obvious. □

5. Interior Points in the Tangent Wedge

The main result of this section is the 'Intrinsic Embedding Theorem,' which states that if W is the Lie wedge of a weakly exponential subsemigroup S then $W - W$ is a Lie algebra and S is also a weakly exponential subsemigroup —in particular, a closed subsemigroup— of the analytic subgroup corresponding to $W - W$, endowed with its intrinsic Lie group topology. Thus for most of the problems we are concerned with we may restrict our attention to the case where the Lie wedge W of a given weakly exponential or exponential subsemigroup S has inner points in the Lie algebra $\mathfrak{g}$. (Recall that W has inner points in $\mathfrak{g}$ if and only if W is generating in $\mathfrak{g}$, i.e., $\mathfrak{g} = W - W$.) In the proof of the 'Intrinsic Embedding Theorem' we use another result of independent interest, the 'Interior Point Theorem' which states (among other things) that the Lie wedge W of a closed subsemigroup S has inner points if $\overline{\exp W}$ has inner points in G.

Our first aim is to show that any wedge W in the Lie algebra $\mathfrak{g}$ contains interior points if its exponential image $\exp W$ contains interior points or if $\overline{\exp W}$ has interior points and W is the Lie wedge of a closed subsemigroup S. This result seems to be of interest also outside the context of exponential semigroups and divisibility.

5.1. LEMMA. *Set $H = \exp \mathfrak{h}$ for a Cartan subalgebra $\mathfrak{h}$ of $\mathfrak{g}$. Suppose that W is a wedge in $\mathfrak{g}$ and that $\mathrm{int}_H(H \cap \exp W) \neq \emptyset$. Then the wedge $\mathfrak{h} \cap W$ has inner points in $\mathfrak{h}$.*

Proof. We first prove that $\exp(W \cap \mathfrak{h})$ has non-empty interior in H. Since $\mathrm{int}_H(H \cap \exp W) \neq \emptyset$, we conclude that also $\mathrm{int}_H(H \cap \exp(\mathrm{REG}\,\mathfrak{g}) \cap \exp W) \neq \emptyset$. (Note that the set $\mathrm{REG}\,\mathfrak{g} = \mathrm{reg}\exp \cap \mathrm{reg}\,\mathfrak{g}$ intersects each Cartan subalgebra in an open and dense subset.) We claim that $\mathrm{int}_H(H \cap \exp(\mathrm{REG}\,\mathfrak{g}) \cap \exp W) \subseteq \exp(W \cap \mathfrak{h})$. Indeed, let $g \in \mathrm{int}_H(H \cap \exp(\mathrm{REG}\,\mathfrak{g}) \cap \exp W)$. Then there exists a point $X \in \mathfrak{h} \cap \mathrm{REG}\,\mathfrak{g}$ and a point $Y \in W$ with $\exp X = \exp Y \in \mathrm{int}_H(H \cap \exp(\mathrm{REG}\,\mathfrak{g}))$. By Lemma 3.5 this implies that $[X, Y] = 0$, so $Y \in W \cap \mathcal{N}(\mathrm{ad}\,X) = W \cap \mathfrak{h}$ and therefore $g \in \exp(W \cap \mathfrak{h})$, which establishes our claim and shows that $\mathrm{int}_H \exp(W \cap \mathfrak{h}) \neq \emptyset$.

Now we endow $\mathfrak{h}$ with the Campbell-Hausdorff multiplication $*$. Then the exponential function $\exp_H : \mathfrak{h} \to H$ is a covering homomorphism and $\exp_H^{-1}(\mathbf{1})$ is a lattice Γ in the center $\mathfrak{z}(\mathfrak{h})$ of $\mathfrak{h}$. For $Z \in \mathfrak{z}(\mathfrak{h})$ and $X \in \mathfrak{h}$ we have $X * Z = X + Z$.

Since $\text{int}_H \exp(W \cap \mathfrak{h}) \neq \emptyset$, the inverse image $(\exp_H)^{-1}\exp(W \cap \mathfrak{h}) = \Gamma + (W \cap \mathfrak{h})$ has interior points in $\mathfrak{h}$. Thus there exists an open subset of $\mathfrak{h}$ which is covered by the countably many closed sets $\gamma + (W \cap \mathfrak{h})$, with $\gamma \in \Gamma$. By Baire's Category Theorem, one of the sets $\gamma + (W \cap \mathfrak{h})$ has inner points in $\mathfrak{h}$. Hence $W \cap \mathfrak{h}$ has inner points in $\mathfrak{h}$. □

5.2. LEMMA. *Set $H = \exp \mathfrak{h}$ for a Cartan subalgebra $\mathfrak{h}$ of $\mathfrak{g}$. Suppose that W is the Lie wedge of a closed subsemigroup S of G and that $\text{int}_H(H \cap \overline{\exp W}) \neq \emptyset$. Then $\text{int}_H(H \cap \exp W) \neq \emptyset$ and therefore the wedge $\mathfrak{h} \cap W$ has inner points in $\mathfrak{h}$.*

Proof. Since $\text{int}_H(H \cap \overline{\exp W}) \neq \emptyset$, we conclude that $\text{int}_H(\exp(\mathfrak{h} \cap \text{REG}\,\mathfrak{g}) \cap \overline{\exp W}) \neq \emptyset$— to see this we only have to recall that REG $\mathfrak{g}$ meets every Cartan subalgebra in an open and dense subset, and that $\exp^{-1} \text{Reg}\, G = \text{REG}\,\mathfrak{g}$ (cf. 3.3(iii)). It therefore suffices to show that $\text{int}_H(\exp(\mathfrak{h} \cap \text{REG}\,\mathfrak{g}) \cap \overline{\exp W}) \subseteq \exp W$.

Let g be a point in $\text{int}_H(\exp(\mathfrak{h} \cap \text{REG}\,\mathfrak{g}) \cap \overline{\exp W})$. Then g is regular and can be written as the limit of a sequence $(\exp Z_n)$ with $Z_n \in W$, and such that also the Γ-limit $\lim_\Gamma \overline{\exp \mathbb{R}^+ \cdot Z_n}$ exists. Since S is closed this limit must be contained in S, thus its Lie wedge $\mathfrak{L}(\lim_\Gamma \overline{\exp \mathbb{R}^+ \cdot Z_n})$ is contained in $W = \mathfrak{L}(S)$. We apply Lemma 3.10 and conclude that $g = \exp X$ for some $X \in W$, that is, $g \in \exp W$. □

5.3. THEOREM. ('The Interior Point Theorem') *Let W be a wedge in the Lie algebra $\mathfrak{g}$ of a connected Lie group G. Then W has interior points in $\mathfrak{g}$ if one of the following conditions holds:*
(a) *The interior of $\exp W$ in G is nonvoid.*
(b) *W is the Lie wedge of a closed subsemigroup S of G and the interior of $\overline{\exp W}$ in G is nonvoid.*

Proof. Since $\exp \text{REG}\,\mathfrak{g}$ is open in G and dense in $\exp \mathfrak{g}$ we can find a point $Z \in \text{REG}\,\mathfrak{g}$ and a neighborhood U_0 of Z in $\mathfrak{g}$ such that $\exp U_0 \subseteq \text{int}\exp W$, in case (a), and $\exp U_0 \subseteq \text{int}\overline{\exp W}$ in case (b). Let $\mathfrak{h}$ be the Cartan algebra generated by Z and let $\mathfrak{e}$ denote a vector space complement of $\mathfrak{h}$ in $\mathfrak{g}$ so that $\mathfrak{g} = \mathfrak{e} \oplus \mathfrak{h}$.

Consider the function $F_Z \colon \mathfrak{g} = \mathfrak{e} \oplus \mathfrak{h} \to \mathfrak{g}$ given by $F_Z(X \oplus Y) = e^{\text{ad}\,X}(Z + Y)$. Note that $F_Z(s \cdot X, t \cdot Y) = Z + t \cdot Y + s \cdot [X, Z] + R(s, t)$, with a remainder $R(s, t)$ whose summands are at least quadratic in s and t, thus $dF_Z(0) = \text{pr}_\mathfrak{h} - (\text{ad}\,Z) \circ \text{pr}_\mathfrak{e}$. Since Z is regular, the restriction of $\text{ad}\,Z$ to $\mathfrak{e} \cong \mathfrak{g}/\mathfrak{h}$ is a nonsingular linear map and therefore $dF_Z(0)$ is nonsingular. Hence there is a compact neighborhood $V_\mathfrak{e}$ of 0 in $\mathfrak{e}$ and a compact neighborhood $V_\mathfrak{h}$ of Z in $\mathfrak{h}$ such that the function

$$V_\mathfrak{e} \times V_\mathfrak{h} \to \mathfrak{g}, \quad (X, Y) \mapsto e^{\text{ad}\,X} Y$$

maps $V_\mathfrak{e} \times V_\mathfrak{h}$ diffeomorphically onto a compact neighborhood

$$U_1 = \bigcup_{X \in V_\mathfrak{e}} \exp(e^{\operatorname{ad} X} V_\mathfrak{h}) = \exp(e^{\operatorname{ad} V_\mathfrak{e}} V_\mathfrak{h})$$

of Z in U_0. Note that U_1 is contained in $\operatorname{int} \exp W$, respectively $\operatorname{int} \overline{\exp W}$.

Now we observe that for every $X \in V_\mathfrak{e}$ the interior of $\exp(e^{\operatorname{ad} X} \mathfrak{h} \cap W)$ with respect to $\exp(e^{\operatorname{ad} X} \mathfrak{h})$ is nonvoid. This is trivial in case (a), it follows from 5.2 in case (b).

Thus by Lemma 5.1, $e^{\operatorname{ad} X} \mathfrak{h} \cap W$ has inner points with respect to $e^{\operatorname{ad} X} \mathfrak{h}$ and therefore $(e^{\operatorname{ad} X} \mathfrak{h} \cap W) - (e^{\operatorname{ad} X} \mathfrak{h} \cap W) = e^{\operatorname{ad} X} \mathfrak{h}$ for each $X \in V_\mathfrak{e}$ implies that

$$U_1 = e^{\operatorname{ad} V_\mathfrak{e}} V_\mathfrak{h} \subseteq \bigcup_{X \in V_\mathfrak{e}} e^{\operatorname{ad} X} \mathfrak{h} = \bigcup_{X \in V_\mathfrak{e}} (e^{\operatorname{ad} X} \mathfrak{h} \cap W) - (e^{\operatorname{ad} X} \mathfrak{h} \cap W) \subseteq W - W.$$

The vector space $W - W$ therefore contains an open set of $\mathfrak{g}$, and thus agrees with $\mathfrak{g}$. The assertion follows. $\square$

5.4. EXERCISE. Give an example showing that the assumption "W is the Lie wedge of a closed subsemigroup" cannot be dispensed with in condition (ii) above.

We now approach the 'Intrinsic Embedding Theorem,' as announced in the beginning of this section. The following Lemma recalls a few basic facts recorded in [27] (p. 382, and p. 377, V.1.10).

5.5. LEMMA. *Suppose that W is a Lie wedge in the Lie algebra $\mathfrak{g}$ of a connected Lie group G. Then the following assertions are equivalent:*
   (i) *$\mathfrak{g}$ is generated by W as a Lie algebra;*
   (ii) *$\exp W$ generates G as a group;*
   (iii) *The subsemigroup S generated by $\exp W$ has nonvoid interior in G.*
   (iv) *The interior of the subsemigroup S generated by $\exp W$ is dense in S.* $\square$

In order to handle the somewhat delicate case of a weakly exponential rather than an exponential subsemigroup we need "many" G-regular points in the analytic subgroup corresponding to the Lie algebra $\langle\langle W \rangle\rangle$ generated by W. These are provided by our next two lemmas.

5.6. LEMMA. *Let A be a not necessarily connected Lie group and suppose that $\varphi: A \to G$ is a continuous and injective homomorphism into a connected Lie group G such that $\varphi(A)$ is dense in G. Then $\varphi^{-1}(\operatorname{Reg} G)$ is open and dense in A.*

Proof. We write $\mathfrak{g} = \mathfrak{L}(G)$. Define a representation ρ of A on $\mathfrak{g}$ by

$$\rho: A \to \operatorname{Aut} \mathfrak{g}, \quad a \mapsto \operatorname{Ad}(a).$$

Let r denote the rank of $\mathfrak{g}$ and r_ρ the rank of the representation ρ. Then the set Reg_ρ of all elements $a \in A$ which are regular with respect to the representation ρ (i.e., such that the nilspace of $\rho(a) - 1$ has the minimal dimension r_ρ) is open and dense in A (cf. BOURBAKI [5] VII, §4, Prop 4.1, p. 33f). Furthermore, $\varphi(A) \cap \mathrm{Reg}\, G \neq \emptyset$, since $\varphi(A)$ must meet the open dense subset $\mathrm{Reg}\, G$ of G. It follows that $r_\rho = r$, hence every regular element in A with respect to the representation ρ is a regular element of G. This establishes the assertion. □

Another lemma will be helpful in the proof of the main result:

5.7. LEMMA. *Suppose that G is a connected Lie group and A a dense analytic subgroup. Write A_{Lie} for the same group A, but endowed with its intrinsic Lie group topology. Then*
  (i) *there is a unique Lie group topology on the group G making it into a Lie group G^* such that A_{Lie} is an open subgroup;*
  (ii) *$\mathrm{Reg}\, G$ is open and dense in G^*.*

Proof. (i) As a dense analytic subgroup, A is normal in G. For $g \in G$ let $I_g: A \to A$ denote the inner automorphism induced by g on A via $I_g(a) = gag^{-1}$. The exponential function $\exp_{A_{\mathrm{Lie}}} = \exp_G |A: \mathfrak{a} \to A_{\mathrm{Lie}}$ maps each sufficiently small open zero neighborhood of $\mathfrak{a}$ homeomorphically onto an open identity neighborhood of A_{Lie}. The relation

$$I_g \circ \exp_{A_{\mathrm{Lie}}} = I_g \circ (\exp_G |\mathfrak{a}) = (\exp_G |\mathfrak{a}) \circ \mathrm{Ad}(g)$$

then shows that $I_g: A_{\mathrm{Lie}} \to A_{\mathrm{Lie}}$ is continuous at $\mathbf{1}$ and thus is continuous. Write $\mathcal{U}$ for the filter of identity neighborhoods of A_{Lie}. We have just seen that $I_g(\mathcal{U}) = \mathcal{U}$, thus the filter $\langle \mathcal{U} \rangle$ generated by $\mathcal{U}$ in G is the filter of a group topology making G into a topological group G^* which induces on A the topology of A_{Lie} and relative to which A_{Lie} is open in G^*. Obviously there is only one such topology.

(ii) We apply Lemma 5.6 to the identity morphism $\varphi: G^* \to G$ and derive the assertion. □

5.8. THEOREM. ('The Intrinsic Embedding Theorem') *Suppose that G is a Lie group and that S is a closed subsemigroup with Lie wedge W. We write A for the analytic group generated in G by $\exp W$ and A_{Lie} for the same group, but endowed with its intrinsic Lie group topology. Further, let T denote the smallest closed subsemigroup of A_{Lie} containing $\exp W$. Then the following assertions hold:*
  (i) *If T is weakly exponential in A_{Lie} then S is weakly exponential in G.*
  (ii) *If S is weakly exponential then $S = T$, and T is weakly exponential in A_{Lie}. Also, $W - W = \mathfrak{a}$.*

Proof. (i) If T is weakly exponential in A_{Lie} then S is weakly exponential in G, since the map $A_{\mathrm{Lie}} \to G$, $a \mapsto a$, is continuous.

(ii) We assume that A is dense in G, this will obviously cause no loss of generality. Then A is normal in G because of $G' \subseteq A$. Suppose that S is weakly exponential in G and construct a Lie group G^* according to the preceding Lemma 5.7. By Lemma 5.5, T has dense interior with respect to A_{Lie} and thus with respect to G^* in view of 5.7(i). Let $s \in G$. Then sT has dense interior in G^* and $s = s1 \in sT$. We conclude that $S = \bigcup_{s \in S} sT$ has dense interior in G^*. Therefore, denoting with cl^* the closure and with int^* the interior in G^*, we conclude from Lemma 5.7(ii) that

$$(1) \qquad S = \text{cl}^*\big(\text{int}^*(\operatorname{Reg} G \cap S)\big).$$

On the other hand we have

$$(2) \qquad \operatorname{Reg} G \cap S \subseteq \exp W \subseteq T$$

by Theorem 3.11. But since T is closed in G^*, then (1) and (2) imply that $S \subseteq T$ and that $\exp W = \exp_{A_{\text{Lie}}} W$ is dense in T with respect to A_{Lie}. Now we have that $S = T$ and that T is weakly exponential in A_{Lie}, as asserted.

It remains to show that $W - W = \mathfrak{a}$. But this follows from the Interior Point Theorem 5.3: Note first that its condition (ii) is satisfied, with T in place of S and A_{Lie} in place of G. Then we conclude that W has inner points in $\mathfrak{L}(A_{\text{Lie}}) = \mathfrak{a}$ and therefore $\mathfrak{a} = W - W$. □

In general the analytic subgroup A generated by a weakly exponential subsemigroup S need not be closed. Here is an example where S is isomorphic with the Master Example $\text{Sl}(2)^+$.

5.9. EXAMPLE. Let $G = \frac{\widetilde{\text{Sl}}(2,\mathbb{R}) \times \mathbb{R}}{\Delta}$ where $\Delta = \{(z^{m+n}, n - m\sqrt{2}) \mid m, n \in \mathbb{Z}\}$, with z a generator of the center of $\widetilde{\text{Sl}}(2, \mathbb{R})$. We identify the Lie algebra $\mathfrak{g}$ of G with $\mathfrak{sl}(2, \mathbb{R}) \times \mathbb{R}$. Now $A = \frac{(\widetilde{\text{Sl}}(2,\mathbb{R}) \times \{0\})\Delta}{\Delta}$ is a dense analytic subgroup of G with Lie algebra $\mathfrak{a} = \mathfrak{sl}(2,\mathbb{R}) \times \{0\}$. The wedge $W = \mathfrak{sl}(2,\mathbb{R})^+ \times \{0\}$ is a Lie semialgebra in $\mathfrak{g}$ and $S := \exp W$ is closed, since $\text{Sl}(2)^+$ is absolutely closed (as we shall prove in V.2.1 below). Thus S is a closed exponential subsemigroup of G but generates the non-closed analytic subgroup A. □

By Theorem 5.8 we know that for a weakly exponential subsemigroup S of a Lie group G we have $S = \overline{\exp W}$ with respect both to the topology of G and the intrinsic topology of A_{Lie}. But we do not yet know, however, whether also the *topologies* induced on S by G and A_{Lie} are the same. Using the familiar arguments from the theory of topological groups it is not difficult to show, even under much more general assumptions, that there exists a dense open subset O of S with respect to the topology induced by G such that the two topologies agree on O.

We shall show now that the two topologies actually coincide, the proof uses arguments well in line with the above ones, but it depends on a result in the final chapter which says that if W is the Lie wedge of a weakly exponential subsemigroup and has inner points in $\mathfrak{g}$ then the set $\operatorname{comp}\mathfrak{g} \cap H(W)$ of compact elements in $H(W)$ form a subset of the largest ideal of $\mathfrak{g}$ which is contained in $H(W)$. Since every ideal of $\mathfrak{a}$ is an ideal also in $\mathfrak{g}$ (as A is dense in G) this implies in particular that condition (coinv) in the following proposition is satisfied for arbitrary weakly exponential semigroups.

5.10. PROPOSITION. *Suppose that in the situation of Theorem 5.8 the following condition is satisfied:*

(coinv) $\operatorname{comp}_G(\mathfrak{g}) \cap W = \operatorname{comp}_G(\mathfrak{g}) \cap H(W)$ *is invariant under all inner automorphisms of* $\mathfrak{g}$.

Then G and A_{Lie} induce the same topology on the semigroup S.

Proof. We first remark that they induce the same topology on the group $S \cap S^{-1}$ of units of S. (In both topologies the group of units is locally compact and connected, thus σ-compact. Hence the standard Baire technique applies and proves the claim.)

Now define J to be the set of all $x \in S$ such that the topologies induced by A_{Lie} and G differ at x, that is to say, such that there exists a sequence (x_n) in S with $x_n \to x$ in G but not in A_{Lie}. Obviously, for such x we may choose (x_n) so that it has no convergent subsequence in A_{Lie}, A_{Lie}-$\lim x_n = \infty$. We have to show that J is empty.

Suppose, to the contrary, that $J \ne \emptyset$. Then J is an ideal since A_{Lie}-$\lim x_n = \infty$ implies A_{Lie}-$\lim sx_n = A_{\mathrm{Lie}}$-$\lim x_n s = \infty$ for all $s \in S$. Since every ideal in S has dense interior with respect to the topology induced by A_{Lie} we see from (1) that there exists a G-regular point $g \in J$. We pick a sequence (g_n) in S with G-$\lim g_n = g$ but A_{Lie}-$\lim g_n = \infty$ and derive a contradiction.

Since g is a regular point in G we may assume that $g = \exp w_0$ and $g_n = \exp w_n$, $n = 1, 2 \ldots$, for suitable $w_0, w_n \in W$. As in the proof of Lemma 3.10 we choose a sequence $a_n \in \mathfrak{a}(g)$ such that $(e^{\operatorname{ad} a_n})$ converges (uniformly on compact subsets) to the identity, and such that $v_n = e^{\operatorname{ad} a_n} w_n \in \mathfrak{h}(g)$ for all n. Note that $e^{\operatorname{ad} a_n} w_n \in \mathfrak{a}$, since $\mathfrak{a}$ is an ideal of $\mathfrak{g}$. For convenience we introduce a euclidean inner product on $\mathfrak{h}(g)$. Furthermore, we endow $\mathfrak{h}(g)$ with the Campbell-Hausdorff multiplication $*$ and write $\kappa \colon (\mathfrak{h}(g), *) \to H(g)$, $h \mapsto \exp h$ for the natural covering morphism.

Write $\mathfrak{k} = H(W) \cap \operatorname{comp}_G(\mathfrak{g}) \cap \mathfrak{h}$ and note that $\mathfrak{k}$ is central in the nilpotent Lie algebra $\mathfrak{h}(g)$, so $\exp \mathfrak{k}$ is a torus group (with respect to both topologies). Then we decompose v_n as $v_n = v_n^{(1)} + v_n^{(2)}$ with $v_n^{(1)} \in \mathfrak{k}$ and $v_n^{(2)} \in \mathfrak{k}^\perp$.

By passing to a suitable subsequence and renaming we enforce that the sequence $(\exp v_n^{(1)})$ converges in $\exp \mathfrak{k}$, hence in A_{Lie}. Since every convergent sequence in $H(g)$ is the κ-image of a convergent sequence in $\mathfrak{h}(g)$ there are elements $z_n^{(1)} \in \ker \kappa \cap \mathfrak{k}$ and $z_n^{(2)} \in \ker \kappa$ such that

$$\lim v_n^{(1)} + z_n^{(1)} = w_0^{(1)}, \quad \text{and} \quad \lim v_n^{(2)} + z_n^{(2)} = w_0 - w_0^{(1)},$$

for some $w_0^{(1)} \in \mathfrak{k} \subseteq H(W)$. We claim that the sequence $(v_n^{(2)})$ is bounded.

Suppose that it is not. After passing to a subsequence and renaming, we may assume that $\lim \|v_n^{(2)}\| = \infty$ and that $v = \lim \frac{1}{\|v_n^{(2)}\|} v_n^{(2)}$ exists. By its definition,

$\|v\| = 1$, so $v \neq 0$. Now $\lim \frac{1}{\|v_n^{(2)}\|}(v_n^{(2)} + z_n^{(2)}) = 0$ and $\mathbb{R} \cdot z_n^{(2)} \subseteq \operatorname{span}_\mathbb{R}(\ker \kappa) \subseteq \operatorname{comp}_G(\mathfrak{g})$. Therefore,

$$v = \lim \frac{1}{\|v_n^{(2)}\|} v_n^{(2)} = \lim \frac{1}{\|v_n^{(2)}\|} z_n^{(2)} \in \operatorname{comp}_G(\mathfrak{g}) \cap \mathfrak{h}(g) \cap \mathfrak{k}^\perp.$$

Also, we recall that by assumption the set $\operatorname{comp}_G(\mathfrak{g}) \cap H(W)$ is a subset of the largest ideal of g contained in $H(W)$, so $e^{-\operatorname{ad} a_n} v_n^{(1)}$ lies in $H(W)$ for all n. Thus

$$\begin{aligned}
v &= \lim \frac{1}{\|v_n^{(2)}\|} v_n^{(2)} \\
&= \lim \frac{1}{\|v_n^{(2)}\|} e^{-\operatorname{ad} a_n}(v_n - v_n^{(1)}) \\
&= \lim \frac{1}{\|v_n^{(2)}\|}(w_n - e^{-\operatorname{ad} a_n} v_n^{(1)}) \in W.
\end{aligned}$$

But since $v \in \operatorname{comp}_G(\mathfrak{g})$ this means that $v \in H(W) \cap \mathfrak{h}(g) \cap \operatorname{comp}_G(\mathfrak{g}) = \mathfrak{k}$, so $v \in \mathfrak{k} \cap \mathfrak{k}^\perp = \{0\}$, a contradiction.

Thus we have shown that $(v_n^{(2)})$ is bounded. Hence the sequence $(\exp v_n) = (\exp v_n^{(1)} \exp v_n^{(2)})$ is bounded, contradicting our supposition that $(\exp w_n)$ and hence $(\exp v_n)$ is unbounded. This finishes the proof. $\square$

In Chapter VI we shall use this proposition to show that in Theorem 5.8 the topologies induced on S by G and A_{Lie} actually coincide.

CHAPTER 3

WEYL GROUPS AND FINITENESS PROPERTIES OF CARTAN SUBALGEBRAS

Let $\mathfrak{e}$ be a compactly embedded proper subalgebra of a real Lie algebra $\mathfrak{g}$, i.e., $\langle e^{\operatorname{ad}\mathfrak{e}} \rangle$ is contained in a compact subgroups of Aut $\mathfrak{g}$. For $x \in \mathfrak{g}$ let $\mathfrak{z}(x)$ denote the centralizer of x in $\mathfrak{g}$. In the process of our later investigations we must consider the set

$$\mathfrak{z}(\mathfrak{e}) = \{x \in \mathfrak{g} \mid (\exists y \in \mathfrak{e} \setminus \{0\}) : [x, y] = 0\} = \bigcup_{0 \neq y \in \mathfrak{e}} \mathfrak{z}(y) = \{x \in \mathfrak{g} \mid \mathfrak{z}(x) \cap \mathfrak{e} \neq \{0\}\}.$$

We will observe later that $\mathfrak{z}(\mathfrak{e})$ is closed in the ZARISKI-topology of $\mathfrak{g}$. In particular, as an algebraic variety, it has a well defined dimension. This entire chapter is devoted to a proof of the estimate $\dim \mathfrak{z}(\mathfrak{e}) \leq \dim \mathfrak{g} - 2$.

In the course of this work we produce several results of independent interest. For example we define a Weyl group $\mathcal{W}(\mathfrak{g}, \mathfrak{h})$ attached to a Cartan subalgebra $\mathfrak{h}$ of an *arbitrary* real or complex Lie algebra $\mathfrak{g}$ and show that is finite.

1. Cartan Dense Ideals

In this section we consider the structure of Lie algebra pairs $(\mathfrak{m}, \mathfrak{g})$ such that $\mathfrak{g}$ is an ideal of $\mathfrak{m}$ and $\mathfrak{m}/\mathfrak{g}$ is abelian. This is readily seen to be equivalent to the condition that $\mathfrak{g}$ is a subspace of the Lie algebra $\mathfrak{m}$ and $[\mathfrak{m}, \mathfrak{m}] \subseteq \mathfrak{g}$.

Some notational conventions first: If $\mathfrak{k}$ is a subalgebra of a Lie algebra, and $\mathfrak{a}$ is any subset of $\mathfrak{g}$, then $\mathfrak{n}(\mathfrak{k}, \mathfrak{a})$ denotes the normalizer of $\mathfrak{k}$ in $\mathfrak{a}$ and $\mathfrak{z}(\mathfrak{k}, \mathfrak{a})$ the centralizer of $\mathfrak{k}$ in $\mathfrak{a}$. For a subset $\mathfrak{h}$ of a Lie algebra $\mathfrak{m}$ we let $\mathfrak{m}^0(\mathfrak{h}) = \{Y \in \mathfrak{m} \mid (\forall X \in \mathfrak{h})(\exists n \in \mathbb{N})(\operatorname{ad} X)^n Y = 0\}$ denote the nilspace of ad $\mathfrak{h}$ on $\mathfrak{m}$. We write $\mathfrak{m}^0(x)$ in place of $\mathfrak{m}^0(\{x\})$.

For a Lie algebra $\mathfrak{g}$, as in II.3.2, we write again reg($\mathfrak{g}$) for the set of regular elements in $\mathfrak{g}$. We recall that one characterization of a regular element x is that $\operatorname{ad}(x)^n$ has maximal rank among all linear operators $\operatorname{ad}(y)$, $y \in \mathfrak{g}$ for $n \geq \dim \mathfrak{g}$.

1.1. LEMMA. *Suppose that $\mathfrak{h} \subseteq \mathfrak{g} \subseteq \mathfrak{m}$ such that*
(a) *$\mathfrak{h}$ is a Cartan algebra of $\mathfrak{g}$,*
(b) *$[\mathfrak{m}, \mathfrak{m}] \subseteq \mathfrak{g}$.*
Then the following conclusions hold:
  (i) *$\mathfrak{m} = \mathfrak{m}^0(\mathfrak{h}) \oplus \mathfrak{g}^+$ is the Fitting decomposition of $\mathfrak{m}$ with respect to ad $\mathfrak{h}$, where $\mathfrak{g}^+$ is the Fitting one-component of ad $\mathfrak{h}$ on $\mathfrak{g}$.*
  (ii) *$\mathfrak{m}^0(\mathfrak{h})$ is a solvable subalgebra of $\mathfrak{m}$ and $[\mathfrak{m}^0(\mathfrak{h}), \mathfrak{m}^0(\mathfrak{h})] \subseteq \mathfrak{h}$.*

(iii) *If $\mathfrak{k}$ is a subalgebra of $\mathfrak{m}^0(\mathfrak{h})$, then the normalizer of $\mathfrak{k}$ in $\mathfrak{m}$ is $\mathfrak{n}(\mathfrak{k}, \mathfrak{m}^0(\mathfrak{h})) \oplus \mathfrak{z}(\mathfrak{k}, \mathfrak{g}^+)$.*

(iv) *If $\mathfrak{k}$ is a vector subspace of $\mathfrak{m}^0(\mathfrak{h})$ containing $\mathfrak{h}$, then the normalizer of $\mathfrak{k}$ in $\mathfrak{m}$ is $\mathfrak{m}^0(\mathfrak{h})$. In particular, $\mathfrak{m}^0(\mathfrak{h})$ is its own normalizer.*

(v) *If $\mathfrak{k}$ is a Cartan algebra of $\mathfrak{m}$ containing $\mathfrak{h}$, then $\mathfrak{k} = \mathfrak{m}^0(\mathfrak{h})$ and $\mathfrak{m} = \mathfrak{k} \oplus \mathfrak{g}^+$.*

(vi) *$\mathfrak{m}^0(\mathfrak{h})$ is nilpotent if and only if $\mathfrak{m}^0(\mathfrak{h})$ is a Cartan algebra of $\mathfrak{m}$ containing $\mathfrak{h}$.*

Proof. (i) In order to simplify notation in this proof we shall abbreviate $\mathfrak{m}^0(\mathfrak{h})$ by $\mathfrak{m}^0$. Let $\mathfrak{m}^+$ denote the Fitting one-component of ad $\mathfrak{h}$ on $\mathfrak{m}$. Then $\mathfrak{m}^+ = [\mathfrak{h}, \mathfrak{m}^+] \subseteq [\mathfrak{m}, \mathfrak{m}] \subseteq \mathfrak{g}$. Then the modular law implies $\mathfrak{g} = \mathfrak{m} \cap \mathfrak{g} = (\mathfrak{m}^0 \oplus \mathfrak{m}^+) \cap \mathfrak{g} = (\mathfrak{m}^0 \cap \mathfrak{g}) \oplus \mathfrak{m}^+$. Now $\mathfrak{m}^0 \cap \mathfrak{g}$ is the nilspace of ad $\mathfrak{h}$ on $\mathfrak{g}$, therefore agrees with $\mathfrak{h}$. Hence $\mathfrak{m}^+$ is the Fitting one-component $\mathfrak{g}^+$ of ad $\mathfrak{h}$ on $\mathfrak{g}$. Thus $\mathfrak{m} = \mathfrak{m}^0 \oplus \mathfrak{g}^+$ is the Fitting decomposition of $\mathfrak{m}$ with respect to ad $\mathfrak{h}$, and $[\mathfrak{m}^0, \mathfrak{g}^+] = \mathfrak{g}^+$. Also we note

(*) $$\mathfrak{g} \cap \mathfrak{m}^0 = \mathfrak{h}.$$

(ii) Using (*) we observe $[\mathfrak{m}^0, \mathfrak{m}^0] \subseteq [\mathfrak{m}, \mathfrak{m}] \cap \mathfrak{m}^0 \subseteq \mathfrak{g} \cap \mathfrak{m}^0 = \mathfrak{h}$. Since $\mathfrak{h}$ is nilpotent, this implies that $\mathfrak{m}^0$ is solvable.

(iii) Let $\mathfrak{k}$ be a subalgebra of $\mathfrak{m}^0$. Let $n \in \mathfrak{m}$ denote an element of the normalizer $\mathfrak{n}(\mathfrak{k}, \mathfrak{m})$ of $\mathfrak{k}$ in $\mathfrak{m}$. Then $n = m^0 + g^+$ with $m^0 \in \mathfrak{m}^0$ and $g^+ \in \mathfrak{g}^+$ by (i). Now $[\mathfrak{k}, g^+] \subseteq [\mathfrak{k}, n] - [\mathfrak{k}, m^0] \subseteq \mathfrak{k} + \mathfrak{m}^0 \in \mathfrak{m}^0$, but also $[\mathfrak{k}, g^+] \in \mathfrak{g}^+$, and thus $\mathfrak{m}^0 \cap \mathfrak{g}^+ = \{0\}$ implies $[\mathfrak{k}, g^+] = \{0\}$, i.e., $g^+ \in \mathfrak{z}(\mathfrak{k}, \mathfrak{g}^+)$. Then $[\mathfrak{k}, m^0] = [\mathfrak{k}, n] \subseteq \mathfrak{k}$, whence m^0 is in the normalizer of $\mathfrak{k}$ in $\mathfrak{m}^0$. Thus $m^0 \in \mathfrak{n}(\mathfrak{k}, \mathfrak{m}^0)$. Hence $\mathfrak{n}(\mathfrak{k}, \mathfrak{m}) \subseteq \mathfrak{n}(\mathfrak{k}, \mathfrak{m}^0) \oplus \mathfrak{z}(\mathfrak{k}, \mathfrak{g}^+)$, and the reverse inclusion is obvious.

(iv) We have $[\mathfrak{m}^0, \mathfrak{m}^0] \subseteq \mathfrak{h}$. Hence every vector subspace $\mathfrak{k}$ with $\mathfrak{h} \subseteq \mathfrak{k} \subseteq \mathfrak{m}^0$ is an ideal in $\mathfrak{m}^0$. Thus $\mathfrak{m}^0 \subseteq \mathfrak{n}(\mathfrak{k}, \mathfrak{m})$. But $\mathfrak{n}(\mathfrak{k}, \mathfrak{m})$ is an $\mathfrak{h}$-module containing $\mathfrak{m}^0$. Therefore its Fitting one-component is $\mathfrak{g}^+ \cap \mathfrak{n}(\mathfrak{k}, \mathfrak{m})$ and we have $\mathfrak{n}(\mathfrak{k}, \mathfrak{m}) = \mathfrak{m}^0 \oplus (\mathfrak{g}^+ \cap \mathfrak{n}(\mathfrak{k}, \mathfrak{m}))$. Let $x \in \text{reg}(\mathfrak{g}) \cap \mathfrak{h}$. Then $\text{ad } x|\mathfrak{g}^+$ is injective. Thus $\mathfrak{g}^+ \cap \mathfrak{n}(\mathfrak{k}, \mathfrak{m}) = [x, \mathfrak{g}^+ \cap \mathfrak{n}(\mathfrak{k}, \mathfrak{m})] \subseteq [x, \mathfrak{n}(\mathfrak{k}, \mathfrak{m})] \subseteq \mathfrak{k}$. Hence $\mathfrak{g}^+ \cap \mathfrak{n}(\mathfrak{k}, \mathfrak{m}) \subseteq \mathfrak{g}^+ \cap \mathfrak{k} = \{0\}$ and consequently $\mathfrak{n}(\mathfrak{k}, \mathfrak{m}) \subseteq \mathfrak{m}^0$.

(v) If $\mathfrak{k}$ is a nilpotent algebra containing $\mathfrak{h}$, then $\mathfrak{k} \subseteq \mathfrak{m}^0$. If $\mathfrak{k}$, in addition is a Cartan algebra of $\mathfrak{m}$, then $\mathfrak{n}(\mathfrak{k}, \mathfrak{m}) = \mathfrak{k}$. By (iv), $\mathfrak{k} = \mathfrak{m}^0$ follows.

(vi) By (iv), the algebra $\mathfrak{m}^0$ is its own normalizer. If it is nilpotent, then it is therefore a Cartan subalgebra. Conversely Cartan algebras are always nilpotent by definition. □

We are particularly interested in those situations arising in conditions (v) and (vi) of Lemma 1.1.

1.2. DEFINITION. A subalgebra $\mathfrak{g}$ of a Lie algebra $\mathfrak{m}$ is said to be *Cartan dense* in $\mathfrak{m}$ if (i) $[\mathfrak{m}, \mathfrak{m}] \subseteq \mathfrak{g}$, and (ii) there exists a Cartan algebra $\mathfrak{h}$ of $\mathfrak{g}$ such that $\mathfrak{m}^0(\mathfrak{h})$ is a Cartan subalgebra of $\mathfrak{m}$. □

If $\mathfrak{m}$ is a metabelian algebra with commutator algebra $\mathfrak{g} = \mathfrak{h} = [\mathfrak{m}, \mathfrak{m}]$, then we have the situation described in Lemma 1.1. An example is the Heisenberg algebra $\mathfrak{m}$ with center $\mathfrak{h}$. In this case the unique Cartan algebra of $\mathfrak{m}$ is $\mathfrak{m}$, containing $\mathfrak{h}$. Another example is the Lie algebra $\mathfrak{m}$ of the group of euclidean motions of the plane $\mathbb{R}^2 \rtimes \mathbb{R}$ with $\mathfrak{g} = \mathfrak{h} = \mathbb{R}^2 \times \{0\}$. In this case the Cartan algebras of $\mathfrak{m}$ are disjoint from $\mathfrak{h}$, so $\mathfrak{g}$ is not Cartan dense in $\mathfrak{m}$.

1.3. LEMMA. *Suppose that $\mathfrak{g} \subseteq \mathfrak{m}$ and $[\mathfrak{m}, \mathfrak{m}] \subseteq \mathfrak{g}$. If $x \in \mathrm{reg}(\mathfrak{g})$, then for $n \geq \dim \mathfrak{m}$*

$$\mathrm{rank}(\mathrm{ad}_\mathfrak{m} x)^n = \mathrm{rank}(\mathrm{ad}_\mathfrak{g} x)^n = \dim \mathfrak{g}^+.$$

Proof. (i) We let $\mathfrak{h} = \mathfrak{g}^0(x)$ denote the Cartan subalgebra generated by x. Then $[\mathfrak{m}^0(\mathfrak{h}), \mathfrak{m}^0(\mathfrak{h})] \subseteq \mathfrak{h}$ by Lemma 1.1(ii) and thus $\mathrm{ad}_\mathfrak{m} x|\mathfrak{m}^0(\mathfrak{h})$ is nilpotent. Let $\mathfrak{g}^+$ be chosen for $\mathfrak{h}$ as in Lemma 1.1. Since $(\mathrm{ad}_\mathfrak{m} x)|\mathfrak{g}^+ = \mathrm{ad}_\mathfrak{g} x|\mathfrak{g}^+$ is invertible, then Lemma 1.1(i) implies that $\mathrm{rank}(\mathrm{ad}_\mathfrak{m} x)^n = \mathrm{rank}(\mathrm{ad}_\mathfrak{g} x)^n$ as asserted. □

1.4. PROPOSITION. *Suppose that $\mathfrak{h} \subseteq \mathfrak{g} \subseteq \mathfrak{m}$ is such that $\mathfrak{h}$ is a Cartan algebra of $\mathfrak{g}$ and $[\mathfrak{m}, \mathfrak{m}] \subseteq \mathfrak{g}$. Then the following statements are equivalent:*
  (i) *$\mathfrak{h}$ is contained in a Cartan algebra of $\mathfrak{m}$.*
  (ii) *$\mathfrak{m}^0(\mathfrak{h})$ is nilpotent.*
  (iii) *$\mathfrak{m} = \mathfrak{g} + \mathfrak{b}$ and $\mathfrak{h} \subseteq \mathfrak{b}$ for some nilpotent subalgebra $\mathfrak{b}$ of $\mathfrak{m}$.*
  (iv) *$\mathrm{reg}(\mathfrak{g}) \subseteq \mathrm{reg}(\mathfrak{m})$.*
  (v) *$\mathrm{reg}(\mathfrak{m}) \cap \mathfrak{g} \neq \emptyset$.*
  (vi) *$\max\{\mathrm{rank}(\mathrm{ad}_\mathfrak{m} x)^n \mid x \in \mathfrak{m}\} = \dim \mathfrak{g}^+$ for $n \geq \dim \mathfrak{m}$.*
  (vii) *$\mathfrak{g}$ is Cartan dense in $\mathfrak{m}$.*

If these conditions are satisfied, $\mathfrak{m}^0(\mathfrak{h})$ is the unique Cartan algebra $\mathfrak{k}$ of $\mathfrak{m}$ containing $\mathfrak{h}$ and $\mathfrak{m} = \mathfrak{k} \oplus \mathfrak{g}^+$.

Proof. (i)⇒(ii): By Lemma 1.1(v) any Cartan algebra $\mathfrak{k}$ of $\mathfrak{m}$ containing $\mathfrak{h}$ agrees with $\mathfrak{m}^0(\mathfrak{h})$. Hence $\mathfrak{m}^0(\mathfrak{h})$ is nilpotent.

(ii)⇒(i): By Lemma 1.1(vi), if the algebra $\mathfrak{m}^0(\mathfrak{h})$ is nilpotent, it is a Cartan algebra.

(ii)⇒(iii): Immediate from Lemma 1.1(i).

(iii)⇒(ii): Since $\mathfrak{b}$ is nilpotent and contains $\mathfrak{h}$, $\mathfrak{b} \subseteq \mathfrak{m}^0(\mathfrak{h})$. The hypothesis together with Lemma 1.1(i) then yield that $\mathfrak{b} = \mathfrak{m}^0(\mathfrak{h})$.

The proof up to this point together with Lemma 1.1(v) shows that if the equivalent conditions (i)-(iii) are satisfied, then $\mathfrak{m}^0(\mathfrak{h})$ is the unique Cartan algebra $\mathfrak{k}$ of $\mathfrak{m}$ containing $\mathfrak{h}$, and that $\mathfrak{m} = \mathfrak{k} \oplus \mathfrak{g}^+$.

(ii)⇒(vi): The fact that $\mathfrak{m}^0(\mathfrak{h})$ is a Cartan subalgebra implies that there exists $x \in \mathfrak{m}^0(\mathfrak{h})$ such that x is regular in $\mathfrak{m}$. Then $\mathrm{ad}_\mathfrak{m} |\mathfrak{m}^0(\mathfrak{h})$ is nilpotent, and then by Lemma 1.1(i), $\mathrm{rank}(\mathrm{ad}_\mathfrak{m} x)^n \leq \dim \mathfrak{g}^+$, for $n \geq \dim \mathfrak{m}$. Since $(\mathrm{ad}_\mathfrak{m} x)^n$ has maximal rank if and only if x is regular, condition (vi) now follows from Lemma 1.3.

(vi)⇒(iv): Under hypothesis (vi), it follows directly from Lemma 1.3 that for a regular element x of $\mathfrak{g}$, the vector space endomorphism $(\mathrm{ad}_\mathfrak{m} x)^n$ of $\mathfrak{m}$ must have maximal rank for $n \geq \dim \mathfrak{m}$, and hence must be regular in $\mathfrak{m}$.

(iv)⇒(v): Trivial.

(v)⇒(vi): If $x \in \text{reg}(\mathfrak{m}) \cap \mathfrak{g}$, then by Lemma 1.3, $\text{rank}(\text{ad}_\mathfrak{m} x)^n \leq \dim \mathfrak{g}^+$, since the regular elements of $\mathfrak{g}$ must have maximal rank. Since $(\text{ad}_\mathfrak{m} x)^n$ must have maximal rank in $\mathfrak{m}$ (since x is regular), it again follows from Lemma 1.3 that it has rank $\dim \mathfrak{g}^+$.

(iv) ⇒(i): Since $\mathfrak{h}$ is a Cartan algebra of $\mathfrak{g}$ there is an $x \in \mathfrak{h} \cap \text{reg}(\mathfrak{g})$. Then $\mathfrak{h} = \mathfrak{g}^0(x)$. By (iv) we know that $x \in \text{reg}(\mathfrak{m})$. Hence $\mathfrak{m}^0(x)$ is the asserted Cartan subalgebra of $\mathfrak{m}$.

(iv)⇔(vii): This is immediate from Lemma 1.1(vi). □

1.5. COROLLARY. *If $\mathfrak{g}$ is Cartan dense in $\mathfrak{m}$, then for all Cartan subalgebras $\mathfrak{h}'$ of $\mathfrak{g}$, $\mathfrak{m}^0(\mathfrak{h}')$ is a Cartan subalgebra of $\mathfrak{m}$ and $\mathfrak{m} = \mathfrak{m}^0(\mathfrak{h}') \oplus \mathfrak{g}^+(\mathfrak{h}')$ is the corresponding Fitting decomposition of $\mathfrak{m}$ with respect to $\mathfrak{m}^0(\mathfrak{h}')$.*

Proof. One notes that conditions (iv)–(vi) of Proposition 1.4 are independent of the Cartan algebra chosen in $\mathfrak{g}$, and hence imply the concluding statement of the Proposition for any Cartan subalgebra of $\mathfrak{g}$. □

1.6. COROLLARY. *If $\mathfrak{g}$ is Cartan dense in $\mathfrak{m}$, and $\mathfrak{g} \subseteq \mathfrak{m}_1 \subseteq \mathfrak{m}$, then $\mathfrak{g}$ is Cartan dense in $\mathfrak{m}_1$.*

Proof. This follows easily from condition (2) of Proposition 1.4, since $\mathfrak{m}_1^0(\mathfrak{h}) \subseteq \mathfrak{m}^0(\mathfrak{h})$. □

2. Algebraic Hulls

Let $\mathfrak{g}$ be a finite-dimensional Lie algebra over the field $\mathbb{F} = \mathbb{R}$ or $\mathbb{F} = \mathbb{C}$.

2.1. DEFINITION. Let $\mathfrak{g} \to \mathfrak{gl}(n, \mathbb{F})$ be a faithful representation of $\mathfrak{g}$ (by ADO's Theorem such representations always exist), identify $\mathfrak{g}$ with its image, and let $\tilde{\mathfrak{g}}$ be the Lie algebra of the Zariski closure of the group generated by $\exp \mathfrak{g}$ in $\text{Gl}(n, \mathbb{F})$. For each subalgebra $\mathfrak{a}$ of $\mathfrak{g}$, let $\tilde{\mathfrak{a}}$ be the Lie algebra of the Zariski closure of the subgroup generated by $\exp \mathfrak{a}$. The assignment $\mathfrak{a} \mapsto \tilde{\mathfrak{a}}$ is called an *algebraic hull operator* on $\mathfrak{g}$. □

For fields of characteristic 0, the algebraic hull operator $\mathfrak{a} \to \tilde{\mathfrak{a}}$ from the subalgebras of $\mathfrak{g}$ to the subalgebras of $\tilde{\mathfrak{g}}$ satisfies, for all subalgebras $\mathfrak{a}, \mathfrak{b}$ and all subspaces $\mathfrak{c}, \mathfrak{d}$ of $\mathfrak{g}$, the following standard properties (see, for example Chapter II.7 of [1] or Chapter VIII of [29]):

(i) $\mathfrak{a} \subseteq \tilde{\mathfrak{a}} = \tilde{\tilde{\mathfrak{a}}}$;
(ii) $\mathfrak{a} \subseteq \mathfrak{b} \Rightarrow \tilde{\mathfrak{a}} \subseteq \tilde{\mathfrak{b}}$;
(iii) $[\mathfrak{a}, \mathfrak{a}] = [\mathfrak{a}, \mathfrak{a}]^\sim = [\tilde{\mathfrak{a}}, \tilde{\mathfrak{a}}]$;
(iv) $[\mathfrak{a}, \mathfrak{c}] \subseteq \mathfrak{d} \Rightarrow [\tilde{\mathfrak{a}}, \mathfrak{c}] \subseteq \mathfrak{d}$;
(v) If $[\mathfrak{a}, \mathfrak{b}] \subseteq \mathfrak{b}$, then $(\tilde{\mathfrak{a}} + \tilde{\mathfrak{b}})^\sim = \tilde{\mathfrak{a}} + \tilde{\mathfrak{b}}$.

2.2. PROPOSITION. *Let $\mathfrak{a} \mapsto \tilde{\mathfrak{a}}$ be an algebraic hull operator from the subalgebras of $\mathfrak{g}$ to the subalgebras of $\tilde{\mathfrak{g}}$, and let $\mathfrak{h}$ be a Cartan subalgebra of $\mathfrak{g}$. Let $\mathfrak{g}^+$ denote the Fitting one-component of $\operatorname{ad}\mathfrak{h}$ on $\mathfrak{g}$. Then $\mathfrak{g}$ is Cartan dense in $\mathfrak{m} := \tilde{\mathfrak{g}}$, $\tilde{\mathfrak{h}} = \mathfrak{m}^0(\mathfrak{h})$, and $\tilde{\mathfrak{g}} = \tilde{\mathfrak{h}} \oplus \mathfrak{g}^+ = \mathfrak{m}^0(\mathfrak{h}) \oplus \mathfrak{g}^+$.*

Proof. We consider the subalgebra $[\mathfrak{g},\mathfrak{g}]$. It follows from property (iii) that it is equal to $[\mathfrak{m},\mathfrak{m}]$ and that $[\mathfrak{g},\mathfrak{g}]^\sim = [\mathfrak{g},\mathfrak{g}]$. Then

$$\mathfrak{g} = \mathfrak{g}^0(\mathfrak{h}) + \mathfrak{g}^+ = \mathfrak{h} + [\mathfrak{h},\mathfrak{g}^+] \subseteq \mathfrak{h} + [\mathfrak{g},\mathfrak{g}],$$

where the second equality follows from the fact $\mathfrak{h}$ is a Cartan subalgebra and properties of the Fitting decomposition.

Since $[\mathfrak{g},\mathfrak{g}]$ is an ideal, we apply property (v) and the preceding equalities and containments to conclude that

$$\mathfrak{m} = \tilde{\mathfrak{g}} = (\mathfrak{h} + [\mathfrak{g},\mathfrak{g}])^\sim \subseteq (\tilde{\mathfrak{h}} + \widetilde{[\mathfrak{g},\mathfrak{g}]})^\sim = \tilde{\mathfrak{h}} + \widetilde{[\mathfrak{g},\mathfrak{g}]} = \tilde{\mathfrak{h}} + [\mathfrak{g},\mathfrak{g}].$$

By two applications of property (iv), $[\tilde{\mathfrak{h}},\tilde{\mathfrak{h}}] \subseteq \tilde{\mathfrak{h}}$, and also by (iv), $[\tilde{\mathfrak{h}},\mathfrak{h}^n] \subseteq \mathfrak{h}^{n+1}$. It follows that $\tilde{\mathfrak{h}}$ is nilpotent, and hence condition (3) of Proposition 1.4 is established. Thus $\mathfrak{g}$ is Cartan dense in $\tilde{\mathfrak{g}}$. The equality $\mathfrak{m}^0(\mathfrak{h}) = \tilde{\mathfrak{h}}$ follows in the same manner as the implication (iii)$\Rightarrow$(ii) of Proposition 1.4. The equality $\tilde{\mathfrak{g}} = \tilde{\mathfrak{h}} \oplus \mathfrak{g}^+$ follows from this one and the conclusion of Proposition 1.4. □

2.3. REMARK. In the setting of Proposition 2.2, one can consider other subalgebras $\mathfrak{m}$ such that $\mathfrak{g} \subseteq \mathfrak{m} \subseteq \tilde{\mathfrak{g}}$, and apply Corollary 1.6. A case of particular interest is the case that $\mathfrak{m}$ is the "splittable hull" of $\mathfrak{g}$, that is, the smallest Lie subalgebra of $\mathfrak{gl}(n,\mathbb{F})$ which contains $\mathfrak{g}$ and also contains the semisimple and nilpotent part in the Jordan decomposition of each of its members. It is a standard fact that $\tilde{\mathfrak{g}}$ is splittable, and hence must contain the splittable hull, and hence by Corollary 1.6 $\mathfrak{g}$ is Cartan dense in $\mathfrak{m}$. □

Before we continue we pause to explain our choice of notation. While for a general Lie group G we shall write the exponential function $\mathfrak{g} \to G$ as $\exp$, for a Banach algebra A with group A^{-1} of units we shall denote the exponential function $A \to A^{-1}$ by $x \mapsto e^x$. Here the exponential function is explicitly given by its power series. In our context, this most frequently applies to $A = \operatorname{End}(V)$ for a finite dimensional real or complex vector space and $A^{-1} = \operatorname{Gl}(V)$.

2.4. COROLLARY. *Let $\mathfrak{g}$ be a subalgebra of a Lie algebra $\mathfrak{m}$ over $\mathbb{F}$. Suppose that either*
  (i) *for each $X \in \mathfrak{m}$, $e^{\operatorname{ad} X}$ belongs to the Zariski closure of the group $\langle e^{\operatorname{ad}\mathfrak{g}} \rangle$,*
or
  (ii) *there exists a connected Lie group M and an exponential map $\exp\colon \mathfrak{m} \to M$ such that the subgroup generated by $\exp\mathfrak{g}$ is dense in M.*
Then $\mathfrak{g}$ is Cartan dense in $\mathfrak{m}$.

Proof. (i) Consider the subalgebras $\operatorname{ad}\mathfrak{g} \subseteq \operatorname{ad}\mathfrak{m}$ in $\mathfrak{gl}(\mathfrak{m})$, and consider the algebraic hull operator $\mathfrak{a} \mapsto \tilde{\mathfrak{a}}$ on the subalgebras of $\operatorname{ad}\mathfrak{g}$. By hypothesis,

$\operatorname{ad} \mathfrak{m} \subseteq \widetilde{\operatorname{ad} \mathfrak{g}}$. Since $\mathfrak{h}$ is a Cartan subalgebra of $\mathfrak{g}$, it follows that $\operatorname{ad} \mathfrak{h}$ is a Cartan subalgebra of $\operatorname{ad} \mathfrak{g}$. Since $\operatorname{ad}: \mathfrak{m} \to \operatorname{ad} \mathfrak{m}$ has kernel the center of $\mathfrak{m}$, one sees directly that $\mathfrak{m}^0(\mathfrak{h})$ is nilpotent if and only if its image under ad is. That the image under ad is nilpotent follows from Proposition 2.2 applied to $\operatorname{ad} \mathfrak{h} \subseteq \operatorname{ad} \mathfrak{g} \subseteq \operatorname{ad} \mathfrak{m} \subseteq \mathfrak{gl}(\mathfrak{m})$. Thus condition (3) of Proposition 1.4 is satisfied, and $\mathfrak{g}$ is Cartan dense in $\mathfrak{m}$.

(ii) Let G be the subgroup generated by $\exp \mathfrak{g}$ in M. We are assuming that G is dense in M. Then

$$\langle e^{\operatorname{ad} \mathfrak{m}} \rangle = \operatorname{Ad}(M) = \overline{\operatorname{Ad} G} = \overline{\langle e^{\operatorname{ad} \mathfrak{g}} \rangle} \subseteq \overline{\langle e^{\operatorname{ad} \mathfrak{g}} \rangle}^{\operatorname{Zar}}.$$

Thus the hypotheses of part (i) are satisfied, and the proof is reduced to that case. □

2.5. THEOREM. *Let $\mathfrak{g}$ be a real Lie algebra and $\mathfrak{h}$ a Cartan subalgebra. Suppose that $\mathfrak{g} \subseteq \mathfrak{gl}(E)$ for a finite dimensional complex vector space E. Let $\widetilde{\mathfrak{g}_\mathbb{C}}$ denote the algebraic hull of $\mathfrak{g}_\mathbb{C} = \mathfrak{g} + i \cdot \mathfrak{g}$ in $\mathfrak{gl}(E)$. Then the following conclusions hold:*

(i) *The algebraic hull $\widetilde{\mathfrak{h}_\mathbb{C}}$ of $\mathfrak{h}_\mathbb{C} = \mathfrak{h} + i \cdot \mathfrak{h}$ in $\mathfrak{gl}(E)$ is the nilspace $\widetilde{\mathfrak{g}_\mathbb{C}}^0$ of $\operatorname{ad} \mathfrak{h}_\mathbb{C}$ in $\widetilde{\mathfrak{g}_\mathbb{C}}$ and is the unique Cartan subalgebra of $\widetilde{\mathfrak{g}_\mathbb{C}}$ containing $\mathfrak{h}_\mathbb{C}$ (or, equivalently, containing $\mathfrak{h}$).*

(ii) *Every root $\lambda \in \mathfrak{h}_\mathbb{C}^*$ of $\mathfrak{g}_\mathbb{C}$ with respect to $\mathfrak{h}_\mathbb{C}$ extends uniquely to a root $\tilde{\lambda} \in (\widetilde{\mathfrak{h}_\mathbb{C}})^*$ of $\widetilde{\mathfrak{g}_\mathbb{C}}$ with respect to $\mathfrak{h}_\mathbb{C}$, and $\lambda \mapsto \tilde{\lambda}: \Lambda \to \tilde{\Lambda}$ is a bijection from the set of roots of $\mathfrak{g}_\mathbb{C}$ with respect to $\mathfrak{h}_\mathbb{C}$ to the set of nonzero roots of $\widetilde{\mathfrak{g}_\mathbb{C}}$ with respect to $\widetilde{\mathfrak{h}_\mathbb{C}}$.*

(iii) *If*

$$\mathfrak{g}_\mathbb{C} = \mathfrak{h}_\mathbb{C} \oplus \bigoplus_{\lambda \in \Lambda} \mathfrak{g}_\mathbb{C}^\lambda$$

is a root space decomposition of $\mathfrak{g}_\mathbb{C}$, then

$$\widetilde{\mathfrak{g}_\mathbb{C}} = \widetilde{\mathfrak{h}_\mathbb{C}} \oplus \bigoplus_{\tilde{\lambda} \in \tilde{\Lambda}} \mathfrak{g}_\mathbb{C}^\lambda$$

is a root space decomposition of $\widetilde{\mathfrak{g}_\mathbb{C}}$.

(iv) $[\widetilde{\mathfrak{g}_\mathbb{C}}, \widetilde{\mathfrak{g}_\mathbb{C}}] = [\mathfrak{g}_\mathbb{C}, \mathfrak{g}_\mathbb{C}] \subseteq \mathfrak{g}_\mathbb{C}$, *and $\mathfrak{g}_\mathbb{C}$ is Cartan dense in $\widetilde{\mathfrak{g}_\mathbb{C}}$.*

Proof. Proposition 1.4 and Proposition 2.2 together prove part (i).

(ii) and (iii): Let $\tilde{\Lambda}$ denote the set of nonzero roots of $\widetilde{\mathfrak{g}_\mathbb{C}}$ with respect to $\widetilde{\mathfrak{h}_\mathbb{C}}$ and $\mathfrak{g}_\mathbb{C} = \mathfrak{h}_\mathbb{C} \oplus \bigoplus_{\lambda \in \Lambda} \mathfrak{g}_\mathbb{C}^\lambda$ the root decomposition. We apply Lemma 1.1(i) with $\mathfrak{g}_\mathbb{C}$ in place of $\mathfrak{g}$, with $\mathfrak{h}_\mathbb{C}$ in place of $\mathfrak{h}$ and with $\widetilde{\mathfrak{g}_\mathbb{C}}$ in place of $\mathfrak{m}$ and find in view of Property (i) that

$$\widetilde{\mathfrak{g}_\mathbb{C}} = \widetilde{\mathfrak{h}_\mathbb{C}} \oplus \widetilde{\mathfrak{g}_\mathbb{C}}^+, \quad \widetilde{\mathfrak{g}_\mathbb{C}}^+ = \bigoplus_{\lambda \in \Lambda} \mathfrak{g}_\mathbb{C}^\lambda$$

is a Fitting decomposition of $\widetilde{\mathfrak{g}_\mathbb{C}}$ for $\operatorname{ad}\widetilde{\mathfrak{h}_\mathbb{C}}$. By Property (iv) of algebraic extensions, all root spaces $\mathfrak{g}_\mathbb{C}^\lambda$ are $\mathfrak{h}_\mathbb{C}$-modules. Then each of these is a direct sum of root spaces of $\widetilde{\mathfrak{g}_\mathbb{C}}$ with respect to $\widetilde{\mathfrak{h}_\mathbb{C}}$. Suppose that $\alpha, \beta \in \widetilde{\Lambda}$ are such that $\alpha|\mathfrak{h}_\mathbb{C} = \beta|\mathfrak{h}_\mathbb{C} \in \Lambda$. We claim that $\alpha = \beta$. Suppose not. Let y_α and y_β be nonzero eigenvectors for $\operatorname{ad}\widetilde{\mathfrak{h}_\mathbb{C}}$ for α, respectively, β, i.e., $[x, y_\alpha] = \alpha(x) \cdot y_\alpha$ and $[x, y_\beta] = \beta(x) \cdot y_\beta$ for all $x \in \widetilde{\mathfrak{h}_\mathbb{C}}$. Note that such eigenvectors do exist. We define $\mathfrak{b} = \mathbb{C} \cdot (y_\alpha + y_\beta)$. If $x \in \mathfrak{h}_\mathbb{C}$ then $\alpha(x) = \beta(x)$ and we compute $[x, y_\alpha + y_\beta] = \alpha(x) \cdot y_\alpha + \beta(x) \cdot y_\beta = \alpha(x) \cdot (y_\alpha + y_\beta) \in \mathfrak{b}$. Thus $[\mathfrak{h}_\mathbb{C}, \mathfrak{b}] \subseteq \mathfrak{b}$. Again by Property (iv) we conclude $[\widetilde{\mathfrak{h}_\mathbb{C}}, \mathfrak{b}] \subseteq \mathfrak{b}$. Thus for each $z \in \widetilde{\mathfrak{h}_\mathbb{C}}$ there is a complex number c such that $[z, y_\alpha + y_\beta] = c \cdot (y_\alpha + y_\beta)$. Now since $\alpha \ne \beta$ we find a $z \in \widetilde{\mathfrak{h}_\mathbb{C}}$ such that $\alpha(z) \ne \beta(z)$. Then $[z, y_\alpha + y_\beta] = \alpha(z) \cdot y_\alpha + \beta(z) \cdot y_\beta$. We conclude

$$(\alpha(z) - c) \cdot y_\alpha + (\beta(z) - c) \cdot y_\beta = 0.$$

Since y_α and y_β are linearly independent we have $\alpha(z) = c = \beta(z)$, a contradiction. This proves the claim and shows that each root space for $\mathfrak{h}_\mathbb{C}$ is also one for $\widetilde{\mathfrak{h}_\mathbb{C}}$. Thus every root λ of $\mathfrak{g}_\mathbb{C}$ with respect to $\mathfrak{h}_\mathbb{C}$ extends uniquely to a root $\widetilde{\lambda}$ of $\widetilde{\mathfrak{h}_\mathbb{C}}$ with respect to $\widetilde{\mathfrak{h}_\mathbb{C}}$, and $\widetilde{\mathfrak{g}_\mathbb{C}}^{\widetilde{\lambda}} = \mathfrak{g}_\mathbb{C}^\lambda$.

(iv): This follows from Property (iii) for algebraic extensions, and from Proposition 2.2. □

3. Generalized Weyl Groups

For a subalgebra $\mathfrak{a}$ of a Lie algebra $\mathfrak{g}$ over $\mathbb{K} = \mathbb{R}$ or $\mathbb{K} = \mathbb{C}$ we want to define a group which deserves the name of the "Weyl group of $\mathfrak{a}$ in $\mathfrak{g}$". This requires a little preparation. We let $\mathbb{G}$ denote the group $\langle e^{\operatorname{ad}\mathfrak{g}} \rangle$ of inner automorphisms of $\mathfrak{g}$ in its intrinsic Lie group topology. Now for a subalgebra $\mathfrak{a}$ of $\mathfrak{g}$ we set

$$\mathbb{N}(\mathfrak{a}, \mathbb{G}) = \{\varphi \in \mathbb{G} \mid \varphi(\mathfrak{a}) = \mathfrak{a}\},$$
$$\mathbb{Z}(\mathfrak{a}, \mathbb{G}) = \{\varphi \in \mathbb{N}(\mathfrak{a}, \mathbb{G}) \mid (\forall a \in \mathfrak{a})\varphi(a) - a \in [\mathfrak{a}, \mathfrak{a}]\}.$$

3.1. DEFINITION. We say that $\mathcal{W}(\mathfrak{a}, \mathfrak{g}) = \mathbb{N}(\mathfrak{a}, \mathbb{G})/\mathbb{Z}(\mathfrak{a}, \mathbb{G})$ is the *generalized Weyl group of $\mathfrak{a}$ in $\mathfrak{g}$*. □

3.2. REMARK. We note that the generalized Weyl group acts in a natural way on $\mathfrak{a}/[\mathfrak{a}, \mathfrak{a}]$ by the formula $(\varphi \cdot \mathbb{Z}(\mathfrak{a}, \mathbb{G}))(x + [\mathfrak{a}, \mathfrak{a}]) = \varphi(x) + [\mathfrak{a}, \mathfrak{a}]$. This yields a well-defined linear action of $\mathcal{W}(\mathfrak{a}, \mathfrak{g})$ on $\mathfrak{a}/[\mathfrak{a}, \mathfrak{a}]$ and this linear representation is faithful on $\mathcal{W}(\mathfrak{a}, \mathfrak{g})$. In particular, if $\mathfrak{a}$ is abelian, then $\mathcal{W}(\mathfrak{a}, \mathfrak{g})$ acts faithfully and linearly on $\mathfrak{a}$. □

Let $\mathfrak{g}$ be a Lie algebra and G *any* connected Lie group with Lie algebra $\mathfrak{g}$ and $\exp: \mathfrak{g} \to G$ its exponential function. Now $\operatorname{Ad}: G \to \operatorname{Aut}\mathfrak{g}$ is given by

$g(\exp x)g^{-1} = \exp \operatorname{Ad}(g)(x)$. The image of Ad is a connected analytic subgroup of $\operatorname{Aut}(\mathfrak{g})$. If $g = \exp y$ then we have $\operatorname{Ad}(\exp y) = e^{\operatorname{ad} y}$. Thus

$$\operatorname{Ad}(G) = \langle e^{\operatorname{ad} \mathfrak{g}} \rangle = \mathbb{G}.$$

Hence $\operatorname{Ad}: G \to \mathbb{G}$ is the unique morphism which on the algebra level induces the homomorphism $\operatorname{ad}: \mathfrak{g} \to \operatorname{ad} \mathfrak{g}$.

We recall that $\ker \operatorname{Ad} = Z(G)$ is the center of G. (Indeed, $\operatorname{Ad} g = 1$ implies $\exp x = \exp \operatorname{Ad}(g)(x) = g(\exp x)g^{-1}$ and since the $\exp x$ generate G, the assertion follows.) We set

$$N(\mathfrak{a}, G) = \{g \in G \mid \operatorname{Ad}(g)(\mathfrak{a}) = \mathfrak{a}\},$$
$$Z(\mathfrak{a}, G) = \{g \in N(\mathfrak{a}, G) \mid (\forall a \in \mathfrak{a}) \operatorname{Ad}(g)(a) - a \in [\mathfrak{a}, \mathfrak{a}]\}.$$

3.3. LEMMA. *The function* $\operatorname{Ad}: G \to \mathbb{G}$ *induces an isomorphism*

$$\Phi: \frac{N(\mathfrak{a}, G)}{Z(\mathfrak{a}, G)} \to \mathcal{W}(\mathfrak{a}, \mathfrak{g}), \quad \Phi(gZ(\mathfrak{a}, G)) = \operatorname{Ad}(g)\mathbb{Z}(\mathfrak{a}, \mathbb{G}).$$

Proof. Firstly, the morphism Φ is well-defined since Ad maps $g \in Z(\mathfrak{a}, G)$ onto $\mathbb{Z}(\mathfrak{a}, \mathbb{G})$ by our definitions. Also, since Ad maps G surjectively onto $\mathbb{G}$, the morphism Φ is surjective. Finally, its kernel consists of those $gZ(\mathfrak{a}, G)$ with $\operatorname{Ad}(g) \in \mathbb{Z}(\mathfrak{a}, \mathbb{G})$, i.e., $g \in Z(\mathfrak{a}, G)$. This shows that the kernel is a singleton and the assertion is proved. □

Let $f: \mathfrak{g}_1 \to \mathfrak{g}_2$ be a surjective morphism of Lie algebras. Let G_j denote the simply connected Lie group with Lie algebra $\mathfrak{g}_j$ and $\exp_j: \mathfrak{g}_j \to G_j$ its exponential function. Then there is a unique morphism $\widetilde{f}: G_1 \to G_2$ such that

(1) $$\widetilde{f} \circ \exp_1 = \exp_2 \circ f.$$

Let $\mathfrak{a}_j$ be a subalgebra of $\mathfrak{g}_j$ such that $f(\mathfrak{a}_1) \subseteq \mathfrak{a}_2$. If $g \in Z(\mathfrak{a}_1, G_1)$, then $\operatorname{Ad}(g)(a) - a \in [\mathfrak{a}_1, \mathfrak{a}_1]$ for all $a \in \mathfrak{a}_1$. Thus $f(\operatorname{Ad}(g)(a)) - f(a) \in [f(\mathfrak{a}_1), f(\mathfrak{a}_1)] \subseteq [\mathfrak{a}_2, \mathfrak{a}_2]$. Also, for all $t \in \mathbb{R}$ we have $\exp_2 t\{f(\operatorname{Ad}(g)(a))\} = \widetilde{f}(\exp_1 t \operatorname{Ad}(g)(a)) = \widetilde{f}(g(\exp_1 ta)g^{-1}) = \widetilde{f}(g)\exp_2 tf(a)\widetilde{f}(g)^{-1} = \exp_2 t \operatorname{Ad}(\widetilde{f}(g))f(a)$. Thus

(2) $$f(\operatorname{Ad}(g)(a)) = \operatorname{Ad}(\widetilde{f}(g))(f(a)).$$

We conclude that $\operatorname{Ad}(\widetilde{f}(g))(f(a)) - f(a) \in [\mathfrak{a}_2, \mathfrak{a}_2]$. Hence

$$\widetilde{f}(Z(\mathfrak{a}_1, G_1)) \subseteq Z(\mathfrak{a}_2, G_2).$$

Further $\operatorname{Ad}(g)(\mathfrak{a}_1) \subseteq \mathfrak{a}_1$ implies $\operatorname{Ad}(\widetilde{f}(g))(f(\mathfrak{a}_1)) = f(\operatorname{Ad}(g)(\mathfrak{a}_1)) \subseteq f(\mathfrak{a}_1) \subseteq \mathfrak{a}_2$.

If we now assume that $\mathfrak{a}_2 = f(\mathfrak{a}_1)$, then $\widetilde{f}(N(\mathfrak{a}_1,\mathfrak{g}_1)) \subseteq N(\mathfrak{a}_2,\mathfrak{g}_2)$. If we use Lemma 3.3 and set $\mathcal{W}(\mathfrak{a}_j,\mathfrak{g}_j) = \frac{N(\mathfrak{a}_j,G_j)}{Z(\mathfrak{a}_j,G_j)}$ then our f induces a morphism

$$\mathcal{W}(f)\colon \mathcal{W}(\mathfrak{a}_1,\mathfrak{g}_1) \to \mathcal{W}(\mathfrak{a}_2,\mathfrak{g}_2), \quad \mathcal{W}(f)(gZ(\mathfrak{a}_1,G_1)) = \widetilde{f}(g)Z(\mathfrak{a}_2,G_2).$$

The image $\operatorname{im} \mathcal{W}(f)$ is

$$\operatorname{im} \mathcal{W}(f) = \frac{\widetilde{f}(N(\mathfrak{a}_1,G_1))Z(\mathfrak{a}_2,G_2)}{Z(\mathfrak{a}_2,G_2)}.$$

If f is surjective then $\widetilde{f}$ is surjective, and if $\mathfrak{a}_1 = f^{-1}(\mathfrak{a}_2)$ then $f(\operatorname{Ad}(g)(\mathfrak{a}_1) = \operatorname{Ad}(\widetilde{f}(g))(\mathfrak{a}_2) = \mathfrak{a}_2 = f(\mathfrak{a}_1)$ iff $\operatorname{Ad}(g)(\mathfrak{a}_1) = \mathfrak{a}_1$ iff $g \in N(\mathfrak{g}_1,\mathfrak{a}_1)$. Thus $\mathcal{W}(f)$ is surjective.

The kernel of $\mathcal{W}(f)$ consists of all $gZ(\mathfrak{a}_1,G_1)$ such that $\widetilde{f}(g) \in Z(\mathfrak{a}_2,G_2)$, i.e., $\operatorname{Ad}(\widetilde{f}(g))(a_2) - a_2 \in [\mathfrak{a}_2,\mathfrak{a}_2]$ for all $a_2 \in \mathfrak{a}_2 = f(\mathfrak{a}_1)$. Pick $a_1 \in \mathfrak{a}_1$ with $f(a_1) = a_2$. Then $f(\operatorname{Ad}(g)(a_1) - a_1) \in [\mathfrak{a}_2,\mathfrak{a}_2] = f([\mathfrak{a}_1,\mathfrak{a}_1])$. This is tantamount to

$$\operatorname{Ad}(g)(a_1) - a_1 \in [\mathfrak{a}_1,\mathfrak{a}_1] + \ker f.$$

Note that the right-hand side collapses to $Z(\mathfrak{a}_1,G_1)$ if $\ker f \subseteq [\mathfrak{a}_1,\mathfrak{a}_1]$, and in this case $\mathcal{W}(f)$ is injective.

We have observed the following facts on the functoriality of the Weyl group.

3.4. LEMMA. *Suppose that $\mathfrak{a}_j \subset \mathfrak{g}_j$ for $j = 1,2$ and that $f\colon \mathfrak{g}_1 \to \mathfrak{g}_2$ is a morphism of Lie algebras such that $f(\mathfrak{a}_1) = \mathfrak{a}_2$. Then f induces a morphism*

$$\mathcal{W}(f)\colon \mathcal{W}(\mathfrak{a}_1,\mathfrak{g}_2) \to \mathcal{W}(\mathfrak{a}_2,\mathfrak{g}_2), \quad \mathcal{W}(f)(gZ(\mathfrak{a}_1,G_1)) = \widetilde{f}(g)Z(\mathfrak{a}_2,G_2).$$

It is surjective iff f is surjective and $\mathfrak{a}_1 = f^{-1}(\mathfrak{a}_2)$. If $\ker f \subseteq [\mathfrak{a}_1,\mathfrak{a}_1]$, then $\mathcal{W}(f)$ is injective. □

As a corollary we obtain

3.5. LEMMA. *Suppose that $\mathfrak{a} \subseteq \mathfrak{g} \subseteq \mathfrak{m}$ are Lie algebras. Then $\mathcal{W}(\mathfrak{a},\mathfrak{g})$ may be identified with a subgroup of $\mathcal{W}(\mathfrak{a},\mathfrak{m})$, and if $\mathcal{W}(\mathfrak{a},\mathfrak{m})$ is finite, then $\mathcal{W}(\mathfrak{a},\mathfrak{g})$ is finite.*

Proof. We apply Lemma 3.4 to the inclusion morphism $f\colon \mathfrak{g} \to \mathfrak{m}$. □

3.6. LEMMA. *Let $\mathfrak{h}$ be a Cartan algebra of $\mathfrak{g}$ and $\mathfrak{a}$ a subalgebra of $\mathfrak{h}$. Let G be any connected Lie group with Lie algebra $\mathfrak{g}$ and let $H = \exp \mathfrak{h}$. Then the following statements are equivalent:*
  (i) $H \subseteq Z(\mathfrak{a},G)$.
  (ii) $\operatorname{Ad}(h)(a) - a \in [\mathfrak{a},\mathfrak{a}]$ *for all* $h \in H$, $a \in \mathfrak{a}$.
  (iii) $(e^{\operatorname{ad} x} - 1)(a) \in [\mathfrak{a},\mathfrak{a}]$ *for all* $x \in \mathfrak{h}$, $a \in \mathfrak{a}$.

(iv) $[\mathfrak{h}, \mathfrak{a}] \subseteq [\mathfrak{a}, \mathfrak{a}]$.
(v) $e^{\operatorname{ad} \mathfrak{h}} \subseteq Z(\mathfrak{a}, \mathbb{G})$.

Proof. (i)⇔(ii) by definition. (ii)⇔(iii) because $\operatorname{Ad}(\exp x) = e^{\operatorname{ad} x}$ and $H = \exp \mathfrak{h}$.

(iii)⇒(iv): For all $t \in \mathbb{R}$ and all $x \in \mathfrak{h}$, $a \in \mathfrak{a}$ we have $e^{t \cdot \operatorname{ad} x} a - a \in [\mathfrak{a}, \mathfrak{a}]$. Now $[x, a] = \lim_{0 \neq t \to 0} \frac{1}{t}(e^{t \cdot \operatorname{ad} x} a - a) \in [\mathfrak{a}, \mathfrak{a}]$.

(iv) ⇒(iii): Let $[x, a] \in [\mathfrak{a}, \mathfrak{a}]$. Then $e^{\operatorname{ad} x} a - a \in \bigoplus_{n=1}^{\infty} (\operatorname{ad} x)^n \mathfrak{a} \subseteq [\mathfrak{a}, \mathfrak{a}]$.

(iii)⇔(v) directly from the definitions. □

3.7. LEMMA. *If $\mathfrak{a} \subseteq \mathfrak{h}$ is such that $[\mathfrak{h}, \mathfrak{a}] \subseteq [\mathfrak{a}, \mathfrak{a}]$ then the identity component $N_0(\mathfrak{a}, \mathbb{G})$ is contained in $Z(\mathfrak{a}, \mathbb{G})$. In particular, $\mathcal{W}(\mathfrak{a}, \mathfrak{g})$ is finite if $N(\mathfrak{a}, \mathbb{G})$ has finitely many components.*

Proof. $N_0(\mathfrak{a}, \mathbb{G})$ is generated by all $e^{t \cdot \operatorname{ad} x}$, $x \in \mathfrak{g}$ such that $e^{t \cdot \operatorname{ad} x} \mathfrak{a} = \mathfrak{a}$. This is tantamount to $[x, \mathfrak{a}] \subseteq \mathfrak{a}$.

Decomposing $\mathfrak{g}$ into its Fitting components $\mathfrak{h} \oplus \mathfrak{g}^+$ with respect to $\mathfrak{h}$, we see that for each $h \in \mathfrak{h}$ we have $[\mathfrak{g}, h] \cap \mathfrak{h} = [\mathfrak{h}, h]$. Thus $[x, a] \in [\mathfrak{h}, \mathfrak{a}]$ for all $a \in \mathfrak{a}$. By hypothesis this implies $[x, \mathfrak{a}] \subseteq [\mathfrak{a}, \mathfrak{a}]$. Hence $e^{t \operatorname{ad} x} \in Z(\mathfrak{a}, \mathbb{G})$ by Lemma 3.6, and this proves the claim. □

The complexification $\mathfrak{g}_\mathbb{C} = \mathbb{C} \otimes \mathfrak{g}$ gives rise to the group $\mathbb{G}_\mathbb{C} = \langle e^{\operatorname{ad} \mathfrak{g}_\mathbb{C}} \rangle$ of complex automorphisms of $\mathfrak{g}_\mathbb{C}$. We may consider $\mathbb{G}$ as embedded into $\mathbb{G}_\mathbb{C}$ since every real automorphism of $\mathfrak{g}$ extends uniquely to a complex automorphism of $\mathfrak{g}_\mathbb{C}$. With this identification, the group $Z(\mathfrak{a}, \mathbb{G})$ becomes identified with $Z(\mathfrak{a}_\mathbb{C}, \mathbb{G}_\mathbb{C}) \cap \mathbb{G}$: Indeed, $\gamma \in Z(\mathfrak{a}, \mathbb{G})$ is equivalent to $(\gamma - 1)(a) \in [\mathfrak{a}, \mathfrak{a}]$ for all $a \in \mathfrak{a}$, and this is equivalent to $(\gamma - 1)(a + i \cdot b) \in [\mathfrak{a}, \mathfrak{a}] + i \cdot [\mathfrak{a}, \mathfrak{a}] = [\mathfrak{a}_\mathbb{C}, \mathfrak{a}_\mathbb{C}]$ for all $a, b \in \mathfrak{a}$. In exactly the same way we have $N(\mathfrak{a}, \mathbb{G}) = N(\mathfrak{a}_\mathbb{C}, \mathbb{G}_\mathbb{C}) \cap \mathbb{G}$. We write

$$\mathcal{W}_\mathbb{C}(\mathfrak{a}_\mathbb{C}, \mathfrak{g}_\mathbb{C}) = \frac{N(\mathfrak{a}_\mathbb{C}, \mathbb{G}_\mathbb{C})}{Z(\mathfrak{a}_\mathbb{C}, \mathbb{G}_\mathbb{C})}.$$

The function

$$\Psi: \mathcal{W}(\mathfrak{a}, \mathfrak{g}) \to \mathcal{W}_\mathbb{C}(\mathfrak{a}_\mathbb{C}, \mathfrak{g}_\mathbb{C}), \quad \Psi(\gamma Z(\mathfrak{a}, \mathbb{G})) = \gamma Z(\mathfrak{a}_\mathbb{C}, \mathbb{G}_\mathbb{C})$$

has the kernel $\frac{Z(\mathfrak{a}_\mathbb{C}, \mathbb{G}_\mathbb{C}) \cap \mathbb{G}}{Z(\mathfrak{a}, \mathbb{G})} = \{1\}$. Thus

3.8. LEMMA. $\mathcal{W}(\mathfrak{a}, \mathfrak{g})$ *can be viewed as a subgroup of* $\mathcal{W}_\mathbb{C}(\mathfrak{a}_\mathbb{C}, \mathfrak{g}_\mathbb{C})$. *In particular, the Weyl group of $\mathfrak{a}$ in $\mathfrak{g}$ is finite if the Weyl group of $\mathfrak{a}_\mathbb{C}$ in $\mathfrak{g}_\mathbb{C}$ (in the sense of complex Lie algebras) is finite.* □

For questions of finiteness we may therefore assume without losing generality that the Lie algebras we consider are complex.

In the following we refer to Theorem 2.5. Thus we let $\mathfrak{g}$ denote a Lie algebra (real or complex) contained in $\mathfrak{gl}(E)$ for a finite dimensional complex vector space E. Let $\widetilde{\mathfrak{g}_\mathbb{C}}$ be the algebraic hull of its complexification.

3.9. THEOREM. *Let $\mathfrak{a}$ be a subalgebra of a real or complex Lie algebra $\mathfrak{g}$ which is contained in a Cartan subalgebra $\mathfrak{h}$ such that $[\mathfrak{a}, \mathfrak{h}] \subseteq [\mathfrak{a}, \mathfrak{a}]$. Then $\mathcal{W}(\mathfrak{a}, \mathfrak{g})$, and in particular $\mathcal{W}(\mathfrak{h}, \mathfrak{g})$ is finite.*

Proof. (i) By ADO's Theorem, we may assume without loss of generality that $\mathfrak{g}$ is a subalgebra of $\mathfrak{gl}(E)$ for a finite dimensional complex vector space E. By Lemma 3.8 we may assume that $\mathfrak{g}$, $\mathfrak{h}$ and $\mathfrak{a}$ are complex subalgebras.

(ii) We form $\widetilde{\mathfrak{g}}$. By Theorem 2.5, $\mathfrak{h}$ is contained in a unique Cartan subalgebra $\widetilde{\mathfrak{h}}$ of $\widetilde{\mathfrak{g}}$ and we have $[\mathfrak{a}, \widetilde{\mathfrak{h}}] \subseteq [\mathfrak{a}, \mathfrak{a}]$ by property (iv) of Definition 2.1. Now we apply Lemma 3.7 to $\widetilde{\mathfrak{g}}$ and the algebraic group $\widetilde{G}$. Then $N(\mathfrak{a}, \widetilde{G})$ has only finitely many components, since it is a Zariski closed subgroup of $\mathrm{Gl}(E)$. Hence $\mathcal{W}(\mathfrak{a}, \widetilde{\mathfrak{g}})$ is finite by Lemma 3.7. Then $\mathcal{W}(\mathfrak{a}, \mathfrak{g})$ is finite by Lemma 3.5. $\square$

4. Intersections of Cartan Subalgebras

We have worked to prove the finiteness of the Weyl group $\mathcal{W}(\mathfrak{a}, \mathfrak{g})$. A finiteness condition of a different type will be the topic of this section. Eventually we shall bring the two lines together in Theorem 4.31. Here we shall consider a complex Lie algebra $\mathfrak{g}$ and fix a Cartan subalgebra $\mathfrak{h}$ of $\mathfrak{g}$. We wish to determine conditions under which $\mathfrak{g}$ is *Cartan finite*, that is, the set $\{\mathfrak{h} \cap \mathfrak{k} \mid \mathfrak{k} \in \mathcal{C}\}$ is finite, where $\mathcal{C}$ is the collection of all Cartan subalgebras of $\mathfrak{g}$. Note that if the intersection of two Cartan algebras contains a regular element x, then the two Cartan algebras are both equal to $\mathfrak{g}^0(x)$. Thus a nontrivial intersection of two Cartan subalgebras must miss the regular elements, and hence be contained in the kernel of some root. A very natural condition is that intersections of Cartan algebras with a fixed Cartan algebra $\mathfrak{h}$ agree with intersections of kernels of roots of $\mathfrak{h}$, and it is a special case of this condition which we seek to elucidate in this section.

First of all we write
$$\mathfrak{g} = \mathfrak{h} \oplus \bigoplus_{\lambda \in \Lambda} \mathfrak{g}^\lambda,$$
where Λ is the set of nonzero roots with respect to $\mathfrak{h}$.

4.1. DEFINITION. (i) For a subset $\Omega \subseteq \Lambda$ let $\Omega^\perp = \{h \in \mathfrak{h} \mid (\forall \omega \in \Omega) \omega(h) = 0\}$. If $x = x_0 + \sum_{\lambda \in \Lambda} x_\lambda \in \mathfrak{g}$, $x_\lambda \in \mathfrak{g}^\lambda$, write $\mathrm{Supp}_\Lambda(x)$ for the set of all roots λ with $x_\lambda \neq 0$.

(ii) For a subset $\mathfrak{a} \subseteq \mathfrak{h}$ let $\mathfrak{a}^{\perp_\Lambda} = \{\lambda \in \Lambda \mid \Lambda(\mathfrak{a}) = \{0\}\}$. $\square$

4.2. LEMMA. (i) *If $\Omega \subseteq \Lambda$ then $\Omega^{\perp \perp_\Lambda} = \Lambda \cap \mathrm{span}(\Omega)$.*
(ii) *If $\mathfrak{a} \subseteq \mathfrak{h}$ then $\mathfrak{a}^{\perp_\Lambda \perp} = \mathrm{span}(\mathfrak{a}) + \Lambda^\perp$.*

Proof. These assertions follow from the duality of finite dimensional vector spaces and the definitions. $\square$

Let $\mathfrak{n}$ denote the nilradical of $\mathfrak{g}$, i.e., the largest nilpotent ideal.

4.3. PROPOSITION. $\Lambda^\perp = \{y \in \mathfrak{h} \mid \operatorname{ad} y \text{ is nilpotent }\} = \mathfrak{h} \cap \mathfrak{n}$.

Proof. It follows from the root space decomposition that $y \in \Lambda^\perp$ iff $\operatorname{ad} y$ is nilpotent. If $y \in \mathfrak{n}$, then $\operatorname{ad} y$ is nilpotent, and thus $y \in \mathfrak{h} \cap \mathfrak{n}$ implies $y \in \Lambda^\perp$. If $y \in \Lambda^\perp$ then $\operatorname{ad} y$ is nilpotent. In order to show that $y \in \mathfrak{n}$ it suffices to show that y is in the radical $\mathfrak{r}$ since $\mathfrak{n}$ is exactly the set of nilpotent elements of $\mathfrak{r}$. Now $(\mathfrak{h} + \mathfrak{r})/\mathfrak{r}$ is a Cartan algebra of the semisimple algebra $\mathfrak{g}/\mathfrak{r}$. The vector space endomorphism $\operatorname{ad}_{\mathfrak{g}/\mathfrak{r}}(y + \mathfrak{r})$ is semisimple. Since $\operatorname{ad} y$ is nilpotent, we conclude that $\operatorname{ad}_{\mathfrak{g}/\mathfrak{r}}(y + \mathfrak{r}) = 0$, i.e., $y + \mathfrak{r}$ is central in $\mathfrak{g}/\mathfrak{r}$. Thus $y \in \mathfrak{r}$. $\square$

4.4. LEMMA. *If $\Omega \subseteq \Lambda$, then there is a regular $x \in \mathfrak{g}$ such that $\Omega = \operatorname{Supp}_\Lambda(x)$.*

Proof. Let $\Omega \subseteq \Lambda$. The set $\mathfrak{v} = \mathfrak{h} \oplus \bigoplus_{\omega \in \Omega} \mathfrak{g}^\omega$ is a vector subspace containing regular elements. Hence $\mathfrak{v} \cap \operatorname{reg}(\mathfrak{g})$ is Zariski dense open in $\mathfrak{v}$. Thus there exists a regular element $x = x_0 + \bigoplus_{\omega \in \Omega} x_\omega$ with $x_0 \neq 0$ and $x_\omega \neq 0$ for $\omega \in \Omega$. Then $\operatorname{Supp}_\Lambda(x) = \Omega$ by definition. $\square$

4.5. LEMMA. *Let $y \in \mathfrak{h}$ and $x \in \mathfrak{g}$.*
(i) *Then the following are equivalent:*
   (1) $x \in \mathfrak{g}^0(y)$,
   (2) $y \in \operatorname{Supp}_\Lambda(x)^\perp$.
   (3) $\operatorname{Supp}_\Lambda(x) \subseteq y^{\perp_\Lambda}$.
(ii) $x \in \mathfrak{g}^0(\operatorname{Supp}_\Lambda(x)^\perp)$.
(iii) *Let $x \in \operatorname{reg}(\mathfrak{g})$. Then if $\mathfrak{r}$ denotes the radical*

$$\mathfrak{h} \cap \mathfrak{g}^0(x) \subseteq \operatorname{Supp}_\Lambda(x)^\perp \subseteq \mathfrak{h} \cap (\mathfrak{g}^0(x) + [\mathfrak{r}, \mathfrak{g}]).$$

Proof. (i) Write $x = x_0 + \sum_{\lambda \in \Lambda} x_\lambda$, $x_\lambda \in \mathfrak{g}^\lambda$ and suppose $x \in \mathfrak{g}^0(y)$. Then there is a n such that $0 = (\operatorname{ad} y)^n(x) = \sum_\lambda (\operatorname{ad} y)^n x_\lambda$. Since also $y \in \mathfrak{h}$, then $(\operatorname{ad} y)^n(x_\lambda) \subseteq \mathfrak{g}^\lambda$. Hence

(1) $\qquad (\operatorname{ad} y)^n(x_\lambda) = 0 \quad \text{for all} \quad \lambda \in \Lambda.$

On the other hand, by the definition of the root spaces

(2) $\qquad (\operatorname{ad} y - \lambda(y))^n(x_\lambda) = 0 \quad \text{for all} \quad \lambda \in \Lambda.$

Thus x_λ is, for all $\lambda \in \Lambda$, a generalized eigenvector for the eigenvalue 0 as well as the eigenvalue $\lambda(y)$ which implies $x_\lambda = 0$ or $\lambda(y) = 0$ and which proves $y \in \operatorname{Supp}_\Lambda(x)^\perp$.

Conversely, if $y \in \operatorname{Supp}_\Lambda(x)^\perp$, then

$$(\operatorname{ad} y)^n(x) = (\operatorname{ad} y)^n(x_0) + \sum_{\lambda \in \Lambda}(\operatorname{ad} y)^n(x_\lambda) = \sum_{\lambda \in \operatorname{Supp}_\Lambda(x)} (\operatorname{ad} y - \lambda(y))^n x_\lambda = 0.$$

Thus $x \in \mathfrak{g}^0(y)$. Thus (1) and (2) are equivalent.

Now (2) means that $\lambda(y) = 0$ iff $x_\lambda \neq 0$. But $\lambda(y) = 0$ is tantamount to $\lambda \in y^{\perp_\Lambda}$ by Definition 4.1(ii). Likewise, $x_\lambda \neq 0$ is equivalent to $\lambda \in \mathrm{Supp}_\Lambda(x)$ by Definition 4.1(i). Thus (2) and (3) are equivalent.

(ii) By (i), the element x is in $\mathfrak{g}^0(y)$ for every $y \in \mathrm{Supp}_\Lambda(x)^\perp$. Thus $x \in \mathfrak{g}^0(\mathrm{Supp}_\Lambda(x)^\perp)$.

(iii) Firstly, let $y \in \mathfrak{h} \cap \mathfrak{g}^0(x)$. Since $x \in \mathrm{reg}(\mathfrak{g})$ we have $\mathfrak{g}^0(x) \in \mathcal{C}$. Then $x \in \mathfrak{g}^0(x) \subseteq \mathfrak{g}^0(y)$. Thus $y \in \mathrm{Supp}_\Lambda(x)^\perp$ by (i).

Secondly, let $y \in \mathrm{Supp}_\Lambda(x)^\perp$. Then $x \in \mathfrak{g}^0(y)$ by (i). Let $\pi\colon \mathfrak{g} \to \mathfrak{g}/[\mathfrak{r},\mathfrak{g}]$ denote the quotient homomorphism onto the quotient algebra modulo $[\mathfrak{r},\mathfrak{g}]$. Then the quotient algebra $\mathfrak{g}/[\mathfrak{r},\mathfrak{g}]$ is reductive. Thus the element $\pi(y)$ is semisimple in $\pi(\mathfrak{g})$, since it is in the Cartan algebra $\pi(\mathfrak{h})$ in the reductive algebra $\pi(\mathfrak{g})$. Then $\pi(x) \in \pi(\mathfrak{g}^0(y)) = \pi(\mathfrak{g})^0(\pi(y)) = \ker \mathrm{ad}_{\pi(\mathfrak{g})} \pi(y)$, i.e., $\pi([x,y]) = [\pi(x),\pi(y)] = 0$. Since $\pi(x)$ is regular in $\pi(\mathfrak{g})$ we have $\pi(\mathfrak{g}^0(x)) = \pi(\mathfrak{g})^0(\pi(x)) = \ker_{\pi(\mathfrak{g})} \mathrm{ad}\,\pi(x)$. Thus $\pi(y) \in \pi(\mathfrak{g}^0(x))$. Hence $y \in \mathfrak{h} \cap (\mathfrak{g}^0(x) + [\mathfrak{r},\mathfrak{g}])$. □

Recall from the beginning of Section 2 that $\mathbb{G} = \langle e^{\mathrm{ad}\,\mathfrak{g}} \rangle$ is the group of inner automorphisms of $\mathfrak{g}$.

4.6. LEMMA. *Let $\mathfrak{g}$ denote a complex Lie algebra and $\mathfrak{h}$ a Cartan subalgebra. Set*

$$\mathfrak{n}(\mathcal{C}) = \bigcap \mathcal{C} = \bigcap_{\gamma \in \mathbb{G}} \gamma(\mathfrak{h}).$$

Then $\mathfrak{n}(\mathcal{C})$ is a nilpotent characteristic ideal which contains all members of the ascending central series of $\mathfrak{g}$ and is contained in $\mathfrak{h} \cap \mathfrak{n}$. Also, $\mathfrak{n}(\mathcal{C})$ is the largest ideal of $\mathfrak{g}$ contained in $\mathfrak{h}$.

Proof. The assertions are straightforward from the definitions. The last assertion follows from the observation that a subspace is an ideal if and only if it is invariant under all inner automorphisms in $\mathbb{G}$. □

4.7. LEMMA. *Suppose that $\mathrm{Supp}_\Lambda(x)^\perp \subseteq \mathfrak{h} \cap \mathfrak{g}^0(x)$ for all regular $x \in \mathcal{O}$ for a set $\mathcal{O}$ of regular elements meeting all Cartan algebras of $\mathfrak{g}$. Then the following conclusions hold:*

(i) $\mathfrak{h} \cap \mathfrak{n} = \Lambda^\perp = \mathfrak{n}(\mathcal{C})$.

(ii) *If $\mathfrak{m}$ is any subalgebra of $\mathfrak{g}$ containing $\mathfrak{h}$, then $\Lambda(\mathfrak{m})^\perp = \mathfrak{n}(\mathcal{C}(\mathfrak{m}))$.*

Proof. (i) The first equality was established in Proposition 4.3, and one inclusion in the second equality in Lemma 4.6. Let $\mathfrak{k} \in \mathcal{C}$. Then $\mathfrak{k} \cap \mathcal{O} \neq \emptyset$. Since $\Lambda^\perp \subseteq \mathrm{Supp}_\Lambda(x)^\perp$ for all x the hypothesis implies

$$\Lambda^\perp \subseteq \bigcap_{x \in \mathcal{O}} \mathfrak{g}^0(x) = \bigcap_{\mathfrak{k} \in \mathcal{C}} \mathfrak{k} = \mathfrak{n}(\mathcal{C}).$$

(ii) We apply (i) with $\mathfrak{m}$ in place of $\mathfrak{g}$ and the set $\mathcal{O}(\mathfrak{m}) = \mathcal{O} \cap \mathfrak{m}$. We check that the hypotheses of (i) are satisfied for $\mathfrak{m}$: Firstly, if $\mathfrak{k} \in \mathcal{C}(\mathfrak{m})$, then $\mathfrak{k}$ is

conjugate in m to $\mathfrak{h}$. Hence $\mathfrak{k} \in \mathcal{C}$. Thus $\mathcal{O} \cap \mathfrak{m} \neq \emptyset$ and thus $\mathcal{O}(\mathfrak{m}) \cap \mathfrak{k} \neq \emptyset$.
Secondly, let $x \in \mathcal{O}(\mathfrak{m})$. Then $\mathfrak{m}^0(x) = \mathfrak{g}^0(x) \cap \mathfrak{m} = \mathfrak{g}^0(x)$ as $\mathfrak{m}$ has maximal rank. Also $(\mathrm{Supp}_\Lambda)_\mathfrak{m}(x) = (\mathrm{Supp}_\Lambda)_\mathfrak{g}(x)$. Then by hypothesis $(\mathrm{Supp}_\Lambda)_\mathfrak{m}(x)^\perp = (\mathrm{Supp}_\Lambda)_\mathfrak{g}(x)^\perp \subseteq \mathfrak{h} \cap \mathfrak{g}^0(x) = \mathfrak{h} \cap \mathfrak{m}^0(x)$.
The assertion now follows from (i). □

4.8. LEMMA. *Suppose that $\mathfrak{g}$ satisfies* Condition (ii) *of Lemma 4.7. Then*

$$\mathfrak{h} \cap \mathfrak{g}^0(x) = \mathrm{Supp}_\Lambda(x)^\perp.$$

Proof. After 4.5 we must show the containment "$\supseteq$". Let $y \in \mathrm{Supp}_\Lambda(x)^\perp$. Then $x \in \mathfrak{g}^0(y)$ by 4.5(i). Since $\mathfrak{h} \subseteq \mathfrak{g}^0(y)$, the algebra $\mathfrak{m} = \mathfrak{g}^0(y)$ is a maximal rank subalgebra. If x is regular in $\mathfrak{g}$ it is regular in $\mathfrak{m}$. Hence $\mathfrak{g}^0(x) \cap \mathfrak{m}$ is a Cartan algebra of $\mathfrak{m}$. Now $\dim(\mathfrak{g}^0(x) \cap \mathfrak{m}) = \mathrm{rank}\,\mathfrak{m} = \mathrm{rank}\,\mathfrak{g} = \dim \mathfrak{g}^0(x)$. Thus $\mathfrak{g}^0(x) \subseteq \mathfrak{m}$. Now $\mathrm{ad}\,y|\mathfrak{m}$ is nilpotent and $y \in \mathfrak{h}$, whence $y \in \mathfrak{h} \cap \mathfrak{n}_\mathfrak{m}$, where $\mathfrak{n}_\mathfrak{m}$ is the nilradical of $\mathfrak{m}$. Recall $\mathfrak{h} \cap \mathfrak{n}_\mathfrak{m} = \Lambda(\mathfrak{m})$ by Proposition 4.3. By 4.7(ii) we have $\Lambda(\mathfrak{m}) = \mathfrak{n}(\mathcal{C}(\mathfrak{m}))$, where $\mathcal{C}(\mathfrak{m})$ is the set of Cartan algebras of $\mathfrak{m}$. Thus $y \subseteq \mathfrak{n}(\mathcal{C}(\mathfrak{m})) \subseteq \mathfrak{g}^0(x)$. Hence $y \in \mathfrak{h} \cap \mathfrak{g}^0(x)$ which is what we had to show. □

4.9. THEOREM. *For a complex Lie algebra $\mathfrak{g}$ and a Cartan subalgebra $\mathfrak{h}$ the following statements are equivalent:*
  (1) $\mathfrak{g}^0(x) \cap \mathfrak{h} = \mathrm{Supp}_\Lambda(x)^\perp$ *for all regular $x \in \mathfrak{g}$.*
  (1′) $\mathfrak{g}^0(x) \cap \mathfrak{h} \supseteq \mathrm{Supp}_\Lambda(x)^\perp$ *for all regular $x \in \mathfrak{g}$.*
  (2) *For all $\mathfrak{m} = \mathfrak{g}^0(y)$ with $y \in \mathfrak{h}$ one has $\Lambda(\mathfrak{m})^\perp = \mathfrak{n}(\mathcal{C}(\mathfrak{m}))$.*
  (3) $[\ker \lambda, \mathfrak{g}^\lambda] = \{0\}$ *for all $\lambda \in \Lambda$.*
  (4) *The image of the representation $\mathrm{ad}\,|\mathfrak{h}: \mathfrak{h} \to \mathfrak{gl}(\mathfrak{g}^\lambda)$ is one-dimensional for each $\lambda \in \Lambda$.*

Each of the preceding equivalent conditions implies each of the following equivalent conditions:
  (ω) *If $\mathfrak{k}$ is a Cartan subalgebra of $\mathfrak{g}$, then $\mathfrak{k} \cap \mathfrak{h} = (\mathfrak{k} \cap \mathfrak{h})^{\perp \wedge \perp}$.*
  (Ω) *If $\mathfrak{k}$ is a Cartan subalgebra of $\mathfrak{g}$, then $\mathfrak{h} \cap \mathfrak{n} = \Lambda^\perp \subseteq \mathfrak{k}$.*

Proof. The equivalence of (1) and (1′) is clear from Lemma 4.5.
From Lemmas 4.7 and 4.8 we have the equivalence of (1) and (2).
(3)$\Rightarrow$(1′): Suppose $y \in \mathrm{Supp}_\Lambda(x)^\perp$ and $x \in \mathrm{reg}\,\mathfrak{g}$. For $\lambda \in \mathrm{Supp}_\Lambda(x)^\perp$ we have $y \in \ker \lambda$. Hence (3) implies $[x, y] = [x_0, y] + \sum_{\lambda \in \mathrm{Supp}_\Lambda(x)}[x_\lambda, y] = [x_0, y] \in [\mathfrak{h}, \mathfrak{h}] \subseteq \Lambda^\perp$. Hence $\lambda([x, y]) = 0$ for all $\lambda \in \Lambda$. Thus $(\mathrm{ad}\,x)^2(y) = [x, [x, y]] = [x_0, [x, y]] + \sum_{\lambda \in \Lambda} [x_\lambda, [x, y]] = [x_0, [x_0, y]] = (\mathrm{ad}\,x_0)^2(y)$. By induction we have $(\mathrm{ad}\,x)^n(y) = (\mathrm{ad}\,x_0)^n(y)$ and this element vanishes for sufficiently large n because x_0 and y are in the nilpotent algebra $\mathfrak{h}$. Thus $y \in \mathfrak{g}^0(x) \cap \mathfrak{h}$ which we claimed.
(2)$\Rightarrow$(3): We set $\mathfrak{m} = \mathfrak{h} \oplus \bigoplus_{n=1}^\infty \mathfrak{g}^{n \cdot \lambda}$. Then $\mathfrak{m}$ is a subalgebra containing $\mathfrak{h}$ and $\mathfrak{g}^\lambda$. Now $\ker \lambda = \ker n \cdot \lambda$ for $n = 2, \ldots$ Thus $\Lambda(\mathfrak{m}) = \{n \cdot \lambda |\, n = 1, 2, \ldots\}$

and $\Lambda(\mathfrak{m})^\perp = \ker \lambda$. By (2) we know that $\ker \lambda = \Lambda(\mathfrak{m})^\perp = \mathfrak{n}(\mathcal{C}(\mathfrak{m}))$, i.e., that $\ker \lambda$ is an ideal in $\mathfrak{m}$. Then $[\ker \lambda, \mathfrak{g}^\lambda] \subseteq \mathfrak{g}^\lambda \cap \ker \lambda = \{0\}$. This proves (3).

(3)$\Rightarrow$(4): Condition (3) implies that $\text{ad}\,|\mathfrak{h}: \mathfrak{h} \to \mathfrak{gl}(\mathfrak{g}^\lambda)$ factors through $\mathfrak{h}/\ker \lambda$. Since the representation is not zero by the definition of a root space, the image is isomorphic to $\mathfrak{h}/\ker \lambda$ and thus is one-dimensional.

(4)$\Rightarrow$(3): If the image of the adjoint representation $\rho: \mathfrak{h} \to \mathfrak{gl}(\mathfrak{g}^\lambda)$ of $\mathfrak{h}$ on $\mathfrak{g}^\lambda$ is one-dimensional, then the kernel $\ker \rho$ is a hyperplane in $\mathfrak{h}$. If $\rho(x) = 0$ then $\lambda(x) = 0$. Thus $\ker \rho \subseteq \ker \lambda$, since $\lambda \neq 0$ we have $\ker \rho = \ker \lambda$ for all $\lambda \in \Lambda$. But this is exactly (3).

(1)$\Rightarrow (\omega)$: We can write $\mathfrak{k}$ as $\mathfrak{g}^0(x)$ for x regular in $\mathfrak{k}$. Then

$$\mathfrak{g}^0(x) \cap \mathfrak{h} = \text{Supp}_\Lambda(x)^\perp = \{h \in \mathfrak{h}: \lambda(h) = 0 \text{ if } x_\lambda \neq 0\}$$
$$= \bigcap\{\ker \lambda: x_\lambda \neq 0\} \supseteq \bigcap\{\ker \lambda: \lambda \in \Lambda, \mathfrak{k} \cap \mathfrak{h} \subseteq \ker \lambda\} = (\mathfrak{h} \cap \mathfrak{k})^{\perp_\Lambda \perp}.$$

The reverse inclusion always obtains.

Finally (Ω) is equivalent to (ω) by 4.2(ii). $\square$

In order to facilitate the formulation of the main consequences we resort to the following nomenclature. The second part of the definition was already discussed in the introduction of this section; it is recorded here for the sake of systematic completeness.

4.10. DEFINITION. Let $\mathfrak{g}$ be a real or complex Lie algebra.
(i) We say that $\mathfrak{g}$ is *Cartan adapted* if $\mathfrak{g}$ (in case $\mathfrak{g}$ is complex), respectively $\mathfrak{g}_\mathbb{C}$ (in case $\mathfrak{g}$ is real) satisfies the equivalent conditions of Theorem 4.9 with respect to one, and thus with respect to each Cartan subalgebra (since they are all conjugate).
(ii) $\mathfrak{g}$ is said to be *Cartan finite* if for any fixed Cartan subalgebra $\mathfrak{h}$ the family of intersections

$$\{\mathfrak{h} \cap \mathfrak{k} \mid \mathfrak{k} \text{ is a Cartan subalgebra of } \mathfrak{g}\}$$

is finite. $\square$

An immediate consequence of Lemma 4.8, Theorem 4.9 and Lemma 4.4 is the following:

4.11. PROPOSITION. *If $\mathfrak{g}$ is a Cartan adapted complex Lie algebra, then*

$$\{\mathfrak{h} \cap \mathfrak{k} \mid \mathfrak{k} \in \mathcal{C}\} = \{\Omega^\perp \mid \Omega \subseteq \Lambda\}.$$

In particular, $|\{\mathfrak{h} \cap \mathfrak{k} : \mathfrak{k} \in \mathcal{C}\}| \leq 2^{|\Lambda|}$. $\square$

4.12. COROLLARY. *Suppose that $\mathfrak{g}$ is Cartan adapted. Then $\mathfrak{g}$ is Cartan finite.*

Proof. We pass to the complexification $\mathfrak{g}_\mathbb{C}$. Then $\mathfrak{h}_\mathbb{C}$ is a Cartan subalgebra in the complex Lie algebra $\mathfrak{g}_\mathbb{C}$ if $\mathfrak{h}$ is a Cartan subalgebra of $\mathfrak{g}$. Thus the set of all $\mathfrak{h}_\mathbb{C} \cap \mathfrak{k}_\mathbb{C}$, as $\mathfrak{k}_\mathbb{C}$ ranges through the set of all Cartan subalgebras of $\mathfrak{g}_\mathbb{C}$, is finite by Proposition 4.11. Then the set of all $\mathfrak{h} \cap \mathfrak{k} = \mathfrak{h}_\mathbb{C} \cap \mathfrak{k}_\mathbb{C} \cap \mathfrak{g}$ is finite. □

We conclude with some examples which show that these results are sharp.

4.13. EXAMPLE. (i) We let $\mathfrak{g}$ denote the Lie algebra of all matrices of the form

$$[c; x, z; u, v, w; p, q] := \begin{pmatrix} c & x & z & u & 0 & 0 & 0 \\ 0 & c & 0 & v & 0 & 0 & 0 \\ 0 & 0 & c & w & 0 & 0 & 0 \\ 0 & 0 & 0 & 0 & 0 & 0 & 0 \\ 0 & 0 & 0 & 0 & x & 0 & p \\ 0 & 0 & 0 & 0 & 0 & z & q \\ 0 & 0 & 0 & 0 & 0 & 0 & 0 \end{pmatrix}, \quad c, x, z, u, v, w, p, q \in \mathbb{C}.$$

Then $\mathfrak{h} := \{[c; x, z; 0, 0, 0; 0, 0] \mid c, x, z \in \mathbb{C}\}$ is an abelian Cartan subalgebra. The set Λ of roots contains three elements λ_j, $j = 1, 2, 3$ given by

$$\lambda_1([c; x, z; 0, 0, 0; 0, 0]) = c, \quad \lambda_2([c; x, z; 0, 0, 0; 0, 0]) = x, \quad \lambda_3([c; x, z; 0, 0, 0; 0, 0]) = z.$$

The corresponding root spaces are

$$\mathfrak{g}^{\lambda_1} = \{[0; 0, 0, ; u, v, w; 0, 0] \mid u, v, w \in \mathbb{C}\},$$
$$\mathfrak{g}^{\lambda_2} = \{[0; 0, 0; 0, 0, 0; p, 0] \mid p \in \mathbb{C}\},$$
$$\mathfrak{g}^{\lambda_3} = \{[0; 0, 0; 0, 0, 0; 0, q] \mid q \in \mathbb{C}\}.$$

The nilradical is $\mathfrak{n} = \{[0; 0, 0; u, v, w; p, q] \mid u, v, w, p, q \in \mathbb{C}\}$. Hence $\mathfrak{h} \cap \mathfrak{n} = \Lambda^\perp = \{0\}$. For each $v \in \mathbb{C}$, the set of all

$$\langle c; x, z \rangle := [c; x, z; xv + z, cv, c; 0, 0], \quad c, x, z \in \mathbb{C}$$

is a Cartan algebra $\mathfrak{h}_v$. Then the intersection $\mathfrak{h} \cap \mathfrak{h}_v$ contains an element $\langle c; x, z \rangle$ iff $c = 0$ and $z = -vx$. Hence the set $\{\mathfrak{h} \cap \mathfrak{k} \mid \mathfrak{k} \in \mathcal{C}\}$ is infinite. One regular element X in $\mathfrak{h}_v$ is given by $X = \langle 1; 0, 0 \rangle = [1; 0, 0; 0, v, 1; 0, 0]$. Then $\mathrm{Supp}_\Lambda(X) = \{\lambda_1\}$. Thus $\mathrm{Supp}_\Lambda(X)^\perp = \{[0; x, z; 0, 0, 0; 0, 0] \mid x, z \in \mathbb{C}\}$ and this set is different from the one-dimensional set $\mathfrak{h} \cap \mathfrak{g}^0(X) = \mathfrak{h} \cap \mathfrak{h}_v$.

(ii) We let $\mathfrak{g}$ denote the Lie algebra of all matrices of the form

$$[c; x, z; u, v, w] := \begin{pmatrix} c & x & z & u \\ 0 & c & 0 & v \\ 0 & 0 & c & w \\ 0 & 0 & 0 & 0 \end{pmatrix}, \quad c, x, z, u, v, w \in \mathbb{C}.$$

This Lie algebra is the Lie algebra of an algebraic subgroup of $\mathrm{Gl}(\mathbb{C}, 4)$. Further, $\mathfrak{h} := \{[c; x, z; 0, 0, 0; 0, 0] \mid c, x, z \in \mathbb{C}\}$ is an abelian Cartan subalgebra. The set Λ of roots contains one element λ, given by $\lambda([c; x, z; 0, 0, 0]) = c$. The corresponding root space is

$$\mathfrak{g}^{\lambda_1} = \{[0; 0, 0; u, v, w] \mid u, v, w \in \mathbb{C}\}.$$

The nilradical is $\mathfrak{n} = \{[0; x, z; u, v, w] \mid x, z, u, v, w \in \mathbb{C}\}$. Hence $\mathfrak{h} \cap \mathfrak{n} = \Lambda^\perp = \{[0; x, z; 0, 0, 0]\}$.

The remaining properties are the same as those of Example (i) above. In Example (ii) one gains the property that $\mathfrak{g} = \tilde{\mathfrak{g}}$ is the Lie algebra of an algebraic group and that the dimension is smaller, and one loses the property that $\mathfrak{h} \cap \mathfrak{n} = \{0\}$. □

This example shows that there are solvable complex Lie algebras (even those of an algebraic group) whose Cartan algebras are abelian such that the family $\{\mathfrak{h} \cap \mathfrak{k} \mid \mathfrak{k} \in \mathcal{C}\}$ is not finite. If one does not insist that the Lie algebra be one of an algebraic group then the examples show that $\mathfrak{h} \cap \mathfrak{n} = \Lambda^\perp = \{0\}$ is possible all the while.

The algebra of Part (ii) in Example 4.13 is (isomorphic to) a maximal rank subalgebra of the algebra in Part (i). Hence neither of the Examples 4.13 satisfies the conditions of Theorem 4.9.

In Example 4.13(i) we have $\mathfrak{n}(\mathcal{C}) = \{0\} = \mathfrak{n} \cap \mathfrak{h}$, in Example 4.13(ii), however, $\mathfrak{n}(\mathcal{C}) = \{0\} \neq \mathfrak{n} \cap \mathfrak{h}$.

There are examples of Lie algebras in which the set $\{\mathfrak{h} \cap \mathfrak{k} : \mathfrak{k} \in \mathcal{C}\}$ is finite, but in which the equivalent conditions in Theorem 4.9 are not satisfied, i.e., $\mathfrak{g}$ is not Cartan adapted. A coherent theory of Cartan finite algebras seems difficult to achieve at this stage however. Thus we have restricted our attention to the general, yet accessible, case of Cartan adapted algebras where the set of intersections is finite and can be precisely computed.

We observe that there is a special important class of Lie algebras which are always Cartan adapted. We recall an important property of certain subalgebras.

4.14. DEFINITION. (i) A subset $\mathfrak{e}$ of a Lie algebra $\mathfrak{g}$ is called *reductively embedded* if the operators $\mathrm{ad}_\mathfrak{g} x$, $x \in \mathfrak{e}$ are semisimple (cf. BOURBAKI [5] I, §6.6).

(ii) We say that $\mathfrak{e}$ is *compactly embedded* if $e^{\mathrm{ad}\,\mathfrak{e}}$ is contained in a compact subgroup of the automorphism group $\mathrm{Aut}\,\mathfrak{g}$ of $\mathfrak{g}$. □

Every compactly embedded subset is semisimply embedded.

We cite a well known proposition (cf., e.g. BOURBAKI, [5], Chap. VII., §2, n°3, Proposition 10) which is significant in this context.

4.15. PROPOSITION. *Let $\mathfrak{a}$ be a reductively embedded commutative subalgebra of a Lie algebra $\mathfrak{g}$. Then the Cartan subalgebras of the centralizer $\mathfrak{z}(\mathfrak{a},\mathfrak{g})$ of $\mathfrak{a}$ in $\mathfrak{g}$ are exactly the Cartan subalgebras of $\mathfrak{g}$ which contain $\mathfrak{a}$.* □

In particular, every reductively embedded abelian vector subspace is contained in a Cartan subalgebra.

4.16. LEMMA. *In a real or complex Lie algebra $\mathfrak{g}$, if one Cartan subalgebra is reductively embedded then all are.*

Proof. Let $\mathfrak{h}$ be reductively embedded in $\mathfrak{g}$ and let $\mathfrak{k}$ be any Cartan subalgebra of $\mathfrak{g}$. If $\mathfrak{g}$ is real, then $\mathfrak{h}_{\mathbb{C}}$ is reductively embedded in $\mathfrak{g}_{\mathbb{C}}$. Since $\mathfrak{k}_{\mathbb{C}}$ is conjugate to $\mathfrak{h}_{\mathbb{C}}$ in $\mathfrak{g}_{\mathbb{C}}$ under inner automorphisms, $\mathfrak{k}_{\mathbb{C}}$ is reductively embedded in $\mathfrak{g}_{\mathbb{C}}$. Then $\mathfrak{k}$ is reductively embedded in $\mathfrak{g}$. □

4.17. PROPOSITION. *Suppose that $\mathfrak{g}$ is a complex Lie algebra whose Cartan algebras are reductively embedded. Then $\mathfrak{g}$ is Cartan adapted and for each Cartan subalgebra $\mathfrak{h}$ we have*
$$\Lambda^{\perp} = \mathfrak{h} \cap \mathfrak{n} = \mathfrak{z} = \mathfrak{n}(\mathcal{C}).$$

Proof. If the Cartan algebra $\mathfrak{h}$ is reductively embedded, then Condition (3) of Theorem 4.9 is satisfied and so $\mathfrak{g}$ is Cartan adapted by Definition 4.10. By 4.3 we know $\Lambda^{\perp} = \mathfrak{h} \cap \mathfrak{n}$. Let $h \in \mathfrak{h} \cap \mathfrak{n}$. Then, firstly, $\operatorname{ad} h$ is nilpotent and, secondly, $\operatorname{ad} h$ is semisimple by hypothesis. Hence $\operatorname{ad} h = 0$, i.e., $h \in \mathfrak{z}$. Trivially, $\mathfrak{z} \subseteq \mathfrak{n}(\mathcal{C}) \subseteq \mathfrak{h} \cap \mathfrak{n}$. □

Example 4.13(i) shows that the hypothesis in Proposition 4.17 is sufficient but not necessary for $\Lambda^{\perp} = \mathfrak{h} \cap \mathfrak{n} = \mathfrak{z} = \mathfrak{n}(\mathcal{C})$ to hold.

4.18. PROPOSITION. *If all Cartan algebras of a real or complex Lie algebra $\mathfrak{g}$ are reductively embedded, then $\mathfrak{g}$ is Cartan finite, i.e., $\{\mathfrak{h} \cap \mathfrak{k} \mid \mathfrak{k} \in \mathcal{C}\}$ is finite for each $\mathfrak{h} \in \mathcal{C}$.*

Proof. Consider first the complex case. By the preceding Proposition 4.17, $\mathfrak{g}$ is Cartan adapted. Then Corollary 4.12 proves the assertion.

For the real case, one notes that the Cartan algebras of the complexification $\mathfrak{g}_{\mathbb{C}}$ of $\mathfrak{g}$ are reductively embedded, and hence we are reduced to the previous case. □

4.19. DEFINITION. Suppose that $\mathfrak{g}_1$ is a subalgebra of a Lie algebra $\mathfrak{g}$. We shall say that a subalgebra $\mathfrak{a}$ of $\mathfrak{g}_1$ is $\mathfrak{g}$-*reductive* if it is reductively embedded in $\mathfrak{g}$ (not merely in $\mathfrak{g}_1$). □

4.20. PROPOSITION. *Let $\mathfrak{g}$ be a Lie algebra.*

(i) *Every $\mathfrak{g}$-reductive abelian subalgebra $\mathfrak{a}$ of $\mathfrak{g}_1$ is contained in a maximal $\mathfrak{g}$-reductive abelian subalgebra of $\mathfrak{g}_1$.*

(ii) *If $\mathfrak{a}$ is a maximal $\mathfrak{g}$-reductive abelian subalgebra of $\mathfrak{g}_1$ then so is $\varphi(\mathfrak{a})$ for every automorphism of $\mathfrak{g}$ leaving $\mathfrak{g}_1$ invariant.*

(iii) *Every $\mathfrak{g}$-reductive abelian subalgebra of $\mathfrak{g}_1$ is contained in the center of a Cartan subalgebra of $\mathfrak{g}_1$. Conversely, every Cartan subalgebra of $\mathfrak{g}_1$ contains a unique maximal $\mathfrak{g}$-reductive abelian subalgebra, which lies in its center.*

(iv) *The intersection of all maximal $\mathfrak{g}$-reductive abelian subalgebras of $\mathfrak{g}$ is the center of $\mathfrak{g}$.*

(v) *The set of all maximal $\mathfrak{g}$-reductive abelian subalgebras of $\mathfrak{g}_1$ is a finite union of conjugacy classes. If all Cartan subalgebras of $\mathfrak{g}_1$ are conjugate then so are all maximal $\mathfrak{g}$-reductive abelian subalgebras.*

(vi) *All maximal $\mathfrak{g}$-reductive abelian subalgebras of $\mathfrak{g}_1$ are isomorphic.*

Proof. (i) and (ii) are straightforward.

(iii) Let $\mathfrak{a}$ be a $\mathfrak{g}$-reductive subalgebra in $\mathfrak{g}_1$. Then $\mathfrak{a}$ is reductive also in $\mathfrak{g}_1$, hence by 4.15 is contained in the center of a Cartan subalgebra of $\mathfrak{g}_1$. The second assertion follows from the fact that two commuting reductively embedded subalgebras of $\mathfrak{h}$ have a reductively embedded sum.

(iv) The center of $\mathfrak{g}$ is reductively embedded, hence contained in every maximal abelian $\mathfrak{g}$-reductive subalgebra. Conversely, let x be an element which is contained in every maximal $\mathfrak{g}$-reductive subalgebra. Then x commutes with every element of each Cartan subalgebra by (iii). However, the Cartan subalgebras of $\mathfrak{g}$ generate $\mathfrak{g}$.

(v) follows from (iii) and the fact that $\mathcal{C}(\mathfrak{g}_1)$ is a union of finitely many conjugacy classes.

(vi) The assertion is true if the ground field is $\mathbb{C}$. Let $\mathfrak{v}$ be a maximal $\mathfrak{g}$-reductive abelian subalgebra of $\mathfrak{g}_1$. Then $\mathfrak{v}_\mathbb{C} \subseteq (\mathfrak{g}_1)_\mathbb{C}$ is $\mathfrak{g}_\mathbb{C}$-reductive and abelian and is, therefore, contained in a maximal $\mathfrak{g}_\mathbb{C}$-reductive abelian subalgebra $\mathfrak{m}$ of $(\mathfrak{g}_1)_\mathbb{C}$. We claim that $\mathfrak{v}_\mathbb{C} = \mathfrak{m}$: once this claim is established then $\dim_\mathbb{R} \mathfrak{v} = \dim_\mathbb{C} \mathfrak{m}$. Since all maximal $\mathfrak{g}_\mathbb{C}$-reductive abelian subalgebras of $(\mathfrak{g}_1)_\mathbb{C}$ are conjugate by (v) (since in a complex Lie algebra all Cartan subalgebras are conjugate), it follows that they have the same dimension and the assertion follows.

Since the abelian algebra $\mathfrak{v}$ is reductively embedded in $\mathfrak{g}_1$ it is contained in the center of a Cartan algebra $\mathfrak{h}$ of $\mathfrak{g}_1$. Then $\mathfrak{h} \subseteq \mathfrak{z}(\mathfrak{v}, \mathfrak{g}_1)$ and thus $\mathfrak{h}_\mathbb{C} \subseteq \mathfrak{z}(\mathfrak{v}_\mathbb{C}, (\mathfrak{g}_1)_\mathbb{C})$. Also $\mathfrak{m} \subseteq \mathfrak{z}(\mathfrak{v}_\mathbb{C}, \mathfrak{g}_\mathbb{C})$ since $\mathfrak{m}$ is abelian. Since $\mathfrak{m}$ is abelian and $\mathfrak{g}_\mathbb{C}$-reductively embedded, there is a Cartan algebra $\mathfrak{k}$ of $(\mathfrak{g}_1)_\mathbb{C}$ containing $\mathfrak{m}$. Since $\mathfrak{k}$ and $\mathfrak{h}_\mathbb{C}$ are conjugate in $(\mathfrak{g}_1)_\mathbb{C}$, then a conjugate $\mathfrak{n}$ of $\mathfrak{m}$ in $(\mathfrak{g}_1)_\mathbb{C}$ contains $\mathfrak{v}_\mathbb{C}$ and is contained in $\mathfrak{h}_\mathbb{C}$. We may assume without losing generality that $\mathfrak{n} = \mathfrak{m}$.

Now let $\kappa: \mathfrak{g}_\mathbb{C} \to \mathfrak{g}_\mathbb{C}$ be the real involution given by $\kappa(x + i\cdot y) = x - i\cdot y$ for $x, y \in \mathfrak{g}$. Then $\kappa(\mathfrak{h}_\mathbb{C}) = \mathfrak{h}_\mathbb{C}$, whence $\kappa(\mathfrak{z}(\mathfrak{h}_\mathbb{C})) = \mathfrak{z}(\mathfrak{h}_\mathbb{C})$. Thus $\mathfrak{m} \subseteq \mathfrak{z}(\mathfrak{h}_\mathbb{C})$ implies $\kappa(\mathfrak{m}) \subseteq \mathfrak{z}(\mathfrak{h}_\mathbb{C})$. Therefore $\mathfrak{m} + \kappa(\mathfrak{m})$ is an abelian subalgebra of $(\mathfrak{g}_1)_\mathbb{C}$ containing $\mathfrak{v}_\mathbb{C}$. Moreover, $\kappa(\mathfrak{m}) + \mathfrak{m}$ is also $\mathfrak{g}_\mathbb{C}$-reductive: for $m \in \mathfrak{m}$ and $m' \in \kappa(\mathfrak{m})$ the operators ad m and ad m' are semisimple and commute, so their sum $\operatorname{ad}(m + m') = \operatorname{ad} m +$

ad m' is semisimple. By the maximality of $\mathfrak{m}$ we conclude $\mathfrak{m} + \kappa(\mathfrak{m}) \subseteq \mathfrak{m}$. This means $\kappa(\mathfrak{m}) = \mathfrak{m}$ and thus $\mathfrak{m} = (\mathfrak{m} \cap \mathfrak{g}) + i \cdot (\mathfrak{m} \cap \mathfrak{g}) = \mathfrak{v} + i \cdot \mathfrak{v} = \mathfrak{v}_{\mathbb{C}}$, as asserted. (We thank K.-H. NEEB for this proof.) □

4.21. DEFINITION. *A subalgebra $\mathfrak{g}_1$ of $\mathfrak{g}$ is called Cartan reductive in $\mathfrak{g}$ if every Cartan subalgebra of $\mathfrak{g}$ meets $\mathfrak{g}_1$ in a reductively embedded subalgebra.*
□

One notices at once two principal sufficient conditions under which a subalgebra is Cartan reductive in a containing algebra:

4.22. PROPOSITION. *A subalgebra $\mathfrak{g}_1$ of a Lie algebra $\mathfrak{g}$ is Cartan reductive if at least one of the following conditions is satisfied:*
  (i) *At least one Cartan subalgebra of $\mathfrak{g}$ is reductively embedded.*
  (ii) *$\mathfrak{g}_1$ is reductively embedded.*
Condition (i) is equivalent to saying that all Cartan subalgebras are reductively embedded. It is satisfied if at least one Cartan subalgebra is compactly embedded. Condition (ii) holds whenever $\mathfrak{g}_1$ is compactly embedded. Both conditions are satisfied if $\mathfrak{g}$ is a compact Lie algebra. □

4.23. LEMMA. *Suppose that $\mathfrak{g}$ is a complex Lie algebra and that $\mathfrak{h} \in \mathcal{C}(\mathfrak{g})$. Let $\mathfrak{v}$ be a $\mathfrak{g}$-reductive subalgebra of $\mathfrak{h}$. Then for any $\mathfrak{k} \in \mathcal{C}(\mathfrak{g})$ we have*

$$\mathfrak{v} \cap \mathfrak{k} = \mathfrak{v} \cap \operatorname{Supp}_\Lambda(\mathfrak{k})^\perp.$$

Proof. First we claim

(1) $\qquad \mathfrak{v} \cap \mathfrak{k} = \mathfrak{v} \cap \mathfrak{z}(\mathfrak{k}), \quad \mathfrak{z}(\mathfrak{k}) = \text{center of } \mathfrak{k}.$

Proof: Since $\mathfrak{v}$ is $\mathfrak{g}$-reductive, $\mathfrak{v} \cap \mathfrak{k} \subseteq \mathfrak{z}(\mathfrak{k})$. Thus $\mathfrak{v} \cap \mathfrak{k} \subseteq \mathfrak{v} \cap \mathfrak{z}(\mathfrak{k})$. The converse, however, is trivial.

If $v \in \mathfrak{v}$ and $x \in \mathfrak{g}$ write $x = x_0 + \sum_{\lambda \in \Lambda} x_\lambda$ according with the root space decomposition. Since $\mathfrak{v}$ is $\mathfrak{g}$-reductive we obtain

(2) $\qquad [v, x] = [v, x_0 + \sum_{\mu \in \Lambda} x_\mu] = \sum_{\mu \in \Lambda} \mu(v) \cdot x_\mu.$

Next we claim $\mathfrak{v} \cap \mathfrak{z}(\mathfrak{k}) \subseteq \operatorname{Supp}_\Lambda(\mathfrak{k})^\perp$.
Proof: Let $v \in \mathfrak{z}(\mathfrak{k})$ and take $\lambda \in \operatorname{Supp}_\Lambda(\mathfrak{k})$. By the definition of $\operatorname{Supp}_\Lambda(\mathfrak{k})$ there is a $y \in \mathfrak{k}$ with $\operatorname{pr}_\lambda(y) \neq 0$. Since $v \in \mathfrak{z}(\mathfrak{k})$, in view of (2) we get $0 = [v, y] = [v, y_0 + \sum_{\mu \in \Lambda} y_\mu] = \sum_{\mu \in \Lambda} \mu(v) \cdot y_\mu$. In particular, $0 = \lambda(v) \cdot y_\lambda = \lambda(v) \cdot \operatorname{pr}_\lambda(y)$. Since $\operatorname{pr}_\lambda(y) \neq 0$ it follows that $\lambda(v) = 0$. This proves the claim.

Finally we claim $\mathfrak{v} \cap \operatorname{Supp}_\Lambda(\mathfrak{k})^\perp \subseteq \mathfrak{z}(\mathfrak{k})$.
Proof: Let $v \in \mathfrak{v} \cap \operatorname{Supp}_\Lambda(\mathfrak{k})^\perp$. Take any $y \in \mathfrak{k}$. By (2) we have $[v, y] = [v, y_0 + \sum_{\mu \in \Lambda} y_\mu] = \sum_{\mu \in \Lambda} \mu(v) \cdot y_\mu$. For each $\lambda \in \Lambda$ we have either $y_\lambda = 0$ or $y_\lambda \neq 0$. In the latter case $\lambda \in \operatorname{Supp}_\Lambda(\mathfrak{k})$. Then by the definition of $\operatorname{Supp}_\Lambda(\mathfrak{k})$ and the choice of v we obtain $\lambda(v) = 0$. Thus $[v, y] = 0$ and therefore $v \in \mathfrak{z}(\mathfrak{k})$, as asserted. □

4.24. PROPOSITION. *Let $\mathfrak{g}$ be a real or complex Lie algebra and $\mathfrak{v}$ a $\mathfrak{g}$-reductive abelian subalgebra of $\mathfrak{g}$. Then the set $\{\mathfrak{v} \cap \mathfrak{k} \mid \mathfrak{k} \in \mathcal{C}(\mathfrak{g})\}$ is finite.*

Proof. If $\mathfrak{g}$ is real, then $\mathfrak{v}_\mathbb{C}$ is $\mathfrak{g}_\mathbb{C}$-reductive in $\mathfrak{g}_\mathbb{C}$. Also $\mathfrak{v} \cap \mathfrak{k} = (\mathfrak{v}_\mathbb{C} \cap \mathfrak{k}_\mathbb{C}) \cap \mathfrak{g}$. It follows that a proof of the assertion for complex Lie algebras implies the conclusion for real Lie algebras. Thus we assume henceforth that $\mathfrak{g}$ is complex.

Since $\mathfrak{v}$ is $\mathfrak{g}$-reductive, by Proposition 4.15 there is a $\mathfrak{h} \in \mathcal{C}$ with $\mathfrak{v} \subseteq \mathfrak{h}$. Lemma 4.23 then implies that $\mathfrak{v} \cap \mathfrak{k} = \mathfrak{v} \cap \operatorname{Supp}_\Lambda(\mathfrak{k})^\perp$ for each $\mathfrak{k} \in \mathcal{C}$ where $\operatorname{Supp}_\Lambda(\mathfrak{k})$ is defined with respect to $\mathfrak{h}$. The set Λ being finite, there are only finitely many subsets of Λ. Thus there are only finitely many subsets of $\mathfrak{v}$ of the form $\mathfrak{v} \cap \Omega^\perp$, and the assertion follows. □

We now merge the results on Weyl groups and on the intersections of Cartan subalgebras and establish finiteness properties for subsets of Cartan algebras under automorphisms and conjugation. We begin by exploiting the finiteness of the Weyl group. Again we recall that $\mathbb{G}$ denotes the group of inner automorphisms of $\mathfrak{g}$.

4.25. NOTATION. Suppose that $\mathfrak{a}$ and $\mathfrak{b}$ are subsets of the Lie algebra $\mathfrak{g}$. Let $\Phi(\mathfrak{a}, \mathfrak{b})$ denote the set of all bijections from $\mathfrak{a}$ to $\mathfrak{b}$ which are restrictions of inner automorphisms of $\mathfrak{g}$. Note that in general $\Phi(\mathfrak{a}, \mathfrak{b})$ is non-empty if and only if $\mathfrak{a}$ and $\mathfrak{b}$ are conjugate under an inner automorphism. If both $\mathfrak{a}$ and $\mathfrak{b}$ are Lie subalgebras, we note also that each member of $\Phi(\mathfrak{a}, \mathfrak{b})$ induces an isomorphism from $\mathfrak{a}/[\mathfrak{a}, \mathfrak{a}] \to \mathfrak{b}/[\mathfrak{b}, \mathfrak{b}]$. □

4.26. PROPOSITION. *Let $\mathfrak{g}$ be a complex or solvable Lie algebra with Cartan subalgebra $\mathfrak{h}$, and let $\mathfrak{a}$ and $\mathfrak{b}$ be subalgebras of $\mathfrak{h}$ such that $[\mathfrak{a}, \mathfrak{h}] \subseteq [\mathfrak{a}, \mathfrak{a}]$ and $[\mathfrak{b}, \mathfrak{h}] \subseteq [\mathfrak{b}, \mathfrak{b}]$. Then if $f \in \Phi(\mathfrak{a}, \mathfrak{b})$, there exists $\alpha + \mathbb{Z}(\mathfrak{h}, \mathfrak{g}) \in \mathcal{W}(\mathfrak{h}, \mathfrak{g})$ such that $\alpha(\mathfrak{a}) = \mathfrak{b}$ and α and f induce the same mappings from $\mathfrak{a}/[\mathfrak{a}, \mathfrak{a}]$ to $\mathfrak{b}/[\mathfrak{b}, \mathfrak{b}]$. In particular, if $\mathfrak{a} = \mathfrak{b}$, then the members of $\mathcal{W}(\mathfrak{a}, \mathfrak{g})$ may be viewed as restrictions of those members of $\mathcal{W}(\mathfrak{h}, \mathfrak{g})$ which carry $\mathfrak{a}$ to $\mathfrak{a}$.*

Proof. By definition of $\Phi(\mathfrak{a}, \mathfrak{b})$ there exists $\varphi \in \mathbb{G}$ such that $\varphi|\mathfrak{a} = f$. Then $\varphi(\mathfrak{h})$ is a Cartan subalgebra containing $\mathfrak{b}$, and

$$[\varphi(\mathfrak{h}), \mathfrak{b}] = [\varphi(\mathfrak{h}), \varphi(\mathfrak{a})] = \varphi[\mathfrak{h}, \mathfrak{a}] \subseteq \varphi[\mathfrak{a}, \mathfrak{a}] = [\mathfrak{b}, \mathfrak{b}].$$

Then the subalgebra $\mathfrak{k} := \{x \mid [x, \mathfrak{b}] \subseteq [\mathfrak{b}, \mathfrak{b}]\}$ contains $\mathfrak{h} \cup \varphi(\mathfrak{h})$. Since $\mathfrak{h}$ and $\varphi(\mathfrak{h})$ are Cartan subalgebras of the complex or solvable algebra $\mathfrak{k}$, there exists an inner automorphism θ of $\mathfrak{k}$ carrying $\varphi(\mathfrak{h})$ to $\mathfrak{h}$, and since θ is inner we can also regard it as an automorphism on $\mathfrak{g}$. It follows from Lemma 3.6(v) that $\theta \in \mathbb{Z}(\mathfrak{b}, \mathbb{G})$. The composition $\alpha := \theta \circ \varphi$ is contained in $\mathbb{N}(\mathfrak{h}, \mathbb{G})$. Since $\theta \in \mathbb{Z}(\mathfrak{b}, \mathbb{G})$, it carries $\mathfrak{b}$ into itself and induces the identity on $\mathfrak{b}/[\mathfrak{b}, \mathfrak{b}]$. Thus $\theta \circ \varphi$ and φ induce the same mappings from $\mathfrak{a}/[\mathfrak{a}, \mathfrak{a}]$ to $\mathfrak{b}/[\mathfrak{b}, \mathfrak{b}]$.

The last assertion now follows in a straightforward manner. □

4.27. THEOREM. *Assume that $\mathfrak{a}$ is a subalgebra of a Cartan subalgebra $\mathfrak{h}$ satisfying $[\mathfrak{a}, \mathfrak{h}] = \{0\}$. Then, if $\alpha \in \mathbb{G}$ and $\alpha(\mathfrak{a}) = \mathfrak{a}' \subseteq \mathfrak{h}$ and $[\mathfrak{a}', \mathfrak{h}] = 0$, the assignment*

$$\theta \colon \mathcal{W}(\mathfrak{a}, \mathfrak{g}) \to \Phi(\mathfrak{a}, \mathfrak{a}'), \quad \theta(\varphi \cdot \mathbb{Z}(\mathfrak{a}, \mathbb{G})) = \alpha\varphi|\mathfrak{a}$$

is a well defined bijection. In particular, $|\Phi(\mathfrak{a}, \mathfrak{a}')| = |\mathcal{W}(\mathfrak{a}, \mathfrak{g})|$ is finite. Furthermore, each member of $\Phi(\mathfrak{a}, \mathfrak{a}')$ can be viewed as a restriction and corestriction of a member of the Weyl group of $\mathfrak{h}$.

Proof. By Remark 3.2 the member of the Weyl group $\varphi \cdot \mathbb{Z}(\mathfrak{a}, \mathbb{G})$ acts on $\mathfrak{a}$ by restricting φ to $\mathfrak{a}$, since $[\mathfrak{a}, \mathfrak{a}] = \{0\}$. Suppose that $\beta \in \mathbb{G}$ is such that $\beta(\mathfrak{a}) = \mathfrak{a}' = \alpha(\mathfrak{a})$. Then $\psi := \alpha^{-1}\beta \in \mathbb{G}$ satisfies $\psi(\mathfrak{a}) = \mathfrak{a}$ and is, therefore in $N(\mathfrak{a}, \mathbb{G})$. If $\beta' \in \mathbb{G}$ satisfies $\beta'(y) = \beta(y)$ for all $y \in \mathfrak{a}$, then $\zeta := \beta^{-1}\beta' \in \mathbb{Z}(\mathfrak{a}, \mathbb{G})$ and $\beta' = \alpha\varphi\zeta$. These remarks show that θ is well defined and bijective. The remainder is now a consequence of Theorem 3.9 and the preceding Proposition 4.26. □

4.28. PROPOSITION. *Let $\mathfrak{g}$ be a real or complex Lie algebra and $\mathfrak{v}$ a $\mathfrak{g}$-reductive abelian subalgebra of $\mathfrak{g}$. Then for any $\mathfrak{h} \in \mathcal{C}(\mathfrak{g})$ containing $\mathfrak{v}$ the set $\{\mathfrak{h} \cap \varphi(\mathfrak{v}) \mid \varphi \in \mathbb{G}\}$ is finite.*

Proof. Let $\mathfrak{m}$ denote the unique maximal $\mathfrak{g}$-reductive subalgebra of $\mathfrak{h}$. Then $\mathfrak{m}$ is abelian and $\mathfrak{v} \subseteq \mathfrak{m}$. By 4.24, the set

$$\mathfrak{J} := \{\mathfrak{a} = \mathfrak{m} \cap \varphi^{-1}(\mathfrak{h}) \mid \varphi \in \mathbb{G}\}$$

is finite as a subset of a finite set. The set $\{\mathfrak{v} \cap \mathfrak{a} \mid \mathfrak{a} \in \mathfrak{J}\}$ is also finite.

By 4.20(iii) we have $[\mathfrak{h}, \mathfrak{a}] = \{0\}$ for all $\mathfrak{a} \in \mathfrak{J}$, so for for each $\mathfrak{a} \in \mathfrak{J}$ and each $\mathfrak{b} \in \mathfrak{J}$, Theorem 4.27 applies and shows that the set $\Phi(\mathfrak{a}, \mathfrak{b})$ is finite (possibly empty). Thus the following set is finite:

$$\mathcal{S} := \{f(\mathfrak{v} \cap \mathfrak{a}) \mid \mathfrak{a} \in \mathfrak{J}, \ f \in \Phi(\mathfrak{a}, \mathfrak{b})) \text{ for some } \mathfrak{b} \in \mathfrak{J}\}.$$

Now consider a fixed inner automorphism $\varphi \in \mathbb{G}$, and let $\mathfrak{a} = \mathfrak{m} \cap \varphi^{-1}(\mathfrak{h}) \in \mathfrak{J}$. We claim that $\mathfrak{b} := \varphi(\mathfrak{a})$ is in $\mathfrak{J}$: Indeed, $\mathfrak{b} = \varphi(\mathfrak{m} \cap \varphi^{-1}(\mathfrak{h})) = \varphi(\mathfrak{m}) \cap \mathfrak{h} \subseteq \mathfrak{h}$ is $\mathfrak{g}$-reductive in $\mathfrak{h}$ and thus contained in $\mathfrak{m}$. Also, $\mathfrak{b} \subseteq \varphi(\mathfrak{m}) \subseteq \varphi(\mathfrak{h})$, and therefore, $\mathfrak{b} \subseteq \mathfrak{m} \cap \varphi(\mathfrak{h})$. Since $\varphi(\mathfrak{m})$ is the unique maximal $\mathfrak{g}$-reductive subalgebra of $\varphi(\mathfrak{h})$ we have $\mathfrak{m} \cap \varphi(\mathfrak{h}) \subseteq \varphi(\mathfrak{m})$. Also, $\mathfrak{m} \cap \varphi(\mathfrak{h}) \subseteq \mathfrak{m} \subseteq \mathfrak{h}$. Thus $\mathfrak{m} \cap \varphi(\mathfrak{h}) \subseteq \varphi(\mathfrak{m}) \cap \mathfrak{h} = \mathfrak{b}$. Thus $\mathfrak{b} = \mathfrak{m} \cap \varphi(\mathfrak{h}) \in \mathfrak{J}$. We set $f = \varphi|\mathfrak{a}$. Then $\mathfrak{h} \cap \varphi(\mathfrak{v}) = \varphi(\mathfrak{v} \cap \varphi^{-1}(\mathfrak{h})) = \varphi(\mathfrak{v} \cap \mathfrak{m} \cap \varphi^{-1}(\mathfrak{h})) = f(\mathfrak{v} \cap \mathfrak{a}) \in \mathcal{S}$. This establishes the finiteness of the set $\{\mathfrak{h} \cap \varphi(\mathfrak{v}) \mid \varphi \in \mathbb{G}\}$. □

4.29. REMARK. Suppose that $\mathfrak{e}$ is a subalgebra of a Lie algebra $\mathfrak{g}$. Let $\mathfrak{g}_{\mathbb{C}}$ denote the complexification of $\mathfrak{g}$ and $\mathfrak{e}_{\mathbb{C}} = \mathfrak{e} + i\mathfrak{e}$. For each $\varphi \in \mathbb{G} = \langle e^{\operatorname{ad} \mathfrak{g}} \rangle$ there is a unique extension $\varphi_{\mathbb{C}} \in \mathbb{G}_{\mathbb{C}} = \langle e^{\operatorname{ad} \mathfrak{g}_{\mathbb{C}}} \rangle$ and the map $\varphi \mapsto \varphi_{\mathbb{C}} \colon \mathbb{G} \to \mathbb{G}_{\mathbb{C}}$ is an embedding. Then for each Cartan subalgebra $\mathfrak{h}$ of $\mathfrak{g}$ the complexification $\mathfrak{h}_{\mathbb{C}}$ is a Cartan subalgebra of $\mathfrak{g}_{\mathbb{C}}$ and $\mathfrak{h} \cap \varphi(\mathfrak{e}) = \mathfrak{h}_{\mathbb{C}} \cap \varphi_{\mathbb{C}}(\mathfrak{e}_{\mathbb{C}}) \cap \mathfrak{g}$ for all $\varphi \in \mathbb{G}$. In particular:

(†) If the set $\{\mathfrak{h}_{\mathbb{C}} \cap \psi(\mathfrak{e}_{\mathbb{C}}) \mid \psi \in \mathbb{G}_{\mathbb{C}}\}$ is finite then the set $\{\mathfrak{h} \cap \varphi(\mathfrak{e}) \mid \varphi \in \mathbb{G}\}$ is finite, too. □

4.30. PROPOSITION. *Let $\mathfrak{e}$ be a Cartan reductive subalgebra of a Lie algebra $\mathfrak{g}$. Further, we assume that $\mathfrak{h}$ is a Cartan subalgebra of $\mathfrak{g}$ which contains a maximal abelian $\mathfrak{g}$-reductive subalgebra $\mathfrak{v}$ of $\mathfrak{e}$. Then the set $\{\mathfrak{h} \cap \varphi(\mathfrak{e}) \mid \varphi \in \mathbb{G}\}$ is finite.*

Proof. Pick an inner automorphism $\varphi \in \mathbb{G}$. Since $\mathfrak{e}$ is Cartan reductive, $\mathfrak{e} \cap \varphi^{-1}(\mathfrak{h}) \subseteq \mathfrak{h}_{\mathfrak{e}}$ for some Cartan subalgebra $\mathfrak{h}_{\mathfrak{e}}$ of $\mathfrak{e}$, by 4.15. Then $\mathfrak{e}_{\mathbb{C}} \cap \varphi_{\mathbb{C}}^{-1}(\mathfrak{h}_{\mathbb{C}}) \subseteq (\mathfrak{h}_{\mathfrak{e}})_{\mathbb{C}}$. Since the Cartan subalgebras of $\mathfrak{e}_{\mathbb{C}}$ are conjugate we find an inner automorphism γ of $\mathfrak{e}_{\mathbb{C}}$ with $(\mathfrak{h}_{\mathfrak{e}})_{\mathbb{C}} \supseteq \gamma(\mathfrak{v}_{\mathbb{C}})$. Now $\gamma(\mathfrak{v}_{\mathbb{C}})$ is a maximal abelian $\mathfrak{g}_{\mathbb{C}}$-reductive subalgebra of $(\mathfrak{h}_{\mathfrak{e}})_{\mathbb{C}}$, and $\mathfrak{e}_{\mathbb{C}} \cap \varphi_{\mathbb{C}}^{-1}(\mathfrak{h}_{\mathbb{C}})$ is $\mathfrak{g}_{\mathbb{C}}$-reductive abelian and contained in $(\mathfrak{h}_{\mathfrak{e}})_{\mathbb{C}}$. Thus $\mathfrak{e}_{\mathbb{C}} \cap \varphi_{\mathbb{C}}^{-1}(\mathfrak{h}_{\mathbb{C}})$ is contained in $\gamma(\mathfrak{v}_{\mathbb{C}})$, by maximality. Therefore,

$$\mathfrak{e}_{\mathbb{C}} \cap \varphi_{\mathbb{C}}^{-1}(\mathfrak{h}_{\mathbb{C}}) \subseteq \gamma(\mathfrak{v}_{\mathbb{C}}) \cap \varphi_{\mathbb{C}}^{-1}(\mathfrak{h}_{\mathbb{C}}) \subseteq \mathfrak{e}_{\mathbb{C}} \cap \varphi_{\mathbb{C}}^{-1}(\mathfrak{h}_{\mathbb{C}}).$$

Hence $\mathfrak{e}_{\mathbb{C}} \cap \varphi_{\mathbb{C}}^{-1}(\mathfrak{h}_{\mathbb{C}}) = \gamma(\mathfrak{v}_{\mathbb{C}}) \cap \varphi_{\mathbb{C}}^{-1}(\mathfrak{h}_{\mathbb{C}})$. It follows that

$$\mathfrak{h}_{\mathbb{C}} \cap \varphi_{\mathbb{C}}(\mathfrak{e}_{\mathbb{C}}) = \varphi_{\mathbb{C}}(\mathfrak{e}_{\mathbb{C}} \cap \varphi_{\mathbb{C}}^{-1}(\mathfrak{h}_{\mathbb{C}})) = \varphi_{\mathbb{C}}(\gamma(\mathfrak{v}_{\mathbb{C}})) \cap \mathfrak{h}_{\mathbb{C}}.$$

By Proposition 4.28 the set

$$\{\mathfrak{h}_{\mathbb{C}} \cap \psi(\mathfrak{v}_{\mathbb{C}}) \mid \psi \in \mathbb{G}_{\mathbb{C}}\}$$

is finite. So the set

$$\{\mathfrak{h}_{\mathbb{C}} \cap \varphi_{\mathbb{C}}(\mathfrak{e}_{\mathbb{C}}) = \mathfrak{h}_{\mathbb{C}} \cap \varphi_{\mathbb{C}}(\gamma(\mathfrak{v}_{\mathbb{C}})) \mid \varphi \in \mathbb{G}\}$$

is finite. Thus by 4.29(†) the assertion follows from the equation $\mathfrak{h} \cap \varphi(\mathfrak{e}) = \mathfrak{h}_{\mathbb{C}} \cap \varphi_{\mathbb{C}}(\mathfrak{e}_{\mathbb{C}}) \cap \mathfrak{g}$. □

4.31. THEOREM. *Let $\mathfrak{g}$ be a real or complex Lie algebra and $\mathfrak{e}$ a Cartan reductive subalgebra of $\mathfrak{g}$. Then for any Cartan subalgebra $\mathfrak{h}$ the set*

$$\{\mathfrak{h} \cap \varphi(\mathfrak{e}) \mid \varphi \in \mathbb{G}\}$$

is finite.

Proof. By Remark 4.29(†) we may assume that the ground field is $\mathbb{C}$. In $\mathfrak{e}$ we pick a maximal abelian $\mathfrak{g}$-reductive subalgebra $\mathfrak{v}$. Then by 4.15 there is a Cartan subalgebra $\mathfrak{h}$ containing $\mathfrak{v}$. Since all Cartan subalgebras of $\mathfrak{g}$ are conjugate, the assertion then follows from 4.30. □

5. Porcupine Varieties

5.1. NOTATION. (i) For a subset $\mathfrak{a} \subseteq \mathfrak{g}$ we write
$$\mathfrak{z}(\mathfrak{a}, \mathfrak{g}) = \{y \in \mathfrak{g} \mid (\forall x \in \mathfrak{a})\, [x, y] = 0\};$$
we abbreviate $\mathfrak{z}(\{x\}, \mathfrak{g})$ by $\mathfrak{z}(x)$. Notice that $\mathfrak{z}(x, \mathfrak{g}) = \ker \operatorname{ad} x$.

(ii) We write $\mathfrak{g}^0(\operatorname{ad} x)$ or shortly $\mathfrak{g}^0(x)$ for the nilspace of ad x; i. e.,
$$\mathfrak{g}^0(x) = \{y \in \mathfrak{g} \mid (\exists n)(\operatorname{ad} x)^n(y) = 0\} = \ker(\operatorname{ad} x)^d,$$
where $d = \dim \mathfrak{g}$. □

Given a fixed subset $\mathfrak{e}$ of a Lie algebra $\mathfrak{g}$, usually a subalgebra, and a class $\mathcal{A}$ of subalgebras we shall consider the union $\mathfrak{U}$ of all elements in the class $\mathcal{A}$ which hit the set $\mathfrak{e}$ non-trivially. This construct we shall refer to as a *Porcupine set*. Not unexpectedly, the geometry of such a set exhibits interesting structural relations between $\mathfrak{e}$ and the class $\mathcal{A}$. In the following definition we introduce three procupine sets.

5.2. DEFINITION. Let $\mathfrak{e}$ be any subset of $\mathfrak{g}$. We set
$\mathfrak{Z}(\mathfrak{e}) = \{x \in \mathfrak{g} \mid (\exists y \in \mathfrak{e} \setminus \{0\}) : [x, y] = 0\},$
$\mathfrak{H}(\mathfrak{e}) = \bigcup \{\mathfrak{h} \mid \mathfrak{h} \text{ is a Cartan subalgebra of } \mathfrak{g} \text{ and } \mathfrak{h} \cap \mathfrak{e} \neq \{0\}\};$
$\mathfrak{N}(\mathfrak{e}) = \{x \in \mathfrak{g} \mid (\exists y \in \mathfrak{e} \setminus \{0\}, n \in \mathbb{N})\, (\operatorname{ad} x)^n(y) = 0\}.$ □

Note that $\mathfrak{Z}(\mathfrak{e}) = \bigcup_{0 \neq y \in \mathfrak{e}} \mathfrak{z}(y) = \{x \in \mathfrak{g} \mid \mathfrak{z}(x) \cap \mathfrak{e} \neq \{0\}\}$ and $\mathfrak{N}(\mathfrak{e}) = \{x \in \mathfrak{g} \mid \mathfrak{g}^0(x) \cap \mathfrak{e} \neq \{0\}\}$.

5.3. LEMMA. *Suppose that $\mathfrak{e}$ is any subset of $\mathfrak{g}$. Then*
 (i) $\mathfrak{Z}(\mathfrak{e}) \subseteq \mathfrak{N}(\mathfrak{e})$.
 (ii) $\mathfrak{H}(\mathfrak{e}) \subseteq \mathfrak{N}(\mathfrak{e})$.
 (iii) $\mathfrak{Z}(\mathfrak{e}) \cap \operatorname{reg}(\mathfrak{g}) \subseteq \mathfrak{H}(\mathfrak{e})$
 (iv) $\mathfrak{N}(\mathfrak{e}) \cap \operatorname{reg}(\mathfrak{g}) = \mathfrak{H}(\mathfrak{e}) \cap \operatorname{reg}(\mathfrak{g})$.
 (v) *$\mathfrak{H}(\mathfrak{e}) \cap \operatorname{reg}(\mathfrak{g})$ is open dense in $\mathfrak{H}(\mathfrak{e})$ with respect to the topology induced by the topology of $\mathfrak{g}$.*
 (vi) *If for every Cartan subalgebra $\mathfrak{h}$ of $\mathfrak{g}$ the intersection $\mathfrak{h} \cap \mathfrak{e}$ is central in $\mathfrak{h}$, i.e., $[\mathfrak{h} \cap \mathfrak{e}, \mathfrak{h}] = \{0\}$, then $\mathfrak{H}(\mathfrak{e}) \subseteq \mathfrak{Z}(\mathfrak{e})$ and $\mathfrak{H}(\mathfrak{e}) \cap \operatorname{reg}(\mathfrak{g}) = \mathfrak{Z}(\mathfrak{e}) \cap \operatorname{reg}(\mathfrak{g})$. Consequently,*
$$\operatorname{Zcl}(\mathfrak{H}(\mathfrak{e})) = \operatorname{Zcl}(\mathfrak{Z}(\mathfrak{e}) \cap \operatorname{reg}(\mathfrak{g})).$$

In particular, $\mathfrak{H}(\mathfrak{e}) \subseteq \mathfrak{Z}(\mathfrak{e})$ and $\operatorname{Zcl}(\mathfrak{H}(\mathfrak{e})) = \operatorname{Zcl}(\mathfrak{Z}(\mathfrak{e}) \cap \operatorname{reg}(\mathfrak{g}))$ if $\mathfrak{e}$ is Cartan reductive in $\mathfrak{g}$.

Proof. (i) is trivial.
(ii) Let $x \in \mathfrak{H}(\mathfrak{e})$. Then there is a Cartan subalgebra $\mathfrak{h}$ of $\mathfrak{g}$ with $x \in \mathfrak{h}$ and $\mathfrak{h} \cap \mathfrak{e} \neq \{0\}$. Now $\mathfrak{h} \subseteq \mathfrak{g}^0(x)$, so $\mathfrak{g}^0(x) \cap \mathfrak{e} \neq \{0\}$, and thus $x \in \mathfrak{N}(\mathfrak{e})$.

(iii) Let $x \in \mathfrak{z}(\mathfrak{e}) \cap \mathrm{reg}(\mathfrak{g})$. Then there is a non-zero $y \in \mathfrak{e}$ with $[x,y] = 0$. Since $x \in \mathrm{reg}(\mathfrak{g})$ the nilspace $\mathfrak{g}^0(x)$ is a Cartan subalgebra of $\mathfrak{g}$; furthermore $y \in \mathfrak{z}(x) \subseteq \mathfrak{g}^0(x)$. Thus $x \in \mathfrak{H}(\mathfrak{e})$.

(iv) Let $x \in \mathfrak{N}(\mathfrak{e}) \cap \mathrm{reg}(\mathfrak{g})$. Then $\mathfrak{g}^0(x) \cap \mathfrak{e} \neq \{0\}$, since $x \in \mathfrak{N}(\mathfrak{e})$, and $\mathfrak{g}^0(x)$ is a Cartan subalgebra, since x is regular. Thus $x \in \mathfrak{H}(\mathfrak{e})$. The assertion follows now from (ii).

(v) The set $\mathrm{reg}(\mathfrak{g})$ is Zariski open hence open in $\mathfrak{g}$. Thus $\mathfrak{H}(\mathfrak{e}) \cap \mathrm{reg}(\mathfrak{g})$ is open in $\mathfrak{H}(\mathfrak{e})$. As is well known, each Cartan subalgebra of $\mathfrak{g}$ meets $\mathrm{reg}(\mathfrak{g})$ in a dense subset. Thus

$$\mathfrak{H}(\mathfrak{e}) \cap \mathrm{reg}(\mathfrak{g}) = \bigcup \{\mathfrak{h} \cap \mathrm{reg}(\mathfrak{g}) \mid \mathfrak{h} \text{ is a Cartan subalgebra of } \mathfrak{g} \text{ with } \mathfrak{h} \cap \mathfrak{e} \neq \{0\}\}$$

is dense in $\mathfrak{H}(\mathfrak{e})$.

(vi) If $\mathfrak{h} \cap \mathfrak{e}$ is central in $\mathfrak{h}$ then $\mathfrak{h} \subseteq \mathfrak{z}(\mathfrak{e} \cap \mathfrak{h}, \mathfrak{g})$, and if $\mathfrak{e} \cap \mathfrak{h} \neq \{0\}$ then $\mathfrak{z}(\mathfrak{e} \cap \mathfrak{h}, \mathfrak{g}) \subseteq \mathfrak{z}(\mathfrak{e})$. Thus, under the hypothesis of (vi) we have $\mathfrak{H}(\mathfrak{e}) \subseteq \mathfrak{z}(\mathfrak{e})$ and therefore $\mathfrak{H}(\mathfrak{e}) \cap \mathrm{reg}(\mathfrak{g}) = \mathfrak{z}(\mathfrak{e}) \cap \mathrm{reg}(\mathfrak{g})$ by (iii).

As a consequence of (v), since the Zariski topology is coarser than the natural one, $\mathfrak{H}(\mathfrak{e}) \cap \mathrm{reg}(\mathfrak{g})$ is Zariski dense in $\mathrm{Zcl}(\mathfrak{H}(\mathfrak{e}))$. The remainder then follows. $\square$

Recall that an element x in a Lie algebra $\mathfrak{g}$ is called *semisimple* if $\mathrm{ad}\,x$ is semisimple.

5.4. PROPOSITION. *Let $\mathfrak{e}$ be a Cartan reductive subalgebra of $\mathfrak{g}$. Then $\mathfrak{H}(\mathfrak{e}) \subseteq \mathfrak{z}(\mathfrak{e})$. If $\mathfrak{e}$ contains at least one nonzero element x which is either regular in $\mathfrak{g}$ or semisimple, then $\mathfrak{H}(\mathfrak{e}) \neq \emptyset$.*

Proof. Suppose that $x \in \mathfrak{H}(\mathfrak{e})$. Then x lies in some Cartan subalgebra $\mathfrak{h}$ of $\mathfrak{g}$ such that there exists a non-zero vector $y \in \mathfrak{h} \cap \mathfrak{e}$. Since $\mathfrak{h} \cap \mathfrak{e}$ is reductively embedded then $[\mathfrak{h} \cap \mathfrak{e}, \mathfrak{h}] = \{0\}$ by 4.20(iii). Thus $[h, y] = \{0\}$ for all $y \in \mathfrak{h} \cap \mathfrak{e}$ and the assertion follows.

If x is regular in $\mathfrak{g}$ or if $\mathrm{ad}\,x$ is semisimple then there is a Cartan subalgebra $\mathfrak{h}$ of $\mathfrak{g}$ such that $x \in \mathfrak{h}$ (for x semisimple this follows from 4.15). If $0 \neq x \in \mathfrak{e}$ then $\mathfrak{h} \subseteq \mathfrak{H}(\mathfrak{e})$. $\square$

5.5. PROPOSITION. *Suppose that $\mathfrak{e}$ is any nonzero vector subspace of a Lie algebra $\mathfrak{g}$, let $\{e_1, e_2, \ldots, e_k\}$ be a basis of $\mathfrak{e}$, and set $d = \dim \mathfrak{g}$. Then*
  (i) $\mathfrak{z}(\mathfrak{e}) = \{x \in \mathfrak{g} \mid (\mathrm{ad}\,x)e_1 \wedge (\mathrm{ad}\,x)e_2 \cdots \wedge (\mathrm{ad}\,x)e_k = 0 \text{ in } \bigwedge^k \mathfrak{g}\}$;
  (ii) $\mathfrak{N}(\mathfrak{e}) = \{x \in \mathfrak{g} \mid (\mathrm{ad}\,x)^d e_1 \wedge (\mathrm{ad}\,x)^d e_2 \wedge \cdots \wedge (\mathrm{ad}\,x)^d e_k = 0 \text{ in } \bigwedge^k \mathfrak{g}\}$.
  (iii) $\mathfrak{z}(\mathfrak{e})$ *and* $\mathfrak{N}(\mathfrak{e})$ *are Zariski closed subsets, so*

(1) $$\mathrm{Zcl}(\mathfrak{H}(\mathfrak{e})) \subseteq \mathfrak{N}(\mathfrak{e}).$$

  (iv) *If* $[\mathfrak{h} \cap \mathfrak{e}, \mathfrak{h}] = \{0\}$, *for each Cartan algebra $\mathfrak{h}$ of $\mathfrak{g}$, then*

(2) $$\mathrm{Zcl}(\mathfrak{z}(\mathfrak{e}) \cap \mathrm{reg}(\mathfrak{g})) = \mathrm{Zcl}(\mathfrak{H}(\mathfrak{e})) \subseteq \mathfrak{z}(\mathfrak{e}).$$

In particular, (2) holds if $\mathfrak{g}$ has abelian Cartan subalgebras.

(v) If $\mathfrak{h} \cap \mathfrak{e} \neq \{0\}$ for at least one Cartan subalgebra $\mathfrak{h}$ of $\mathfrak{g}$ and if the variety $\mathfrak{N}(\mathfrak{e})$ is irreducible, then

(3) $$\operatorname{Zcl}(\mathfrak{H}(\mathfrak{e})) = \mathfrak{N}(\mathfrak{e}).$$

(vi) Suppose that
- (a) $\mathfrak{h} \cap \mathfrak{e} \neq \{0\}$ for at least one Cartan subalgebra $\mathfrak{h}$ of $\mathfrak{g}$,
- (b) the variety $\mathfrak{N}(\mathfrak{e})$ is irreducible,
- (c) $[\mathfrak{h} \cap \mathfrak{e}, \mathfrak{h}] = \{0\}$ for all Cartan subalgebras $\mathfrak{h}$ of $\mathfrak{g}$.

Then we have

(4) $$\mathfrak{z}(\mathfrak{e}) = \operatorname{Zcl}(\mathfrak{H}(\mathfrak{e})) = \mathfrak{N}(\mathfrak{e}).$$

Proof. Since $\mathfrak{z}(x, \mathfrak{g}) = \ker \operatorname{ad} x$ and $\mathfrak{g}^0(x) = \ker(\operatorname{ad} x)^d$ the equations asserted in (i) and (ii) follow from the equivalence of the following five statements (a)–(e):
- (a) $\mathfrak{z}(x, \mathfrak{g}) \cap \mathfrak{e} \neq \{0\}$, respectively, $\mathfrak{g}^0(x) \cap \mathfrak{e} \neq \{0\}$;
- (b) $(\ker \operatorname{ad} x) \cap \mathfrak{e} \neq \{0\}$, respectively, $\ker((\operatorname{ad}(x))^d) \cap \mathfrak{e} \neq \{0\}$;
- (c) $(\operatorname{ad} x)|\mathfrak{e}$, respectively, $(\operatorname{ad} x)^d|\mathfrak{e}$ is not injective;
- (d) $\{(\operatorname{ad} x)e_1, \ldots, (\operatorname{ad} x)e_k\}$, respectively, $\{(\operatorname{ad} x)^d e_1, \ldots, (\operatorname{ad} x)^d e_k\}$ is a linearly dependent set;
- (e) in the k-th exterior power $\bigwedge^k \mathfrak{g}$ of $\mathfrak{g}$ we have $(\operatorname{ad} x)e_1 \wedge \cdots \wedge (\operatorname{ad} x)e_k = 0$, respectively, $(\operatorname{ad} x)^d e_1 \wedge \cdots \wedge (\operatorname{ad} x)^d e_k = 0$.

The elements of $\mathfrak{z}(\mathfrak{e})$, respectively, $\mathfrak{N}(\mathfrak{e})$ are thus recognized as those satisfying a homogeneous algebraic equation of degree k, respectively, kd. They therefore form Zariski closed subsets of $\mathfrak{g}$. Relations (1) and (2) now follow from Lemma 5.3(ii) and (vi).

Suppose now that $\mathfrak{h} \cap \mathfrak{e} \neq \{0\}$ for some Cartan subalgebra $\mathfrak{h}$ and that the variety $\mathfrak{N}(\mathfrak{e})$ is irreducible. In order to prove (v) and (vi) we notice that

$$\mathfrak{H}(\mathfrak{e}) \neq \emptyset \Leftrightarrow (\exists \mathfrak{h}: \mathfrak{h} \text{ is a Cartan subalgebra of } \mathfrak{g} \text{ and } \mathfrak{h} \cap \mathfrak{e} \neq \{0\}),$$

so the intersection $\mathfrak{N}(\mathfrak{e}) \cap \operatorname{reg}(\mathfrak{g})$ is not empty, by Lemma 5.3(iv, v). Since $\mathfrak{N}(\mathfrak{e})$ is irreducible, any of its non-empty Zariski open subsets is Zariski dense in it. Thus by Lemma 5.3(iv)

$$\mathfrak{N}(\mathfrak{e}) = \operatorname{Zcl}(\mathfrak{N}(\mathfrak{e}) \cap \operatorname{reg}(\mathfrak{g})) \subseteq \operatorname{Zcl}(\mathfrak{H}(\mathfrak{e})),$$

together with (1) this implies (3). If also $[\mathfrak{h} \cap \mathfrak{e}, \mathfrak{h}] = \{0\}$, for all Cartan subalgebras $\mathfrak{h}$ in $\mathfrak{g}$, then Lemma 5.3(i) and (3) together show (4). □

In [46], p. 416, 5.5 we exhibit examples showing that neither $\mathfrak{z}(\mathfrak{e})$ nor $\mathfrak{N}(\mathfrak{e})$ need be irreducible. The preceding proposition justifies the term *porcupine variety* in place of procupine set in the case of $\mathfrak{z}(\mathfrak{e})$ and $\mathfrak{N}(\mathfrak{e})$

5.6. LEMMA. *If $x \in \mathfrak{e}$ is a nonzero central element of a maximal rank subalgebra $\mathfrak{m}$ of $\mathfrak{g}$, then $\mathfrak{m} \subseteq \overline{\mathfrak{H}(\mathfrak{e})} \subseteq \mathrm{Zcl}(\mathfrak{H}(\mathfrak{e}))$.*

Proof. The center of $\mathfrak{m}$, hence x, is contained in every Cartan subalgebra of $\mathfrak{m}$. The union of all Cartan subalgebras of $\mathfrak{m}$ contains $\mathrm{reg}(\mathfrak{m})$. Since $\mathfrak{m}$ is of maximal rank, $\mathrm{reg}(\mathfrak{m}) \subseteq \mathfrak{H}(\mathfrak{e})$. It follows that $\mathfrak{m} = \overline{\mathrm{reg}(\mathfrak{m})} \subseteq \overline{\mathfrak{H}(\mathfrak{e})} \subseteq \mathrm{Zcl}(\mathfrak{H}(\mathfrak{e}))$ since the natural topology is finer than the Zariski topology. □

5.7. THEOREM. *Suppose that $\mathfrak{e}$ is a nonzero vector subspace of a Lie algebra $\mathfrak{g}$.*
  (i) *If $\mathfrak{e}$ is a Cartan reductive subalgebra of $\mathfrak{g}$ containing an $x \neq 0$ such that $\mathrm{ad}\, x$ is semisimple then $\emptyset \neq \mathrm{Zcl}(\mathfrak{H}(\mathfrak{e})) \subseteq \mathfrak{z}(\mathfrak{e}) \subseteq \mathfrak{N}(\mathfrak{e})$.*
  (ii) *If $\mathfrak{e}$ is reductively embedded then $\mathfrak{z}(\mathfrak{e}) = \mathrm{Zcl}(\mathfrak{H}(\mathfrak{e})) \subseteq \mathfrak{N}(\mathfrak{e})$.*

Proof. (i) In view of 5.4, all hypotheses for 5.5(iv) are satisfied and the assertion follows.
  (ii) Take an element $z \in \mathfrak{z}(\mathfrak{e})$. Then there exists a non-zero $x \in \mathfrak{z}(z, \mathfrak{g}) \cap \mathfrak{e}$. By hypothesis $\mathrm{ad}\, x$ is semisimple and thus by Proposition 4.15 $\mathfrak{z}(x, \mathfrak{g})$ is a maximal rank subalgebra containing x in its center. Then by Lemma 5.6 $\mathfrak{z}(x, \mathfrak{g}) \subseteq \mathrm{Zcl}(\mathfrak{H}(\mathfrak{e}))$. Thus $\mathfrak{z}(\mathfrak{e}) \subseteq \mathrm{Zcl}(\mathfrak{H}(\mathfrak{e}))$. This together with (i) above proves our claim. □

Note that if $\mathfrak{e}$ is Cartan reductive and $0 \neq x \in \mathfrak{e} \cap \mathfrak{h}$ for some Cartan subalgebra $\mathfrak{h}$ then $\mathrm{ad}\, x$ is semisimple.

5.8. PROPOSITION. *If $\mathfrak{e}$ is a nonzero subalgebra of $\mathfrak{g}$ then every nilpotent element (i.e., every element x for which $\mathrm{ad}\, x$ is nilpotent) is contained in $\mathfrak{N}(\mathfrak{e})$. In particular, $\mathfrak{N}(\mathfrak{e})$ contains the nilradical $\mathfrak{n}$ of $\mathfrak{g}$. Similarly, $\mathfrak{z}(\mathfrak{e})$ always contains the center of $\mathfrak{g}$.*

Proof. If x is nilpotent, then $\mathfrak{g}^0(x) = \mathfrak{g}$, and if x is central, then $\mathfrak{z}(x, \mathfrak{g}) = \mathfrak{g}$. Thus $\mathfrak{g}^0(x) \cap \mathfrak{e} \neq \{0\}$, respectively $\mathfrak{z}(x, \mathfrak{g}) \cap \mathfrak{e} \neq \{0\}$. □

5.9. EXAMPLES. (i) Let $\mathfrak{g} = \mathbb{R} \cdot u + \mathbb{R} \cdot x + \mathbb{R} \cdot y$ with $[u, x] = y$, $[u, y] = -x$ and $[x, y] = 0$. Set $\mathfrak{e} = \mathbb{R} \cdot x$. Then $\mathfrak{e}$ is ideal free, i.e., does not contain nonzero ideals of $\mathfrak{g}$. Further, $\mathfrak{H}(\mathfrak{e}) = \emptyset$, $\mathfrak{z}(\mathfrak{e}) = \mathfrak{N}(\mathfrak{e}) = \mathbb{R} \cdot x + \mathbb{R} \cdot y$.

(ii) Let $\mathfrak{g} = \mathfrak{sl}(2, \mathbb{K})$, $\mathbb{K} = \mathbb{R}, \mathbb{C}$. Choose $h, p, q \in \mathfrak{g}$ with $[h, p] = 2p$, $[h, q] = -2q$, $[p, q] = h$.
  (a) Set $\mathfrak{e} = \mathbb{K} \cdot p$. Then $\mathfrak{e}$ is ideal free and $\mathfrak{H}(\mathfrak{e}) = \emptyset$, $\mathfrak{z}(\mathfrak{e}) = \mathfrak{e}$ and $\mathfrak{N}(\mathfrak{e})$ is the set $\mathcal{N}$ of all nilpotent elements, i.e., the zero set of the Cartan-Killing form.
  (b) Set $\mathfrak{e} = \mathbb{K} \cdot h + \mathbb{K} \cdot p$. Then $\mathfrak{e}$ is ideal free and $\mathfrak{H}(\mathfrak{e}) = \mathfrak{e} \setminus (\mathbb{K} \cdot p \setminus \{0\})$, while $\mathfrak{z}(\mathfrak{e}) = \mathfrak{e}$ is as in (a) above, and $\mathfrak{N}(\mathfrak{e}) = \mathcal{N} \cup \mathfrak{e}$.

(iii) Let $\mathfrak{g} = \mathbb{K} \cdot x + \mathbb{K} \cdot y + \mathbb{K} \cdot z_1 + \mathbb{K} \cdot z_2$ with $[x, y] = y + z_1$, $[x, z_j] = j \cdot z_j$, $[y, z_j] = z_{j+1}$ with $z_3 = 0$ and all other brackets zero. Set $\mathfrak{e} = \mathbb{K} \cdot z_1$. Then $\mathfrak{e}$ is ideal free, $\mathfrak{H}(\mathfrak{e}) = \emptyset$, $\mathfrak{z}(\mathfrak{e}) = \mathbb{K} \cdot z_1 + \mathbb{K} \cdot z_2$ and $\mathfrak{N}(\mathfrak{e}) = \mathbb{K} \cdot y + \mathbb{K} \cdot z_1 + \mathbb{K} \cdot z_2$. □

6. Porcupine Varieties are Proper

6.1. NOTATION. Let $\mathfrak{h}$ be Cartan subalgebra of a Lie algebra $\mathfrak{g}$ and let $\mathfrak{e}$ be any subalgebra of $\mathfrak{g}$.
  (i) We let $\Sigma(\mathfrak{h})$ denote the space of vector subspaces of $\mathfrak{h}$, endowed with the usual topology and set $\Delta\colon \mathbb{G} \to \Sigma(\mathfrak{h})$, $\Delta(\varphi) = \mathfrak{h} \cap \varphi(\mathfrak{e})$.
  (ii) We denote by $\Sigma^v(\mathfrak{h})$ the set $\Sigma(\mathfrak{h})$ with the topology consisting of all open lower sets. Here a subset $\mathsf{U} \subseteq \Sigma(\mathfrak{h})$ is called a *lower set* if $\mathfrak{a} \subseteq \mathfrak{b} \in \mathsf{U}$ implies $\mathfrak{a} \in \mathsf{U}$.
  (iii) Similarily, we let $\mathbb{R}^v$ denote the set $\mathbb{R}$ with the topology $\{\mathbb{R}, \emptyset\} \cup \{]-\infty, a[\mid a \in \mathbb{R}\}$.
  (iv) We define $\delta\colon \mathbb{G} \to \mathbb{R}^v$, $\delta(\varphi) = \dim \Delta(\varphi) = \dim(\mathfrak{h} \cap \varphi(\mathfrak{e}))$.
  (v) We write $\mathcal{E} := \langle e^{\operatorname{ad} \mathfrak{e}} \rangle$. □

6.2. LEMMA. (i) $\Delta\colon \mathbb{G} \to \Sigma^v(\mathfrak{h})$ *is continuous.*
  (ii) $\delta\colon \mathbb{G} \to \mathbb{R}^v$ *is continuous.*
  (iii) *Let* $m = \min \delta(\mathbb{G})$. *Then* $\mathcal{O} := \delta^{-1}(m)$ *is open in* $\mathbb{G}$ *and* $\Delta\colon \mathbb{G} \to \Sigma(\mathfrak{h})$ *is continuous on* $\mathcal{O}$.
  (iv) $\mathcal{O}$ *is Zariski-open and thus open dense in* $\mathbb{G}$.
  (v) *Suppose that* $\Delta(\mathbb{G})$ *is finite. Then* $\mathcal{O}$ *is the finite disjoint union of non-void open sets* $\mathcal{O}_1 \cup \cdots \cup \mathcal{O}_p$ *such that* $\Delta\colon \mathcal{O} \to \Sigma(\mathfrak{h})$ *is constant on each* $\mathcal{O}_j$.

Proof. (i) Suppose that $\varphi = \lim \varphi_n$ in $\mathbb{G}$. Clearly $\varphi \mapsto \varphi(\mathfrak{e}) \colon \mathbb{G} \to \Sigma(\mathfrak{g})$ is continuous, and thus $\varphi(\mathfrak{e}) = \lim_{\Sigma} \varphi_n(\mathfrak{e})$. Let $n(j)$ denote a subsequence of natural numbers such that $\mathfrak{a} = \lim_{\Sigma} \Delta(\varphi_{n(j)})$ exists. If $x \in \mathfrak{a}$, then $x = \lim x_j$ with $x_j \in \Delta(\varphi_{n(j)}) \subseteq \mathfrak{h}$. Then $x \in \mathfrak{h}$. Also $x \in \lim_{\Sigma} \varphi_{n(j)}(\mathfrak{e}) = \varphi(\mathfrak{e})$. Hence $x \in \mathfrak{h} \cap \varphi(\mathfrak{e}) = \Delta(\varphi)$. Thus $\mathfrak{a} \subseteq \Delta(\varphi)$. The compactness of $\Sigma(\mathfrak{h})$ implies now that every $\Sigma^v(\mathfrak{h})$-neighborhood of $\Delta(\varphi)$ eventually contains $\Delta(\varphi_n)$. This proves the claim.

(ii) The function $\dim \colon \Sigma(\mathfrak{h}) \to \mathbb{R}$ is continuous by the definition of $\Sigma(\mathfrak{h})$. Thus $\dim \colon \Sigma^v(\mathfrak{h}) \to \mathbb{R}^v$ is continuous, too.

(iii) We notice that $\{m\}$ is open in $\delta(\mathbb{G}) \cap \mathbb{R}^v$. Hence $\mathcal{O}$ is open in $\mathbb{G}$ by (ii). Since the topologies of $\Sigma(\mathfrak{h})$ and of $\Sigma^v(\mathfrak{h})$ induce the same topology on $(\dim)^{-1}(m)$ the remainder follows.

(iv) The set $\mathcal{R}_k \subseteq \operatorname{Hom}(\mathfrak{e}, \mathfrak{e})$ of vector space endomorphisms ψ of $\mathfrak{e}$ of rank k is Zariski open in the Zariski closed set $\mathcal{R}_{\leq k}$ of all ψ whose rank is $\leq k$. This is seen at once by inspecting the coefficients of the characteristic polynomial of ψ.

Let x be a regular element of $\mathfrak{h}$ and write $n = \dim \mathfrak{g}$. Then $\mathfrak{h} = \mathfrak{g}^0(x) = \ker(\operatorname{ad} x)^n$, hence m is the minimum of the dimensions of

$$\ker(\operatorname{ad} x)^n \cap \varphi(\mathfrak{e}) = \mathfrak{h} \cap \varphi(\mathfrak{e}), \quad \varphi \in \mathbb{G}$$

and therefore $(\operatorname{ad} x)^n \circ \varphi$ has at most rank $p - m$. Thus

$$\{(\operatorname{ad} x)^n \circ \varphi \mid \varphi \in \mathbb{G}\} \subseteq \mathcal{R}_{\leq p-m}.$$

Next we remark that for an element $\varphi \in \mathbb{G}$ the following assertions are equivalent:
  (a) $\varphi \in \mathcal{O}$
  (b) $\ker(\operatorname{ad} x)^n \cap \varphi(\mathfrak{e}) = \mathfrak{g}^0(x) \cap \varphi(\mathfrak{e}) = \mathfrak{h} \cap \varphi(\mathfrak{e})$ is m-dimensional
  (c) $(\operatorname{ad} x)^n (\varphi(\mathfrak{e}))$ has dimension $p - m$
  (d) $(\operatorname{ad} x)^n \circ \varphi \in \mathcal{R}_{p-m}$.
Since $\mathcal{R}_{p-m}$ is open in $\mathcal{R}_{\leq p-m}$ this implies that $\mathcal{O}$ is open in $\mathbb{G}$.
  (v) By hypothesis, the set $\Delta(\mathcal{O}) \subseteq \Delta(\mathbb{G})$ is finite. Since by (iii) the restriction $\Delta | \mathcal{O}$ is continuous with respect to the usual toplogy of $\Sigma(\mathfrak{h})$, the assertion follows. □

We now show that actually $\Delta(\mathcal{O}) = \{0\}$.

6.3. LEMMA. *Suppose that $\mathfrak{h}$ is a Cartan subalgebra a Lie algebra $\mathfrak{g}$. Furthermore, let $\mathfrak{e}$ be a Cartan reductive subalgebra of $\mathfrak{g}$, not containing any nonzero ideal of $\mathfrak{g}$. Then there exists an inner automorphism $\psi \in \mathbb{G}$ with $\mathfrak{h} \cap \psi(\mathfrak{e}) = \{0\}$. Moreover the set of all such inner automorphisms is Zariski-open in $\mathbb{G}$.*

Proof. As in 6.2 we let $\mathcal{O} = \{\varphi \in \mathbb{G} \mid \dim(\mathfrak{h} \cap \varphi(\mathfrak{e})) = m\}$, where m is the minimum dimension of the sets $\mathfrak{h} \cap \varphi(\mathfrak{e})$, $\varphi \in \mathbb{G}$. We fix an element $\psi \in \mathcal{O}$ and define for every subset $\mathcal{X} \subseteq \mathbb{G}$ the vector space $\mathfrak{e}(\mathcal{X}) = \bigcap_{\varphi \in \mathcal{X}} \varphi\psi(\mathfrak{e})$. If $\mathcal{U}$ ranges through the neighborhoods of $\mathbf{1}$ in $\mathbb{G}$, then the spaces $\mathfrak{e}(\mathcal{U})$ form an updirected family of vector subspaces of $\mathfrak{h}$, so for dimensional reasons there must be a $\mathcal{U}_1$ such that $\mathcal{U} \subseteq \mathcal{U}_1$ implies $\mathfrak{e}(\mathcal{U}) = \mathfrak{e}(\mathcal{U}_1)$.

By Theorem 4.31, the set $\Delta(\mathbb{G}) = \{\mathfrak{h} \cap \varphi(\mathfrak{e}) \mid \varphi \in \mathbb{G}\}$ is finite, hence by 6.2(v) there exists an open symmetric identity neighborhood $\mathcal{V}$ in $\mathbb{G}$, so small that $\mathcal{V}\mathcal{V} \subseteq \mathcal{U}_1$, and such that Δ is constant on $\mathcal{V}\psi$, that is, $\mathfrak{h} \cap \varphi\psi(\mathfrak{e}) = \mathfrak{h} \cap \psi(\mathfrak{e})$ for all $\varphi \in \mathcal{V}$. Then

$$\mathfrak{h} \cap \mathfrak{e}(\mathcal{V}) = \bigcap_{\varphi \in \mathcal{V}} (\mathfrak{h} \cap \varphi\psi(\mathfrak{e})) = \mathfrak{h} \cap \psi(\mathfrak{e}).$$

On the other hand, $\varphi \in \mathcal{V}$ implies $\varphi(\mathfrak{e}(\mathcal{V})) \supseteq \mathfrak{e}(\mathcal{V}\mathcal{V}) \supseteq \mathfrak{e}(\mathcal{U}_1) = \mathfrak{e}(\mathcal{V})$. Since $\mathcal{V} = \mathcal{V}^{-1}$ we therefore have $\varphi(\mathfrak{e}(\mathcal{V})) = \mathfrak{e}(\mathcal{V})$ for all $\varphi \in \mathcal{V}$. But $\mathcal{V}$ generates $\mathbb{G}$, so the set $\mathfrak{e}(\mathcal{V})$ is invariant under $\mathbb{G}$ and is therefore an ideal contained in $\psi(\mathfrak{e})$. Since ψ is an automorphism and $\mathfrak{e}$ does not contain any non-zero ideal of $\mathfrak{g}$ it follows that $\mathfrak{e}(\mathcal{V}) = \{0\}$ and we get $\mathfrak{h} \cap \psi(\mathfrak{e}) = \mathfrak{h} \cap \mathfrak{e}(\mathcal{V}) = \{0\}$.

Thus we have shown that $\mathfrak{h} \cap \psi(\mathfrak{e}) = \{0\}$ for all ψ in the Zariski-open set $\mathcal{O}$, which establishes the assertion. □

6.4. THEOREM. *Let $\mathfrak{g}$ be a Lie algebra with a Cartan reductive subalgebra $\mathfrak{e}$ which does not contain any nonzero ideal of $\mathfrak{g}$. Then the following assertions hold:*
  (i) *In each conjugacy class of Cartan subalgebras of $\mathfrak{g}$ there is a member $\mathfrak{h}$ which satisfies $\mathfrak{h} \cap \mathfrak{e} = \{0\}$.*
  (ii) $\dim \mathfrak{g} - \dim \mathfrak{e} \geq \operatorname{rank} \mathfrak{g}$.

(iii) $\mathfrak{N}(\mathfrak{e}) \neq \mathfrak{g}$. A fortiori, $\mathfrak{z}(\mathfrak{e}) \neq \mathfrak{g}$ and $\operatorname{Zcl}(\mathfrak{H}(\mathfrak{e})) \neq \mathfrak{g}$.

Proof. (i) Since $\mathfrak{e}$ is Cartan reductive in $\mathfrak{g}$, Theorem 4.31 applies and shows that $\{\mathfrak{h} \cap \varphi(\mathfrak{e}) \mid \varphi \in \mathbb{G}\}$ is finite for any Cartan subalgebra $\mathfrak{h}$. Now Lemma 6.3 applies and shows that for each Cartan subalgebra $\mathfrak{h}$ there is an inner automorphism ψ such that $\mathfrak{h} \cap \psi(\mathfrak{e}) = \{0\}$. This proves (i).

Assertion (ii) is an obvious consequence of (i).

(iii) Firstly, $\mathfrak{z}(\mathfrak{e}) \subseteq \mathfrak{N}(\mathfrak{e})$ and $\mathfrak{H}(\mathfrak{e}) \subseteq \mathfrak{N}(\mathfrak{e})$ by 5.3(i) and 5.3(ii). Hence the second assertion is a consequence of the first. We prove the first one. Let $\mathfrak{h}$ be a Cartan subalgebra in $\mathfrak{g}$ such that $\mathfrak{h} \cap \mathfrak{e} = \{0\}$, as guaranteed by (i), and let $x \in \operatorname{reg}(\mathfrak{g}) \cap \mathfrak{h}$. Then $\mathfrak{h} = \mathfrak{g}^0(x)$. Suppose that $\mathfrak{N}(\mathfrak{e}) = \mathfrak{g}$. Then $x \in \mathfrak{N}(\mathfrak{e})$ and thus $\mathfrak{g}^0(x) \cap \mathfrak{e} \neq \{0\}$ by the definition of $\mathfrak{N}(\mathfrak{e})$ in 5.2. But then $\mathfrak{h} \cap \mathfrak{e} = \mathfrak{g}^0(x) \cap \mathfrak{e} \neq \{0\}$, a contradiction. □

6.5. THEOREM. *Suppose that $\mathfrak{g}$ is a Lie algebra and $\mathfrak{e}$ a reductively embedded subalgebra not containing any nonzero ideal of $\mathfrak{g}$. Then in every conjugacy class of Cartan subalgebras there is a member $\mathfrak{h}$ with $\mathfrak{h} \cap \mathfrak{e} = \{0\}$.*

As a consequence, $\dim \mathfrak{g} - \dim \mathfrak{e} \geq \operatorname{rank} \mathfrak{g}$ *and* $\mathfrak{N}(\mathfrak{e}) \neq \mathfrak{g}$. *A fortiori,* $\mathfrak{z}(\mathfrak{e}) \neq \mathfrak{g}$ *and* $\operatorname{Zcl}(\mathfrak{H}(\mathfrak{e})) \neq \mathfrak{g}$.

Proof. If $\mathfrak{e}$ is reductively embedded then it is Cartan reductive in $\mathfrak{g}$. Thus the assertion follows from Theorem 6.4. □

Note that every compactly embedded subalgebra is reductively embedded.

7. The Codimension of Porcupine Varieties

We want information on the codimension of $\mathfrak{z}(\mathfrak{e})$ under the circumstances of Theorem 6.4. For this purpose we consider a subalgebra $\mathfrak{e}$ of a Lie algebra $\mathfrak{g}$ such that $[\mathfrak{h} \cap \mathfrak{e}, \mathfrak{h}] = \{0\}$ for every Cartan subalgebra $\mathfrak{h}$ of $\mathfrak{g}$. This condition is satisfied if $\mathfrak{e}$ is Cartan reductive in $\mathfrak{g}$, i.e., $\mathfrak{h} \cap \mathfrak{e}$ is $\mathfrak{g}$-reductive for each $\mathfrak{h} \in \mathcal{C}(\mathfrak{g})$. Then

$$\operatorname{Zcl}(\mathfrak{z}(\mathfrak{e}) \cap \operatorname{reg}(\mathfrak{g})) = \operatorname{Zcl}(\mathfrak{H}(\mathfrak{e})) \subseteq \mathfrak{z}(\mathfrak{e})$$

by 5.5(iv). If $\mathfrak{e}$ is even reductively embedded, then equality holds by Theorem 5.7(ii).

Now the group $\mathcal{E} = \langle e^{\operatorname{ad} \mathfrak{e}} \rangle$ of inner automorphisms generated by $\mathfrak{e}$ acts naturally on the set $\mathcal{I} := \{\mathfrak{h} \cap \mathfrak{e} \mid \mathfrak{h} \in \mathcal{C}, \ \mathfrak{h} \cap \mathfrak{e} \neq \{0\}\}$. Under this action the set $\mathcal{M} := \{\mathfrak{h} \cap \mathfrak{e} \mid \mathfrak{h} \cap \mathfrak{e} \text{ is maximal in } \mathcal{I}\}$ is invariant. Let F denote a system of representatives for the orbits of $\mathcal{I}$.

7.1. LEMMA. *Suppose that $\mathfrak{e}$ is a Cartan reductive subalgebra of a Lie algebra $\mathfrak{g}$. Then* $\mathfrak{H}(\mathfrak{e}) \subseteq \bigcup_{\mathfrak{a} \in F} \mathcal{E} \cdot \mathfrak{z}(\mathfrak{a}, \mathfrak{g}) = \bigcup_{\varphi \in \mathcal{E}, \, \mathfrak{a} \in F} \mathfrak{z}(\varphi(\mathfrak{a}), \mathfrak{g}) \subseteq \overline{\mathfrak{H}(\mathfrak{e})} \subseteq \mathfrak{z}(\mathfrak{e})$.

Proof. For every automorphism φ of $\mathfrak{g}$ we have $\mathfrak{z}(\varphi(\mathfrak{a}), \mathfrak{g}) = \varphi(\mathfrak{z}(\mathfrak{a}, \mathfrak{g}))$. This also establishes the equality $\bigcup_{\mathfrak{a} \in F} \mathcal{E} \cdot \mathfrak{z}(\mathfrak{a}, \mathfrak{g}) = \bigcup_{\varphi \in \mathcal{E}, \, \mathfrak{a} \in F} \mathfrak{z}(\varphi(\mathfrak{a}), \mathfrak{g})$.

Next we let $x \in \mathfrak{H}(\mathfrak{e})$. Then, by the definition of $\mathfrak{H}(\mathfrak{e})$, there is a Cartan subalgebra $\mathfrak{h}$ of $\mathfrak{g}$ with $x \in \mathfrak{h}$ and such that $\mathfrak{h} \cap \mathfrak{e} \neq \{0\}$. Hence by the definition of F there is a $\gamma \in \mathcal{E}$ and an $\mathfrak{a} \in F$ such that $\mathfrak{h} \cap \mathfrak{e} = \gamma(\mathfrak{a})$. Since $\gamma(\mathfrak{a})$ is $\mathfrak{g}$-reductive this implies that $\mathfrak{h} \subseteq \mathfrak{z}(\gamma(\mathfrak{a}), \mathfrak{g}) = \gamma(\mathfrak{z}(\mathfrak{a}, \mathfrak{g}))$, by 4.15. Thus $\mathfrak{H}(\mathfrak{e}) \subseteq \bigcup_{\mathfrak{a} \in F} \mathcal{E} \cdot \mathfrak{z}(\mathfrak{a}, \mathfrak{g})$.

Now let $\varphi \in \mathbb{G}$. Since $\varphi(\mathfrak{a})$ is $\mathfrak{g}$-reductive by hypothesis, 4.15 implies that the Cartan subalgebras of the centralizer $\mathfrak{z}(\varphi(\mathfrak{a}), \mathfrak{g})$ of $\varphi(\mathfrak{a})$ in $\mathfrak{g}$ are exactly the Cartan subalgebras of $\mathfrak{g}$ which contain $\varphi(\mathfrak{a})$. Thus $\mathfrak{H}(\mathfrak{e})$ contains all Cartan subalgebras of $\mathfrak{z}(\varphi(\mathfrak{a}), \mathfrak{g})$. But in any Lie algebra the union of the Cartan subalgebras is dense, so $\mathfrak{z}(\varphi(\mathfrak{a}), \mathfrak{g}) \subseteq \overline{\mathfrak{H}(\mathfrak{e})}$, for each $\varphi \in \mathbb{G}$.

It remains to show the last inclusion of the Lemma. But this is trivial, since we know already that $\mathrm{Zcl}(\mathfrak{H}(\mathfrak{e})) \subseteq \mathfrak{z}(\mathfrak{e})$ and since Zariski closed subsets are closed also in the standard topology. □

7.2. LEMMA. *Let $\mathfrak{g}$ be a Lie algebra with a Cartan reductive subalgebra $\mathfrak{e}$. Then $\mathcal{I}/\mathcal{E}$ is finite and thus any cross section F is finite.*

Proof. Let $\mathfrak{v}$ be a maximal abelian $\mathfrak{g}$-reductive subalgebra of $\mathfrak{e}$. Then there is an $\mathfrak{h} \in \mathcal{C}$ with $\mathfrak{v} \subseteq \mathfrak{h}$; by maximality $\mathfrak{h} \cap \mathfrak{e} = \mathfrak{v}$. Thus $\mathcal{M}$ is the set of all maximal $\mathfrak{g}$-reductive abelian subalgebras of $\mathfrak{e}$. Every such is contained in a Cartan subalgebra of $\mathfrak{e}$ and the function $\mathfrak{k} \mapsto \mathfrak{m}(\mathfrak{k})$ which assigns to a Cartan subalgebra $\mathfrak{k}$ of $\mathfrak{e}$ the unique maximal abelian reductively embedded subalgebra contained in $\mathfrak{k}$ maps the set $\mathcal{C}(\mathfrak{e})$ of Cartan subalgebras of $\mathfrak{e}$ surjectively onto $\mathcal{M}$ in an $\mathcal{E}$-equivariant fashion. Since the set of $\mathcal{E}$-orbits in $\mathcal{C}(\mathfrak{e})$ is finite, then the set $\mathcal{M}/\mathcal{E}$ of $\mathcal{E}$-orbits is finite. Let F_0 denote a cross section for these orbits.

Now let $\mathfrak{a} \in F_0$. Then by 4.24, the set $\mathcal{S}_\mathfrak{a} := \{\mathfrak{k} \cap \mathfrak{a} \mid \mathfrak{k} \in \mathcal{C}(\mathfrak{g})\}$ is finite. Hence the set $F_1 := \bigcup_{\mathfrak{a} \in F_0} \mathcal{S}_\mathfrak{a}$ is finite. If $\mathfrak{h} \in \mathcal{C}(\mathfrak{g})$ then $\mathfrak{h} \cap \mathfrak{e}$ is contained in some element $\mathfrak{v} \in \mathcal{M}$, so there is a $\gamma \in \mathcal{E}$ and an $\mathfrak{a} \in F_0$ such that $\mathfrak{h} \cap \mathfrak{e} \subseteq \gamma(\mathfrak{a})$, i.e., $\gamma^{-1}(\mathfrak{h} \cap \mathfrak{e}) \subseteq \mathfrak{a}$. Hence $\gamma^{-1}(\mathfrak{h} \cap \mathfrak{e}) \in F_1$ and thus $\mathfrak{h} \cap \mathfrak{e} \in \gamma(F_1) \subseteq \mathcal{E} \cdot F_1$. This implies that $\mathcal{I}/\mathcal{E}$ is finite. In particular, any cross section F for the $\mathcal{E}$-orbits of $\mathcal{I}$ is finite. □

It is now our task to determine the nature of each of the finitely many sets $\mathcal{E} \cdot \mathfrak{z}(\mathfrak{a}, \mathfrak{g})$, $\mathfrak{a} \in F$. We shall see that each of these sets is largely determined by the subset

$$\mathfrak{z}_\mathfrak{a}(\mathfrak{e}) := \mathcal{E} \cdot \mathfrak{z}(\mathfrak{a}, \mathfrak{g}) \cap \mathrm{reg}\,\mathfrak{g}$$

of $\mathfrak{g}$-regular points in $\mathcal{E} \cdot \mathfrak{z}(\mathfrak{a}, \mathfrak{g})$ which will turn out to be a submanifold.

Before we go into this discussion we pause briefly to illustrate some geometric aspects of the variety $\mathfrak{z}(\mathfrak{e})$ which are particularly significant in the case of $\dim \mathfrak{z}(\mathfrak{e}) = \dim \mathfrak{g} - 1$. In the following $\mathfrak{g}$ shall denote a finite dimensional Lie algebra, $\mathfrak{a}$ an abelian subalgebra and $\mathfrak{e}$ a subalgebra containing $\mathfrak{a}$. We set $\mathfrak{s} := \mathfrak{z}(\mathfrak{a}, \mathfrak{g}) + \mathfrak{e}$.

7.3. PROPOSITION. *The following conditions are equivalent:*
  (i) *The vector subspace $\mathfrak{s}$ is a subalgebra of $\mathfrak{g}$,*
  (ii) $[\mathfrak{e}, \mathfrak{s}] \subseteq \mathfrak{s}$;
  (iii) $[\mathfrak{z}(\mathfrak{a}, \mathfrak{g}), \mathfrak{e}] \subseteq \mathfrak{s}$;
  (iv) $\mathcal{E} \cdot \mathfrak{z}(\mathfrak{a}, \mathfrak{g}) \subseteq \mathfrak{s}$.
Also, each of these conditions implies
  (v) $\mathrm{Zcl}(\mathfrak{Z}_a(\mathfrak{e})) \subseteq \mathfrak{s}$.
Furthermore, if we assume that $\mathfrak{z}(\mathfrak{a}, \mathfrak{g})$ contains $\mathfrak{g}$-regular points then (v) is equivalent to (iv).

Proof. Trivially, (i)⇒(ii)⇒(iii). It is straightforward that (iii)⇒(i).

The implication (iv)⇒(iii) follows by differentiation, since $\mathcal{E}$ is the group generated by $e^{\mathrm{ad}\,\mathfrak{e}}$. To see (ii)⇒(iv), pick $x \in \mathfrak{e}$ and $s \in \mathfrak{s}$. Then

$$e^{\mathrm{ad}\,x} s = s + [x, s] + \frac{1}{2}[x, [x, s]] + \cdots \in \mathfrak{s},$$

by (ii).

The implication (iv)⇒(v) is obvious, since the vector subspace $\mathfrak{s}$ is Zariski closed.

Now we suppose that $\mathfrak{z}(\mathfrak{a}, \mathfrak{g})$ contains $\mathfrak{g}$-regular points. Then $\mathrm{reg}(\mathfrak{g}) \cap \mathfrak{z}(\mathfrak{a}, \mathfrak{g})$ is dense in $\mathfrak{z}(\mathfrak{a}, \mathfrak{g})$ and thus $\mathfrak{Z}_a(\mathfrak{e}) = \mathrm{reg}(\mathfrak{g}) \cap \mathcal{E} \cdot \mathfrak{z}(\mathfrak{a}, \mathfrak{g})$ is dense in $\mathcal{E} \cdot \mathfrak{z}(\mathfrak{a}, \mathfrak{g})$. Hence $\mathcal{E} \cdot \mathfrak{z}(\mathfrak{a}, \mathfrak{g}) \subseteq \mathrm{Zcl}(\mathfrak{Z}_a(\mathfrak{e}))$. Thus (v) implies (iv). □

7.4. PROPOSITION. $\mathfrak{z}(\mathfrak{a}, \mathfrak{g})$ *contains $\mathfrak{g}$-regular points if $\mathfrak{a}$ is contained in a Cartan subalgebra of $\mathfrak{g}$, and this is the case whenever $\mathfrak{a}$ is reductively embedded.*

Proof. If $\mathfrak{a}$ is reductively embedded then $\mathfrak{a}$ is contained in a Cartan algebra, by Proposition 4.15. □

The prime examples illustrating these circumstance were discussed in [46], p. 416, 5.5.

We assume in the following that $\mathfrak{a}$ is a reductively embedded subalgebra of $\mathfrak{g}$ contained in the Cartan algebra $\mathfrak{h}$. Then $\mathfrak{z}(\mathfrak{a}, \mathfrak{g})$ is a maximal rank subalgebra containing $\mathfrak{h}$. Moreover, it follows from 4.15 that $\mathbb{R} \cdot a$ is contained in the center of $\mathfrak{h}$ for every $a \in \mathfrak{a}$, thus $\mathfrak{a}$ is central in $\mathfrak{h}$, i.e, $[\mathfrak{a}, \mathfrak{h}] = \{0\}$.

7.5. LEMMA. *Let $\mathfrak{h}$ be a Cartan subalgebra of a Lie algebra $\mathfrak{g}$. Further, let $\mathfrak{e}$ be a subalgebra of $\mathfrak{g}$ and $\mathfrak{a}$ a $\mathfrak{g}$-reductive subalgebra of $\mathfrak{h} \cap \mathfrak{e}$. Then for any regular element $h \in \mathfrak{h} \cap \mathrm{reg}(\mathfrak{g})$ we have*
  (i) $(\mathrm{ad}\,h)^{-1}\mathfrak{z}(\mathfrak{a}, \mathfrak{g}) \subseteq \mathfrak{z}(\mathfrak{a}, \mathfrak{g})$;
  (ii) $\dim(\mathfrak{z}(\mathfrak{a}, \mathfrak{g}) + [\mathfrak{e}, h]) = \dim(\mathfrak{z}(\mathfrak{a}, \mathfrak{g}) + \mathfrak{e})$.

Proof. (i) Suppose that $[x, h] \in \mathfrak{z}(\mathfrak{a}, \mathfrak{g})$ for some $x \in \mathfrak{g}$. Then (by 4.15) for every $a \in \mathfrak{a}$ we have $[h, a] = 0$ and therefore $0 = [a, [x, h]] = -[h, [a, x]]$. Since h is regular in $\mathfrak{g}$ we have $\ker \mathrm{ad}\,h \subseteq \mathfrak{h}$. Thus $[a, x] \in \ker \mathrm{ad}\,h \subseteq \mathfrak{h}$. But then $(\mathrm{ad}\,a)^2 x = [a, [a, x]] \in [\mathfrak{a}, \mathfrak{h}] = \{0\}$. Hence x lies in the nilspace of $\mathrm{ad}\,a$ which coincides with the kernel of ad a, since ad a is semisimple. Thus $[a, x] = 0$ for all $a \in \mathfrak{a}$, and therefore $x \in \mathfrak{z}(\mathfrak{a}, \mathfrak{g})$.

(ii) Since $\mathfrak{a}$ is reductively embedded in $\mathfrak{g}$, the $\mathfrak{a}$-module $\mathfrak{e}$ is semisimple, hence can be decomposed into a direct sum $\mathfrak{e} = \mathfrak{z}(\mathfrak{a}, \mathfrak{e}) \oplus \mathfrak{b}$. Then $[\mathfrak{e}, h] = [\mathfrak{z}(\mathfrak{a}, \mathfrak{e}), h] + [\mathfrak{b}, h] \subseteq \mathfrak{z}(\mathfrak{a}, \mathfrak{g}) + [\mathfrak{b}, h]$.

We conclude from (i) that

$$\operatorname{ad} h^{-1}\bigl(\mathfrak{z}(\mathfrak{a}, \mathfrak{g})\bigr) \cap \mathfrak{b} \subseteq \mathfrak{z}(\mathfrak{a}, \mathfrak{g}) \cap \mathfrak{b} = \mathfrak{z}(\mathfrak{a}, \mathfrak{e}) \cap \mathfrak{b} = \{0\},$$

so $\mathfrak{z}(\mathfrak{a}, \mathfrak{g}) \cap [\mathfrak{b}, h] = \{0\}$. Now

$$\mathfrak{z}(\mathfrak{a}, \mathfrak{g}) + [\mathfrak{e}, h] \subseteq \mathfrak{z}(\mathfrak{a}, \mathfrak{g}) + \mathfrak{z}(\mathfrak{a}, \mathfrak{e}) + [\mathfrak{b}, h] = \mathfrak{z}(\mathfrak{a}, \mathfrak{g}) \oplus [\mathfrak{b}, h] \subseteq \mathfrak{z}(\mathfrak{a}, \mathfrak{g}) + [\mathfrak{e}, h],$$

and therefore $\mathfrak{z}(\mathfrak{a}, \mathfrak{g}) + [\mathfrak{e}, h] = \mathfrak{z}(\mathfrak{a}, \mathfrak{g}) \oplus [\mathfrak{b}, h]$. Also, $\ker(\operatorname{ad} h) \subseteq \mathfrak{h}$, hence $\ker(\operatorname{ad} h) \cap \mathfrak{b} \subseteq \mathfrak{h} \cap \mathfrak{b} = (\mathfrak{h} \cap \mathfrak{e}) \cap \mathfrak{b} \subseteq \mathfrak{z}(\mathfrak{a}, \mathfrak{e}) \cap \mathfrak{b} = \{0\}$. Thus $\dim[\mathfrak{b}, h] = \dim \mathfrak{b} = \dim \mathfrak{e} - \dim \mathfrak{z}(\mathfrak{a}, \mathfrak{e})$ and we get $\dim(\mathfrak{z}(\mathfrak{a}, \mathfrak{g}) + [\mathfrak{e}, h]) = \dim \mathfrak{z}(\mathfrak{a}, \mathfrak{g}) + \dim[\mathfrak{b}, h] = \dim \mathfrak{z}(\mathfrak{a}, \mathfrak{g}) + \dim \mathfrak{b} = \dim \mathfrak{z}(\mathfrak{a}, \mathfrak{g}) + \dim \mathfrak{e} - \dim \mathfrak{z}(\mathfrak{a}, \mathfrak{e})$. But $\mathfrak{z}(\mathfrak{a}, \mathfrak{e}) = \mathfrak{z}(\mathfrak{a}, \mathfrak{g}) \cap \mathfrak{e}$ whence $\dim \mathfrak{z}(\mathfrak{a}, \mathfrak{g}) + \dim \mathfrak{e} - \dim \mathfrak{z}(\mathfrak{a}, \mathfrak{e}) = \dim(\mathfrak{z}(\mathfrak{a}, \mathfrak{g}) + \mathfrak{e})$. This establishes the assertion. □

Note that every subalgebra of $\mathfrak{h} \cap \mathfrak{e}$ is automatically reductively embedded in $\mathfrak{g}$ if $\mathfrak{e}$ is assumed to be Cartan reductive.

7.6. NOTATION. An $\mathfrak{a}$-submodule $\mathfrak{v}$ of $\mathfrak{g}$ is called a *zero module*, if $[\mathfrak{a}, \mathfrak{v}] = 0$. □

As before we write $\mathcal{E} = \langle e^{\operatorname{ad} \mathfrak{e}} \rangle$ for the set of all inner automorphisms of $\mathfrak{e}$.

7.7. LEMMA. *Suppose that $\mathfrak{h}$ is a Cartan subalgebra of $\mathfrak{g}$ and $\mathfrak{e}$ a subalgebra of $\mathfrak{g}$, and let $\mathfrak{a} \subseteq \mathfrak{h} \cap \mathfrak{e}$ be a reductively embedded subalgebra. For $h \in \mathfrak{h}$ we define*

$$\eta_h: \mathfrak{e} \times \mathfrak{z}(\mathfrak{a}, \mathfrak{g}) \to \mathcal{E} \cdot \mathfrak{z}(\mathfrak{a}, \mathfrak{g}), \quad \eta_h(x, y) = e^{\operatorname{ad} x}(h + y).$$

(i) *The derivative of η_h at $(0, 0)$ is*

$$d\eta_h(0, 0): \mathfrak{e} \oplus \mathfrak{z}(\mathfrak{a}, \mathfrak{g}) \to \mathfrak{g}, \quad d\eta_h(0, 0)(x, y) = y + [x, h].$$

(ii) $\operatorname{im}\bigl(d\eta_h(0, 0)\bigr) = \mathfrak{z}(\mathfrak{a}, \mathfrak{g}) + [\mathfrak{e}, h]$. *Further, if $h \in \mathfrak{h} \cap \operatorname{reg}(\mathfrak{g})$ then*

$$\operatorname{rank}\bigl(d\eta_h(0, 0)\bigr) = d(\mathfrak{a}) := \dim(\mathfrak{z}(\mathfrak{a}, \mathfrak{g}) + \mathfrak{e}).$$

(iii) *If $\mathfrak{g} \neq \mathfrak{z}(\mathfrak{a}, \mathfrak{g}) + \mathfrak{e}$, then the $\mathfrak{a}$-module $\frac{\mathfrak{g}}{\mathfrak{z}(\mathfrak{a},\mathfrak{g})+\mathfrak{e}}$ does not contain a zero submodule.*

(iv) *If $\mathfrak{a} \neq \{0\}$ is a compactly embedded abelian subalgebra of $\mathfrak{g}$ and contained in $\mathfrak{e}$ then*

$$\dim \mathfrak{g} - \operatorname{rank}\bigl(d\eta_h(0, 0)\bigr) \equiv 0 \bmod 2.$$

(v) *The conclusion of* (iv) *remains intact for non-compactly embedded nonzero abelian subalgebras* $\mathfrak{a}$ *if for each of the roots* λ *of* $\mathfrak{g}_\mathbb{C}$ *with respect to* $\mathfrak{h}_\mathbb{C}$ *which do not vanish on* $\mathfrak{a}$, *the set* $\lambda(\mathfrak{a}) \subseteq \mathbb{C}$ *is not contained in* $\mathbb{R}$.

Proof. (i) We have $e^{\operatorname{ad} x}(h+y) = h + y + [x,h] + \{[x,y] + (\operatorname{ad} x)^2 (h+y) + \cdots\}$ where the summands in $\{\cdots\}$ which are homogeneous in (x,y) are of degree 2 or more. The assertion follows.

(ii) The assertions on the image and the rank of $d\eta_h(0,0)$ follow from (i) and Lemma 7.5(ii).

(iii) The largest zero-submodule of the $\mathfrak{a}$-module $\mathfrak{g}$ is exactly $\mathfrak{z}(\mathfrak{a},\mathfrak{g})$. Since $\mathfrak{a}$ is reductively embedded, $\mathfrak{g} = (\mathfrak{z}(\mathfrak{a},\mathfrak{g}) + \mathfrak{e}) \oplus \mathfrak{m}$ with an $\mathfrak{a}$-submodule $\mathfrak{m}$ which is either $\{0\}$ or is an $\mathfrak{a}$-module without zero-submodule. Note $\mathfrak{m} \cong \frac{\mathfrak{g}}{\mathfrak{z}(\mathfrak{a},\mathfrak{g}) + \mathfrak{e}}$.

(iv), (v) If for each root λ of $\mathfrak{g}_\mathbb{C}$ with respect to $\mathfrak{h}_\mathbb{C}$ the set $\lambda(\mathfrak{a}) \subseteq \mathbb{C}$ is not contained in $\mathbb{R}$, then each $\mathfrak{a}$-submodule of $\mathfrak{g}_\mathbb{C}^\lambda \cap \mathfrak{g}$ has even dimension. Hence $\dim \mathfrak{m}$ is even. This applies, in particular, to the situation that $\mathfrak{a}$ is compactly embedded. □

7.8. REMARK. Before formulating our next lemma, we recall from BOURBAKI [6], n°5.1.8 (p. 36), that the *local dimension* $\dim_x$ of a manifold at a point x is defined to be the dimension of any chart at x. (The global dimension of the manifold is defined as the supremum of the local dimensions.) □

The dimension of a Zariski closed set is the dimension of the subspace of regular points, which is an immersed manifold.

7.9. LEMMA. (i) *Suppose that* $\mathfrak{a}$ *is a nonzero reductively embedded subalgebra of* $\mathfrak{g}$ *contained in* $\mathfrak{h} \cap \mathfrak{e}$. *Then the set*

$$\mathfrak{Z}_\mathfrak{a}(\mathfrak{e}) = \mathcal{E} \cdot \mathfrak{z}(\mathfrak{a},\mathfrak{g}) \cap \operatorname{reg}(\mathfrak{g})$$

of $\mathfrak{g}$-*regular points in* $\mathcal{E} \cdot \mathfrak{z}(\mathfrak{a},\mathfrak{g})$ *is an immersed manifold and is open dense in* $\mathcal{E} \cdot \mathfrak{z}(\mathfrak{a},\mathfrak{g})$.

(ii) *If* $\mathfrak{a}$ *is compactly embedded, then* $\dim \mathfrak{g} - \dim_x \mathfrak{Z}_\mathfrak{a}(\mathfrak{e})$ *is even at any point* $x \in \mathfrak{Z}_\mathfrak{a}(\mathfrak{e})$.

Note that, in particular, $\mathfrak{a}$ is compactly embedded if $\mathfrak{h}$ is compactly embedded.

Proof. (i) Since it contains the Cartan algebra $\mathfrak{h}$, the subalgebra $\mathfrak{z}(\mathfrak{a},\mathfrak{g})$ is of maximal rank; hence $\operatorname{reg}(\mathfrak{g}) \cap \mathfrak{z}(\mathfrak{a},\mathfrak{g})$ is open and dense in $\mathfrak{z}(\mathfrak{a},\mathfrak{g})$. But $\operatorname{reg}(\mathfrak{g})$ is invariant under $\mathbb{G}$, and thus, in particular, under $\mathcal{E}$. It follows readily that $\mathfrak{Z}_\mathfrak{a}(\mathfrak{e}) = \mathcal{E} \cdot (\operatorname{reg}(\mathfrak{g}) \cap \mathfrak{z}(\mathfrak{a},\mathfrak{g}))$ is an open dense subset of $\mathcal{E} \cdot \mathfrak{z}(\mathfrak{a},\mathfrak{g})$.

Every regular point h of $\mathfrak{z}(\mathfrak{a},\mathfrak{g})$ is contained in a Cartan subalgebra of $\mathfrak{g}$ containing $\mathfrak{a}$. We may assume that $h \in \mathfrak{h}$. By Lemma 7.5 the function $h \mapsto \operatorname{rank}(d\eta_h(0,0))$ takes a constant value $d(\mathfrak{a}) = \dim(\mathfrak{z}(\mathfrak{a},\mathfrak{g}) + \mathfrak{e})$ on $\mathfrak{h} \cap \operatorname{reg}(\mathfrak{g})$ and thus is locally constant on $\operatorname{reg}(\mathfrak{g})$. The analytic function $\mathcal{E} \times (\operatorname{reg}(\mathfrak{g}) \cap \mathfrak{z}(\mathfrak{a},\mathfrak{g})) \to \mathfrak{Z}_\mathfrak{a}(\mathfrak{e})$ is therefore a subimmersion by the "Théorème du rang constant" (see BOURBAKI, N., [6] n° 5.10.6, p.53) and thus its image is an immersed manifold of dimension $d(\mathfrak{a})$ on any component. (We note that we have not ruled out possible self-intersections. A subimmersion is the composition of a submersion onto a quotient manifold followed by an immersion, and the dimension of the image of such a subimmersion is the dimension of the quotient manifold.)

(ii) If $\mathfrak{a}$ is compactly embedded, then $\dim \mathfrak{g} - d(\mathfrak{a})$ is even by 7.7(iv). □

7.10. REMARK. We note that $\operatorname{reg}(\mathfrak{g}) \cap \mathfrak{z}(\mathfrak{a}, \mathfrak{g})$ has countably many components, and thus $\mathfrak{z}_\mathfrak{a}(\mathfrak{e})$ has countably many components. □

Now we want to compute the dimension of $\mathfrak{z}(\mathfrak{e}) \cap \operatorname{reg}(\mathfrak{g})$ for a Cartan reductive subalgebra $\mathfrak{e}$. For this purpose we recall the finite system F of representatives for the $\mathcal{E}$-orbits of $\mathcal{I}$ in 7.2 and the definition of the natural number $d(\mathfrak{a})$ in Lemma 7.7(ii):

$$d(\mathfrak{a}) := \dim(\mathfrak{z}(\mathfrak{a}, \mathfrak{g}) + \mathfrak{e}), \quad \mathfrak{a} \in F.$$

By Theorem 6.4 we know that the dimension of $\mathfrak{z}(\mathfrak{e})$, which contains $\operatorname{Zcl}(\mathfrak{H}(\mathfrak{e}))$ is at most $\dim \mathfrak{g} - 1$ whenever $\mathfrak{e}$ does not contain any nontrivial ideals. The following Theorem gives a formula for the dimension of the variety $\operatorname{Zcl}(\mathfrak{z}(\mathfrak{e}) \cap \operatorname{reg}(\mathfrak{g})) = \operatorname{Zcl}(\mathfrak{H}(\mathfrak{e})) \subseteq \mathfrak{z}(\mathfrak{e})$.

7.11. THEOREM. *Let $\mathfrak{g}$ be a Lie algebra and $\mathfrak{e}$ a Cartan reductive subalgebra of $\mathfrak{g}$ such that $\mathfrak{H}(\mathfrak{e}) \neq \emptyset$. Then the dimension of the Zariski closed set $\operatorname{Zcl}(\mathfrak{H}(\mathfrak{e})) = \operatorname{Zcl}(\mathfrak{z}(\mathfrak{e}) \cap \operatorname{reg}(\mathfrak{g}))$ is*

$$\dim \operatorname{Zcl}(\mathfrak{z}(\mathfrak{e} \cap \operatorname{reg}(\mathfrak{g}))) = \max\{d(\mathfrak{a}) = \dim(\mathfrak{z}(\mathfrak{a}, \mathfrak{g}) + \mathfrak{e}) \mid \mathfrak{a} \in F\}.$$

Proof. By 7.1 we know that

$$\mathfrak{H}(\mathfrak{e}) \subseteq \bigcup_{\mathfrak{a} \in F} \mathcal{E} \cdot \mathfrak{z}(\mathfrak{a}, \mathfrak{g}) = \bigcup_{\varphi \in \mathcal{E}, \mathfrak{a} \in F} \mathfrak{z}(\varphi(\mathfrak{a}), \mathfrak{g}) \subseteq \mathfrak{z}(\mathfrak{e}).$$

By Lemma 5.3(vi) we have $\operatorname{Zcl}(\mathfrak{z}(\mathfrak{e}) \cap \operatorname{reg}(\mathfrak{g})) = \operatorname{Zcl}(\mathfrak{H}(\mathfrak{e}))$. Now $\mathfrak{z}(\mathfrak{e}) \cap \operatorname{reg}(\mathfrak{g})$ is Zariski open and dense in $\operatorname{Zcl}(\mathfrak{H}(\mathfrak{e}))$ and we have

$$\mathfrak{z}(\mathfrak{e}) \cap \operatorname{reg}(\mathfrak{g}) = \bigcup_{\mathfrak{a} \in F} \mathfrak{z}_\mathfrak{a}(\mathfrak{e}).$$

The dimension d of $\operatorname{Zcl}(\mathfrak{H}(\mathfrak{e}))$ is the maximum of the dimensions of the irreducible components of $\operatorname{Zcl}(\mathfrak{H}(\mathfrak{e}))$. There is an open subset U of $\mathfrak{g}$ such that $U \cap \operatorname{Zcl}(\mathfrak{z}(\mathfrak{e} \cap \operatorname{reg}(\mathfrak{g}))) = U \cap \operatorname{Zcl}(\mathfrak{H}(\mathfrak{e}))$ is a manifold of dimension d. Since $\mathfrak{z}(\mathfrak{e}) \cap \operatorname{reg}(\mathfrak{g})$ is open dense in $\operatorname{Zcl}(\mathfrak{H}(\mathfrak{e}))$ we may assume that $U \subseteq \operatorname{reg}(\mathfrak{g})$, whence

$$U \cap \mathfrak{z}(\mathfrak{e}) = U \cap \operatorname{Zcl}(\mathfrak{H}(\mathfrak{e})) = \bigcup_{\mathfrak{a} \in F} U \cap \mathfrak{z}_\mathfrak{a}(\mathfrak{e}).$$

The intersection on the left hand side is a manifold of dimension d, and it is no loss of generality to assume that it is an open d-cell. Concerning the union on the right hand side we notice that for each $\mathfrak{a} \in F$ the set $\mathfrak{Z}_\mathfrak{a}(\mathfrak{e})$ is a subimmersed manifold with countably many components in a real vector space. Thus there exists a σ-compact manifold $M_\mathfrak{a}$ and a bijection $f_\mathfrak{a}\colon M_\mathfrak{a} \to U \cap \mathfrak{Z}_\mathfrak{a}(\mathfrak{e})$, which is a regular local diffeomorphism. For each $\mathfrak{a}$ we can find a sequence of compact $d(\mathfrak{a})$-cells $C_n(\mathfrak{a})$ whose union is $M_\mathfrak{a}$. Thus the d-cell $U \cap \mathfrak{Z}(\mathfrak{e})$ is the countable union of the $d(\mathfrak{a})$-cells $f_\mathfrak{a}(C_n(\mathfrak{a}))$, $\mathfrak{a} \in F$, $n = 1, 2, \ldots$ By the Baire category theorem, one of these, say $f_\mathfrak{b}(C_m(\mathfrak{b}))$, has inner points. Hence $d = d(\mathfrak{b})$. It follows that

$$d = d(\mathfrak{b}) = \max\{d(\mathfrak{a}) \mid \mathfrak{a} \in F\}.$$

This concludes the proof. □

Under more special circumstances we can do better, namely, whenever Lemma 7.9(ii) applies. A variation of Definition 4.21 helps to formulate this in a uniform fashion:

7.12. DEFINITION. A subalgebra $\mathfrak{g}_1$ of $\mathfrak{g}$ is called *Cartan compact* in $\mathfrak{g}$ if every Cartan subalgebra of $\mathfrak{g}$ meets $\mathfrak{g}_1$ in a compactly embedded subalgebra. □

One notices at once two principal sufficient conditions under which a subalgebra is Cartan compact in a containing algebra:

7.13. PROPOSITION. *A subalgebra $\mathfrak{g}_1$ of a Lie algebra $\mathfrak{g}$ is Cartan compact in $\mathfrak{g}$ if at least one of the following two conditions is satisfied:*
  (i) *All Cartan subalgebras of $\mathfrak{g}$ meeting $\mathfrak{g}_1$ nontrivially are compactly embedded.*
  (ii) *$\mathfrak{g}_1$ is compactly embedded.* □

7.14. THEOREM. *Suppose that $\mathfrak{e}$ is a Cartan compact proper subalgebra of a Lie algebra $\mathfrak{g}$ with $\mathfrak{H}(\mathfrak{e}) \neq \emptyset$ and suppose that $\mathfrak{e}$ does not contain any nonzero ideals of $\mathfrak{g}$. Then $\dim \operatorname{Zcl}(\mathfrak{H}(\mathfrak{e})) \leq \dim \mathfrak{g} - 2$.*

Proof. Since $\mathfrak{e}$ is Cartan compact in $\mathfrak{g}$ all $\mathfrak{a} \in F$ are compactly embedded in $\mathfrak{g}$ and thus the associated strata $\mathfrak{Z}_\mathfrak{a}(\mathfrak{e})$ have even codimension, by 7.7(iv). Also, all of the intersections $\mathfrak{h} \cap \mathfrak{e}$, $\mathfrak{h} \in \mathcal{C}$ are reductively embedded and thus $\mathfrak{e}$ is Cartan reductive, so Theorem 7.11 applies and shows that $\operatorname{Zcl}(\mathfrak{H}(\mathfrak{e}))$ has even codimension in $\mathfrak{g}$. Since $\mathfrak{e}$ does not contain any nonzero ideals we may also invoke Theorem 6.4, and deduce that $\dim \operatorname{Zcl}(\mathfrak{H}(\mathfrak{e})) \leq \dim \mathfrak{N}(\mathfrak{e}) \leq \dim \mathfrak{g} - 1$. The assertion follows. □

The following consequence is the principal result of this chapter and will be used in a key argument of the following chapter (5.9).

7.15. THEOREM. *Suppose that $\mathfrak{e}$ is a compactly embedded proper subalgebra of a Lie algebra $\mathfrak{g}$, and that it does not contain any nonzero ideals of $\mathfrak{g}$. Then $\dim \mathfrak{z}(\mathfrak{e}) \leq \dim \mathfrak{g} - 2$.*

Proof. Since the algebra $\mathfrak{e}$ is compactly embedded, it is reductively embedded and thus $\mathrm{Zcl}(\mathfrak{H}(\mathfrak{e})) = \mathfrak{z}(\mathfrak{e})$ by Theorem 5.7(ii). If $\mathfrak{e} = \{0\}$ then $\mathfrak{z}(\mathfrak{e})$ is empty, hence has dimension -1, which establishes the assertion. If $\mathfrak{e} \neq \{0\}$ then Theorem 7.14 applies and shows that $\dim \mathfrak{z}(\mathfrak{e}) = \dim \mathrm{Zcl}(\mathfrak{H}(\mathfrak{e})) \leq \dim \mathfrak{g} - 2$. □

7.16. COROLLARY. *Suppose that $\mathfrak{e}$ is a proper subalgebra of a compact Lie algebra $\mathfrak{g}$ and that $\mathfrak{e}$ does not contain any nonzero ideals of $\mathfrak{g}$. Then $\dim \mathfrak{z}(\mathfrak{e}) \leq \dim \mathfrak{g} - 2$.* □

We point out that porcupine varieties do not behave agreeably under the passing to quotient algebras. This is quickly observed by considering the degenerate case that $\mathfrak{e}$ is the center of the Lie algebra $\mathfrak{g}$ and by taking the quotient map $p: \mathfrak{g} \to \mathfrak{g}/\mathfrak{e}$. Then $\mathfrak{z}(\mathfrak{e}) = \mathfrak{g}$. But if $\mathfrak{g}/\mathfrak{e}$ is center free (e.g., if $\mathfrak{g}$ is a compact Lie algebra), then $\mathfrak{z}(p(\mathfrak{e})) = \mathfrak{z}(\{1\}) = \emptyset$. Thus we cannot expect $p(\mathfrak{z}(\mathfrak{e})) = \mathfrak{z}(p(\mathfrak{e}))$ in general.

CHAPTER 4

LIE SEMIALGEBRAS

1. Semialgebras Revisited

The following definitions are standard (cf. [27], IV.1.2, p. 284 and II.2.8, p. 86, II.4.11, p. 130).

1.1. DEFINITION. A subset S of a Campbell-Hausdorff neighborhood B is called a *local semigroup* with respect to B if it contains 0 and satisfies $S*S\cap B \subseteq S$. □

1.2. DEFINITION. (i) A wedge W in a finite dimensional real Lie algebra $\mathfrak{g}$ is called a *Lie semialgebra* if there is a Campbell-Hausdorff neighborhood B of 0 in $\mathfrak{g}$ such that $(W \cap B) * (W \cap B) \subseteq W$.
 (ii) A wedge W in $\mathfrak{g}$ is called a *trivial Lie semialgebra* if $[\mathfrak{g},\mathfrak{g}] \subseteq W$.
 (iii) We say W is a *half-space semialgebra* if W is a half-space, i.e., $\mathfrak{g} = W \cup -W$. □

From [27] we recall that every invariant wedge is a Lie semialgebra. A half-space W in a Lie algebra $\mathfrak{g}$ is a half-space semialgebra if and only if its edge is a Lie algebra. If W is a wedge in $\mathfrak{g}$ and $[\mathfrak{g},\mathfrak{g}] \subseteq W$ then W is invariant.

1.3. NOTATION. Let W be a wedge in some finite-dimensional vector space V, and let $p \in W$. Following [27] we shall use the following notations:
 (i) $L_p(W) := \overline{W - \mathbb{R}\cdot p}$, this is the subtangent wedge of W at p;
 (ii) $T_p(W) := H(L_p(W))$, this is the tangent space of W at p.
 (iii) $C^1(W) := \{x \in W \mid W \text{ has exactly } one \text{ support hyperplane at } x \text{ in } W - W\}$
   for the set of all C^1-points of W with respect to $W - W$.
 (iv) The algebraic interior $\operatorname{int}_{alg} W$ is the interior of W in $W - W$, and the algebraic boundary $\partial_{alg} W$ is $W \setminus \operatorname{int}_{alg} W$.

Note that if p is a C^1-point of W then $T_p(W)$ is a hyperplane in $W-W$ and $L_p(W)$ is a half-space in $W - W$. Moreover, we have $L_{t\cdot p}(W) = L_p(W)$ and $T_{t\cdot p}(W) = T_p(W)$, for every positive real t. □

Recall that for any two vectors x, y in a Campbell-Hausdorff neighborhood B of a Lie algebra $\mathfrak{g}$ we have

$$\left.\frac{d}{dt}\right|_{t=0} (p * t \cdot w) = g(\operatorname{ad} p) w, \quad \text{with} \quad g(Z) = 1 + \frac{1}{2} Z + \sum_{n \in \mathbb{N}} \frac{b_{2n}}{(2n)!} Z^{2n},$$

where the numbers b_n in the power series expansion of g are the Bernoulli numbers. See [27], p.83ff.

Our first proposition is essentially a 'pointwise' version of the well-known characterization of semialgebras (Theorem II.2.14, in [27], p.89; cf. Proposition 1.5 below): it characterizes, in terms of the function g as well as in terms of Lie brackets, those points of $C^1(W)$ where the tangent hyperplane actually 'locally confines' some local semigroup (defined in a suitable zero neighborhood in $\mathfrak{g}$) with Lie wedge W. (The proof arguments are adapted from [27].)

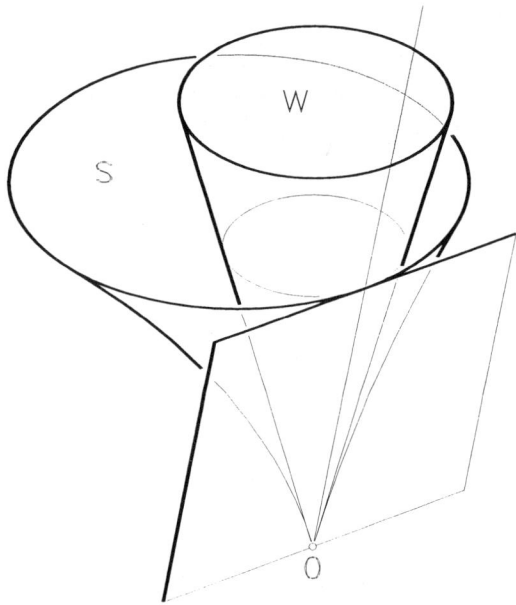

FIGURE 1. Confinement of local semigroups. The local semigroup S in this picture does not cross the Tangent plane $T_p(W)$, that is, it is contained in the tangent object $L_p(W)$ of W at the point p.

1.4. PROPOSITION. *Let B be an open Campbell-Hausdorff neighborhood in a Lie algebra $\mathfrak{g}$, and suppose that S is a local semigroup with respect to the Campbell-Hausdorff multiplication, subordinated to B and satisfying $0 \in S$. We set $W = L(S)$. For a fixed element $p \in W$, consider the following conditions:*
  (a) *$S \subseteq L_p(W)$;*
  (b) *for every $x \in \mathbb{R}^+ \cdot p \cap B$ there is a zero-neighborhood V_x in $\mathfrak{g}$ such that $x * (V_x \cap W) \subseteq L_p(W)$;*

(c) $g(\operatorname{ad} x)(W) \subseteq L_p(W)$, for all $x \in \mathbb{R}^+ \cdot p \cap B$;
(d) $g(\operatorname{ad} x) L_p(W) \subseteq L_p(W)$ for all $x \in \mathbb{R}^+ \cdot p \cap B$;
(e) $[p, T_p] \subseteq T_p$.

Then (a)$\Rightarrow$(b)$\Rightarrow$(c)$\Leftrightarrow$(d)$\Rightarrow$(e). If also int $W \neq \emptyset$ and $p \in C^1(W)$ then (d)$\Leftrightarrow$(e). Furthermore, condition (c) implies that

  (a') there is an open neighborhood C of 0 in B and a local semigroup $S_C \subseteq C$, subordinated to C, such that $L(S_C) = W$ and $S_C \subseteq L_p(W)$.

Proof. The implication (a)$\Rightarrow$(b) is obvious.

(b)$\Rightarrow$(c): Let $w \in W \cap B$ and consider the differential equation

(*) $$\dot{x}(t) = g(\operatorname{ad} x(t))w, \quad x(0) = x.$$

This equation has the solution $x(t) = x * (t \cdot w)$, where t lies in an open interval I about 0; we choose I maximal (cf. [27], p. 83 ff.). By hypothesis (b) we know that $x(t) \in L_p(W)$ for all sufficiently small non-negative values of t. Thus

$$\dot{x}(0) = \lim_{t \to +0} \frac{1}{t}(x(t) - x) \in \overline{L_p(W) - \mathbb{R} \cdot p} = L_p(W).$$

On the other hand we see from equation (*) above that $\dot{x}(0) = g(\operatorname{ad} x(0))(w) = g(\operatorname{ad} x)(w)$, so $g(\operatorname{ad} x)(w) \in L_p(W)$, which establishes (c).

(c)$\Leftrightarrow$(d): Assume that (c) holds. For any vector space automorphism Φ of $\mathfrak{g}$ satisfying $\Phi(W) \subseteq L_p(W)$ we have

$$\Phi(L_p(W)) = \Phi(\overline{W + \mathbb{R} \cdot p}) = \overline{\Phi(W) + \mathbb{R} \cdot \Phi(p)} \subseteq L_p(W).$$

Applying this observation with $\Phi = g(\operatorname{ad} x)$ we derive (d). The converse is obvious.

(d)$\Rightarrow$(e): For all $x \in \mathbb{R} \cdot p \cap B$ the map $\Phi_x := g(\operatorname{ad} x)$ is a vector space automorphism (since $p \in B$), so, as in the proof of (c)$\Leftrightarrow$(d), we have

$$\Phi_x(L_p(W)) = \Phi_x(\overline{W + \mathbb{R} \cdot p}) = \overline{\Phi_x(W) + \mathbb{R} \cdot \Phi_x(p)} = \overline{\Phi_x(W) + \mathbb{R} \cdot p} \subseteq L_p(W).$$

Since Φ_x is an automorphism of $\mathfrak{g}$, it carries the edge T_p of the wedge $L_p(W)$ into itself, and hence $\Phi_x(T_p) = T_p$ (for dimensional reasons). Thus $g(\operatorname{ad} x)q \in T_p$ for all $q \in T_p$ and all $x \in \mathbb{R} \cdot p \cap B$. Applying Lemma II.2.12 of [27] (p. 87) with $T = \operatorname{ad} x$ we conclude that $[T_p, p] \subseteq T_p$, so (e) holds.

Now we suppose that $\operatorname{int}(W) \neq \emptyset$ and that $p \in C^1(W)$.

(e)$\Rightarrow$(d): Condition (e) implies that for all $x \in \mathbb{R} \cdot p \cap B$ the vector subspace T_p is invariant under $\operatorname{ad} x$, and hence under the map $\Phi_x = g(\operatorname{ad} x)$. Since W has interior points $L_p(W)$ is a half space bounded by T_p. It follows that Φ_x maps $L_p(W)$ either onto $L_p(W)$ or onto $-L_p(W)$. Since Φ_x depends continuously on x, and since Φ_0 is the identity on $\mathfrak{g}$, we must have $g(\operatorname{ad} x) L_p(W) \subseteq L_p(W)$ for all $x \in \mathbb{R} \cdot p \cap B$. Thus condition (d) is satisfied.

Finally, we suppose that condition (c) holds.

Let $u: [0,T] \to W$ denote a bounded measurable function and suppose that
$$\dot{x}(t) = g(\operatorname{ad} x(t))u(t), \quad x(0) = p, \quad x(t) \in B.$$
Then by condition (c) we have
$$x(t) = \int_0^T g(\operatorname{ad} x(s))u(s)ds \in L_p(W)$$
for all $t \in [0,T]$. Denote with $\Sigma(W)$ the set of all points $x(t)$ obtained in this fashion. Then by the Confluence Theorem ([27], Theorem IV.3.19 on p. 314) there is an open zero neighborhood $C \subseteq B$ such that $S_C = C \cap \Sigma(W)$ is a local semigroup with respect to C, and satisfying $W = \mathcal{L}(S_C)$. Looking at the above integral formula for $x(t)$ we also see that $\overline{S_C} \subseteq L_p(W)$. □

For the next Proposition we recall that for every subset X and any given Campbell-Hausdorff neighborhood B in a Lie algebra $\mathfrak{g}$ there exists a smallest local semigroup with respect to B containing $B \cap X$. (cf. [27], p. 285, Proposition IV.1.6.)

1.5. PROPOSITION. *For a Lie wedge W in a Lie algebra $\mathfrak{g}$ the following assertions are equivalent:*
  (i) *W is a semialgebra;*
  (ii) *$[p, T_p] \subseteq T_p$ for all $p \in W$;*
  (iii) *$[p, T_p] \subseteq T_p$ for all C^1-points p of W.*
Let B be a Campbell-Hausdorff neighborhood in $\mathfrak{g}$ and write S_0 for the local semigroup with respect to B generated by $B \cap W$. If $\operatorname{int} W \neq \emptyset$ then conditions (i), (ii), (iii) are also equivalent to the following:
  (iv) *$L(S_0 \setminus W) = \emptyset$.*

Proof. The equivalence of the first three conditions is standard information ([27], p.89). Their equivalence with the last one follows from the fact that $L(S_0 \setminus W) = \emptyset$ if and only if $S_0 \setminus W$ does not cluster at 0. □

The following result illuminates these observations:

1.6. PROPOSITION. *Let S_0 be a closed local semigroup in the Lie algebra $\mathfrak{g}$, subordinated to a Campbell-Hausdorff neighborhood B_0 with Lie wedge $L(S_0) = W$ and assume that W has nonvoid interior. Then for every $p \in W \setminus L(S_0 \setminus W)$ we have $[p, T_p] \subseteq T_p$.*

Proof. Suppose that $p \in W$ such that $[p, T_p] \not\subseteq T_p$. Then condition (b) of Proposition 1.4 must fail to be true *for every* Campbell-Hausdorff neighborhood $B \subseteq B_0$. Thus we can find a sequence (λ_n) of positive reals with $\lim \lambda_n = 0$ and such that for every index $n \in \mathbb{N}$ and every zero neighborhood V the set $((\lambda_n \cdot p) * (V \cap W)) \setminus W \supseteq (\lambda_n \cdot p) * (V \cap W) \setminus L_p(W)$ is not void—so we can choose elements $v_n \in W$ with
$$\lim v_n = 0, \quad \lim \lambda_n^{-1} \cdot ((\lambda_n \cdot p) * v_n) = p, \quad \text{and } (\lambda_n \cdot p) * v_n \notin W.$$
But this means that p is a subtangent vector of $S_0 \setminus W$, and hence that $p \in L(S_0 \setminus W)$. The assertion follows. □

2. Dispersion of Weakly Exponential Semigroups

2.1. DEFINITION. A wedge W in the Lie algebra $\mathfrak{g}$ of a Lie group G with exponential function $\exp\colon \mathfrak{g} \to G$ is said to *disperse in G* if we can find an open 0-neighborhood B in $\mathfrak{g}$ such that $B \cap W = B \cap \exp^{-1}(\exp W)$. □

Intuitively speaking, W disperses in G if every 'small' element in the exponential image of W can be reached as the exponential image of a 'small' element in W. We shall see in this section that the Lie wedge of a weakly exponential subsemigroup disperses if and only if it is a semialgebra. Our first Lemma also pertains to a slightly more general situation.

2.2. LEMMA. *Let S be a closed submonoid of a connected Lie group G and suppose that the tangent wedge $W = \mathfrak{L}(S)$ of S satisfies*

$$B \cap W = B \cap \exp^{-1} S \quad \text{for some 0-neighborhood } B.$$

Then S is locally divisible and W is a semialgebra.

Proof. Note that the relation $B \cap W = B \cap \exp^{-1} S$ holds for any smaller neighborhood as well, so we may assume that B is a Campbell-Hausdorff neighborhood on which $\exp$ is injective. Write $\exp_B$ for the restriction of the exponential function to B. Then

$$\exp_B^{-1}(\exp B \cap S) = B \cap \exp_B^{-1}(S) = \{b \in B \mid \exp_B(b) \in S\} = B \cap \exp^{-1} S =$$
$$= B \cap W,$$

thus $(\exp B) \cap S = \exp(B \cap W)$ is divisible. Since the exponential images of the Campbell-Hausdorff neighborhoods form a neighborhood basis at 1 this implies the assertion. □

2.3. LEMMA. *Let $\exp\colon \mathfrak{g} \to G$ be the exponential function of a Lie group and let B be an open Campbell-Hausdorff neighborhood in $\mathfrak{g}$. Consider a weakly exponential semigroup S in G. Then*
  (i) $B \cap \exp^{-1} \exp(\operatorname{int}_{alg} W)$ *is dense in* $B \cap \exp^{-1} S$.
  (ii) $B \cap \exp^{-1}(\exp W)$ *is dense in* $B \cap \exp^{-1} S$.
  (iii) $L((\exp^{-1} S) \setminus W) = L(\exp^{-1}(\exp W) \setminus W)$.

Proof. (i) By hypothesis, $S = \overline{\exp W}$. Pick an element $x \in B \cap \exp^{-1} S$. Then $\exp$ is regular at x, so for every neighborhood V of x the image $\exp V$ is a neighborhood of $\exp x \in S$. Since $S = \overline{\exp W}$ and $\operatorname{int}_{alg} W$ is dense in W we see that $\exp V$ meets $\exp(\operatorname{int}_{alg} W)$. Hence the set $V \cap \exp^{-1} \exp(\operatorname{int}_{alg} W)$ is nonvoid. Thus $B \cap \exp^{-1} \exp(\operatorname{int}_{alg} W)$ is dense in $B \cap \exp^{-1} S$.

Assertion (ii) clearly follows from (i); in view of the equations

$$L((\exp^{-1} S) \setminus W) = L(B \cap ((\exp^{-1} S) \setminus W)),$$
$$L(\exp^{-1}(\exp W) \setminus W) = L(B \cap ((\exp^{-1}(\exp W)) \setminus W)),$$

assertion (iii) is a consequence of (ii). □

2.4. PROPOSITION. *If S is a closed weakly exponential subsemigroup of G then the following statements are equivalent:*
 (i) *The tangent wedge $W = \mathfrak{L}(S)$ disperses in G.*
 (ii) *W is a semialgebra.*

Proof. (i)$\Rightarrow$(ii): Since W disperses in G, we may assume that there is a Campbell-Hausdorff neighborhood B with $B \cap \exp^{-1} \exp W \subseteq W$. Now Lemma 2.3(ii) implies that $B \cap \exp^{-1} S = \overline{B \cap \exp^{-1} \exp W} \subseteq W$, and thus Lemma 2.2 proves the assertion.

(ii)$\Rightarrow$(i) This is a consequence of Theorem III.9 in NEEB's paper [69]. $\square$

The following proposition will be very crucial for the subsequent discussion.

2.5. PROPOSITION. *Assume that $\exp\colon \mathfrak{g} \to G$ is the exponential function of a Lie group. Let W be the tangent wedge of a closed weakly exponential subsemigroup S of G. Then the following statements are equivalent:*
 (i) *The set $L(\exp^{-1}(\exp W) \setminus W)$ is nowhere dense in the algebraic boundary $\partial_{alg}(W)$ of W.*
 (i') *The set $L(\exp^{-1}(\exp W) \setminus W)$ is nowhere dense in the set $C^1(W)$ of C^1-points of the algebraic boundary $\partial_{alg}(W)$ of W.*
 (ii) *W is a semialgebra.*
 (iii) *$L(\exp^{-1}(\exp W) \setminus W) = \emptyset$.*

Proof. The implications (iii)$\Rightarrow$(i) and (i)$\Rightarrow$(i') are trivial.

(i')$\Rightarrow$(ii): From Lemma 2.3(iii) we know that there is a Campbell-Hausdorff neighborhood B in $\mathfrak{g}$ such that $L(\exp^{-1}(\exp W) \setminus W) = L(B \cap ((\exp^{-1} S) \setminus W))$. Thus if (i') is satisfied, then by Lemma 1.6(ii) the relation $[p, T_p] \subseteq T_p$ holds for a dense subset of $C^1(W)$. Using the Approximation Theorem I.3.21 on p. 42 of [27] we conclude that this relation holds for all points of $C^1(W)$. Then by Proposition 1.5 the wedge W is a semialgebra.

(ii)$\Rightarrow$(iii): By Proposition 2.4, the present condition (ii) means that W disperses in G. By Definition 2.1 this means that there is a Campbell-Hausdorff neighborhood B such that $B \cap \exp^{-1}(\exp W) \setminus W = \emptyset$. It now follows that $L(\exp^{-1}(\exp W) \setminus W) = \emptyset$. $\square$

3. More about Subtangent Vectors

In this section we shall show that to every non-zero subtangent vector x of the set $\exp^{-1}(\exp W) \setminus W$ at 0 we can find a non-zero nearly exp-compact element y in the edge of W which commutes with x. Thus any failure of W to disperse in G is reflected by the existence of elements with special properties in the edge of W.

By definition, $p \in \exp^{-1}(\exp W) \setminus W$ if $p \notin W$ but $\exp p = \exp w$ for some $w \in W$. The relation ρ, introduced by our next definition in a rather technical way, is exactly the subtangent set of the set of all pairs $(p, \lambda \cdot w)$ with $\exp p = \exp w$ and $p \notin W$, $w \in W$, $\lambda > 0$.

3.1. DEFINITION. Let $\exp\colon \mathfrak{g} \to G$ be the exponential function of a Lie group and W a wedge in $\mathfrak{g}$. We define a closed binary relation $\rho \subseteq \mathfrak{g} \times \mathfrak{g}$ by stipulating $x\,\rho\,y$ if and only if there exist sequences (p_n, w_n) and (r_n, s_n) with the following properties (a)–(d):
  (a) $(p_n, w_n) \in (\mathfrak{g} \setminus W) \times W$, and $(r_n, s_n) \in \mathbb{R}^+ \times \mathbb{R}^+$;
  (b) $\exp p_n = \exp w_n$ for all $n \in \mathbb{N}$;
  (c) $\lim p_n = 0$;
  (d) $\lim(r_n \cdot p_n, s_n \cdot w_n) = (x, y)$. □

If $x\,\rho\,y$ then by (b) and the definition of subtangent vectors we have that $x \in L(\exp^{-1}(\exp W) \setminus W)$; furthermore $y \in W$. Thus ρ is actually a relation between the subtangent vectors of $\exp^{-1}(\exp W) \setminus W$ and certain elements of W.

3.2. LEMMA. *Let $\exp\colon \mathfrak{g} \to G$ be the exponential function of a Lie group and W a wedge in $\mathfrak{g}$. Suppose that $x\,\rho\,y$. Then*
  (i) $[x, y] = 0$;
  (ii) $y \in W \cap \overline{\mathrm{comp}_G(\mathfrak{g})} \cap H(L(\exp^{-1}(\exp W) \setminus W))$.
  (iii) $\rho \subseteq L(\exp^{-1}(\exp W) \setminus W) \times \{W \cap \overline{\mathrm{comp}_G(\mathfrak{g})} \cap H(L(\exp^{-1}(\exp W) \setminus W))\}$.
If, in addition, W is the Lie wedge of a global subsemigroup of G then
  (iv) $\rho \subseteq L(\exp^{-1}(\exp W) \setminus W) \times \overline{(\mathrm{comp}_G(\mathfrak{g}) \cap H(W))}$.

Proof. We first introduce a Euclidean norm $\|\cdot\|$ on the Lie algebra $\mathfrak{g}$ such that the closed unit ball B is a Campbell-Hausdorff neighborhood. Next we choose sequences (p_n, w_n) and (r_n, s_n) satisfying conditions (a)–(d) of 3.1 and such that $p_n \in \mathrm{int}\,B$ for all indices n.

Proof of (i): By Lemma II.3.5 we have $\exp(w_n - p_n) = 1$ and $[p_n, w_n] = 0$; therefore
$$[x, y] = \lim[r_n \cdot p_n, s_n \cdot w_n] = \lim r_n s_n \cdot [p_n, w_n] = 0,$$
which establishes (i).

Proof of (ii): Since the exponential function is injective on B and $\exp(w_n - p_n) = 1$, the differences $w_n - p_n$ must stay outside of B. Because of $\lim p_n = 0$, the w_n's cannot cluster at 0, hence the sequence (s_n) is bounded. We conclude that $\lim s_n \cdot p_n = 0$ and therefore $y = \lim s_n \cdot w_n = \lim s_n \cdot (w_n - p_n) \in \overline{\mathrm{comp}_G(\mathfrak{g})} \cap W$.

It remains to show that $\pm y \in L(\exp^{-1}(\exp W) \setminus W)$. Since $w_n - p_n \notin B$ and p_n lies in the interior of B, we see that the ray $p_n - w_n + \mathbb{R}^+ \cdot w_n$ intersects B in a line segment, which is non-trivial since it contains p_n. Clearly, $p_n - w_n + \mathbb{R}^+ \cdot w_n \subseteq \exp^{-1} \exp W$, thus the point
$$q_n := p_n \pm (\|p_n\|^{\frac{1}{2}} s_n) \cdot w_n = p_n - w_n + (1 \pm \|p_n\|^{\frac{1}{2}} s_n) \cdot w_n$$
lies in $\exp^{-1} \exp W \setminus W$ whenever $\|p_n\|^{\frac{1}{2}} s_n < 1$. (We have $p_n \neq 0$ since $p_n \notin W$.) But $p_n \to 0$ and (s_n) is bounded, so $\lim \|p_n\|^{\frac{1}{2}} s_n = 0$, and thus $q_n \in \exp^{-1} \exp W$

for all sufficiently large indices n. Also,

$$\lim q_n = \pm \lim \|p_n\|^{\frac{1}{2}} \lim(s_n \cdot w_n) = 0,$$
$$\lim \|p_n\|^{-\frac{1}{2}} \cdot q_n = \lim \|p_n\|^{-\frac{1}{2}} \cdot p_n \pm \lim s_n \cdot w_n = 0 \pm \lim s_n \cdot w_n = \pm y,$$

and we conclude that $\pm y \in L(\exp^{-1}(\exp W) \setminus W)$. This gives us the information $y \in H\bigl(L(\exp^{-1}(\exp W) \setminus W)\bigr)$ and thus completes the proof of (ii).

Conclusion (iii) is just a reformulation of (ii) in view of the remark on the domain of ρ made right after the definition.

Proof of (iv): Suppose now that W is the Lie wedge of a global subsemigroup of G. Then $L(\exp^{-1} \exp W) \subseteq W$ and thus $L(\exp^{-1}(\exp W) \setminus W) \subseteq W$. The last assertion now follows from (iii). □

3.3. PROPOSITION. *Let W be the tangent wedge of a closed subsemigroup S of a connected Lie group G. Then $H\bigl(L(\exp^{-1}(\exp W) \setminus W)\bigr) \subseteq H(W)$ and each nonzero $x \in L(\exp^{-1}(\exp W) \setminus W)$ is ρ-related to a nonzero element*

$$y \in W \cap \overline{\mathrm{comp}_G(\mathfrak{g})} \cap H\bigl(L(\exp^{-1}(\exp W) \setminus W)\bigr).$$

(In particular, $y \in H(W) \cap \overline{\mathrm{comp}_G(\mathfrak{g})}$.)

Proof. As in the proof of 3.2, we introduce a Euclidean norm on $\mathfrak{g}$ such that the unit ball B is a Campbell-Hausdorff neighborhood. By the definition of subtangent vectors there are sequences (p_n) in $B \cap \exp^{-1}(\exp W) \setminus W$ and (r_n) in $\mathbb{R}^+$ with $\lim p_n = 0$ and $\lim r_n \cdot p_n = x$. Since $p_n \in \exp^{-1}(\exp W) \setminus W$ we also find elements $w_n \in W$ such that $\exp w_n = \exp p_n$, that is, $\exp(w_n - p_n) = \mathbf{1}$, by Lemma II.3.5 (note that the elements of B are exp-regular). The exponential function is injective in B and $p_n \notin W$, so the differences $w_n - p_n$ lie outside of B, and hence the points w_n cannot cluster at 0. After passing to a subsequence, if necessary, and renaming the subsequence we may assume that $y = \lim \|w_n\|^{-1} \cdot w_n$ exists. Then $\|y\| = 1$ and so $y \neq 0$. But by the definition of ρ we have $x \, \rho \, y$, and Lemma 3.2 now implies $y \in \overline{\mathrm{comp}_G(\mathfrak{g})} \cap H\bigl(L(\exp^{-1}(\exp W) \setminus W)\bigr)$. □

As an immediate consequence we obtain the following sufficient condition for the Lie wedge of a weakly exponential subsemigroup in a Lie group to be a semialgebra:

3.4. COROLLARY. *Let W be the tangent wedge of a closed subsemigroup S of a connected Lie group G. Suppose that*

$$\overline{\mathrm{comp}_G(\mathfrak{g})} \cap H\bigl(L(\exp^{-1}(\exp W) \setminus W)\bigr) \subseteq \{0\}.$$

Then W disperses in G. If, in addition, S is weakly exponential then W is a semialgebra.

In particular, these statements hold if the edge of W contains no nonzero nearly compact element at all, $\overline{\text{comp}_G(\mathfrak{g})} \cap H(W) = \{0\}$.

Proof. Notice that for any subset A of $\mathfrak{g}$ with $0 \notin A$ the relation $\overline{L(A)} \subseteq \{0\}$ is tantamount to $L(A) = \emptyset$. Thus by Proposition 3.3 the condition $\overline{\text{comp}_G(\mathfrak{g})} \cap H\bigl(L(\exp^{-1}(\exp W) \setminus W)\bigr) \subseteq \{0\}$ implies $L(\exp^{-1}(\exp W) \setminus W) = \emptyset$.

By Proposition 2.4 the Lie wedge of a weakly exponential subsemigroup is a semialgebra if and only if it disperses. □

3.5. LEMMA. *The tangent wedge W of a weakly exponential subsemigroup is a semialgebra whenever one of the following conditions is satisfied:*
  (i) *W is pointed;*
  (ii) *S lies in a Lie group G which has no non-trivial compact subgroups.*

Proof. In case (i) we have $H(W) = \{0\}$, in case (ii) we have $\overline{\text{comp}_G(\mathfrak{g})} = \{0\}$. Thus in each case the intersection in 3.4 must be trivial and the assertion follows. □

3.6. REMARK. In particular we get the known result ([36], [27], p.462, Corollary V.6.11) that the Lie wedge of a divisible subsemigroup is a semialgebra *if it is pointed.* □

4. Reductions by Factoring Normal Subgroups

4.1. LEMMA. *Let S be a closed subsemigroup of a Lie group G and let N be a closed normal subgroup of G with $N \subseteq S$. Then the quotient morphism $\kappa: G \to G/N$ maps S onto a closed subsemigroup S_N of G/N and the induced map $\mathfrak{L}(\kappa): \mathfrak{g} \to \mathfrak{L}(G/N) = \mathfrak{g}/\mathfrak{n}$ maps $W = \mathfrak{L}(S)$ surjectively onto $W_N = \mathfrak{L}(S_N)$.*

In other words: Any one-parameter subsemigroup of S_N can be lifted to a one-parameter subsemigroup of S.

Proof. We must prove that $W/\mathfrak{n} = \mathfrak{L}(\kappa)(W) = W_N$. Since $\mathfrak{n} \subseteq H(W)$ the quotient $W/\mathfrak{n}$ is a closed wedge in $\mathfrak{g}/\mathfrak{n}$ and it therefore suffices to show that $W_N \setminus H(W) \subseteq W/\mathfrak{n}$. Thus let $x \in W_N \setminus H(W_N)$.

Then $\exp_{G/N} \mathbb{R} \cdot x$ cannot be relatively compact, for otherwise $\overline{\exp_{G/N} \mathbb{R}^+ \cdot x}$ would be a compact subgroup of S_N and hence $x \in H(W_N)$. Thus the map $\psi: \mathbb{R} \to \exp \mathbb{R} \cdot x, r \mapsto \exp_{G/N} r \cdot x$, is an isomorphism of topological groups. Write G_x for the identity component of $\kappa^{-1}(\exp \mathbb{R} \cdot x)$. Define $\varphi: G_x \to \mathbb{R}, g \mapsto \psi^{-1}\kappa(g)$. Then φ is surjective; it is an open mapping since the quotient morphism κ is open.

Put $T = S \cap G_x$. Then T is a closed semigroup and $\varphi(T) = \mathbb{R}^+$. We are finished if we can show that there is an $x' \in \mathfrak{L}(T)$ with $\varphi(\exp \mathbb{R}^+ \cdot x') = \mathbb{R}^+$. For $n \in \mathbb{N}$ we find an element $t_n \in T$ with $\varphi(t_n) = \frac{1}{n!}$. Then $\langle t_n, t_n^{-1} \rangle$ is a closed cyclic subgroup A_n such that $A_n \cap T$ maps onto $\frac{1}{n!}\cdot \mathbb{N}$. In the compact space of all closed subgroups of G_x (cf. II.3.7(vi)) there is a subsequence $A_{n(m)}$ converging to a closed abelian and divisible subgroup A. For any positive rational r there is an index m_0 with $A_{n(m)} \cap T \cap \varphi^{-1}(r) \neq \emptyset$ for all $m > m_0$. Thus $A \cap T \cap \varphi^{-1}(r) \neq \emptyset$ for every positive rational r and it follows that $A \cap T$ maps onto $\mathbb{R}^+$ under φ. If $a \in A \cap \varphi^{-1}(0)$ then there are elements $t_m \in T \cap A_{n(m)}$ with $t_m \to a$, so $a \in T$. Thus $A \cap \ker \varphi \subseteq T$.

Let $g \in A$ with $\varphi(g) > 0$. Then there is a $t \in T$ with $\varphi(t) = \varphi(g)$, so we conclude that $\varphi(gt^{-1}) = 0$, $gt^{-1} \in \ker \varphi \subseteq T$, and therefore $g \in Tt \subseteq T$. Thus T is a half space semigroup and $\mathfrak{L}(T)$ is a half-space semialgebra. Now any $x' \in \mathfrak{L}(T)$ with $x' \notin H(\mathfrak{L}(T))$ satisfies the requirements. □

4.2. PROPOSITION. *Let S be a closed subsemigroup of a Lie group G and let N be a closed normal subgroup of G with $N \subseteq S$.*
  (i) *The quotient morphism $\kappa: G \to G/N$ maps S onto a closed subsemigroup S_N of G/N. Further, $\mathfrak{L}(\kappa)$ maps the tangent wedge W of S onto the tangent wedge W_N of S_N.*
  (ii) *The following statements (a),(b) are equivalent:*
     (a) *W is a semialgebra.*
     (b) *W_N is a semialgebra.*
  (iii) *Of the following statements (c) implies (d):*
     (c) *S is weakly exponential [respectively, exponential].*
     (d) *S_N is weakly exponential [respectively, exponential].*

Proof. (i) The first part of the assertion follows from the general properties of quotient maps. As to the second, clearly, $\mathfrak{L}(\kappa)$ maps W into W_N and, by Lemma 4.1, also onto W_N.

Assertion (ii) follows from the fact that $\mathfrak{n} := \mathfrak{L}(N) \subseteq W$ and Proposition II.4.1 of [27], p. 126.

(iii) Straightforward. □

4.3. EXERCISE. (a) Show that W disperses in G if and only if W_N disperses in G/N. (Note that an element $x \in \mathfrak{g}$ lies in W if and only if $x + \mathfrak{n} \subseteq W$.) (b) Deduce from Proposition 3.4 that the Lie wedge of any closed subsemigroup of a nilpotent Lie group (or more generally a Lie group with a unique maximal compact subgroup which is central) disperses. (c) Deduce that the Lie wedge of a weakly exponential subsemigroup of a nilpotent Lie group is a semialgebra. (d) Give an example of a non-dispersing wedge in the Lie algebra of an abelian Lie group. (Such a wedge cannot be the tangent object of a closed subsemigroup.) □

4.4. PROPOSITION. *Let S be a closed subsemigroup of a Lie group G and let N be a closed normal subgroup of G with $N \subseteq S$.*

Then the quotient morphism $\kappa \colon G \to G/N$ maps S onto a closed subsemigroup S_N of G/N. Suppose that the following hypotheses are satisfied:
   (a) *S_N is exponential.*
   (b) *N is exponential.*
   (c) *$\mathfrak{g}$ is a vector space sum $\mathfrak{c} + \mathfrak{n}$ such that for any $x \in \mathfrak{c}$ and $y \in \mathfrak{n}$ we have $[x, y] \in \mathbb{R} \cdot y$.*

Then S is exponential.

Proof. Let $s \in S$. We know from 4.1 that $\mathcal{L}(S_N) = W/\mathfrak{n}$, $W = \mathcal{L}(S)$. By (a) there is a $w \in W$ such that $sN = (\exp w)N$, i.e., $s = (\exp w)n$ for some $n \in N$. By (c) we write $w = w_\mathfrak{c} + w_\mathfrak{n}$ with $w_\mathfrak{c} \in \mathfrak{c}$ and $w_\mathfrak{n} \in \mathfrak{n}$. Then $w_\mathfrak{c} = w - w_\mathfrak{n} \in W + \mathfrak{n} \subseteq W$. Also, $[w_\mathfrak{c}, w_\mathfrak{n}] \in \mathbb{R} \cdot w_\mathfrak{n}$ by (c) and this implies $\exp w \in (\exp w_\mathfrak{c})N$. Hence $s = (\exp w_\mathfrak{c})m$ with $m \in N$. By (b) there is a $y \in \mathfrak{n}$ with $m = \exp y$. Then, by (c) again, there is an $r \in \mathbb{R}$ such that $s = (\exp w_\mathfrak{c})(\exp y) = \exp(w_\mathfrak{c} + r \cdot y)$. Since $w_\mathfrak{c} + r \cdot y \in W + \mathfrak{n} = W$, the Lemma is proved. $\square$

4.5. EXERCISE. Verify that Proposition 4.4 remains intact if in all places the word *exponential* is replaced by *weakly exponential*. $\square$

4.6. EXERCISE. Note that Condition (c) of Proposition 4.4 is satisfied if $\mathfrak{g} = \mathfrak{z}(\mathfrak{n}, \mathfrak{g}) + \mathfrak{n}$, where $\mathfrak{z}(\mathfrak{n}, \mathfrak{g})$ is the centralizer of $\mathfrak{n}$ in $\mathfrak{g}$. Show that this condition is automatically satisfied if $\mathfrak{g}$ is a compact Lie algebra. $\square$

4.7. DEFINITION. Recall that a Lie semialgebra is called *reduced* if it has inner points and does not contain any nonzero ideals. Following this usage we say that a closed subsemigroup of a connected Lie group G is *reduced in G*, or *reduced* for short, if its interior in G is non-empty and if it does not contain any normal subgroups $\neq \{\mathbf{1}\}$ of G. $\square$

Note that by Theorem II.5.3(ii) the Lie wedge of any reduced weakly exponential subsemigroup of a connected Lie group G is generating in the Lie algebra $\mathfrak{g}$.

4.8. PROPOSITION. *Let S be a weakly exponential submonoid of a Lie group G and let A be the analytic subgroup generated in G by $\exp W$. Denote by A_{Lie} the same group but endowed with its intrinsic Lie group topology. Suppose that N is the union of all closed normal subgroups of A_{Lie} which are contained in S. Then the following assertions hold:*
   (i) *N is a closed normal subgroup of A_{Lie} and the quotient map $\kappa \colon A_{\text{Lie}} \to A_{\text{Lie}}/N$ maps S onto a reduced weakly exponential subsemigroup S_N of A_{Lie}/N.*
   (ii) *The tangent wedge of S_N is the image of the tangent wedge of S under the induced quotient map $\mathfrak{a} \to \mathfrak{a}/\mathfrak{n}$ between the respective Lie algebras.*

(iii) W is a semialgebra if and only if $W_N := W/\mathfrak{n} \cong \mathcal{L}(S_N)$ is a semialgebra.

(iv) If S is divisible then so is $\kappa(S)$.

Proof. (i) By Theorem II.5.8 S is weakly exponential in A_{Lie} and $\mathfrak{a} = \mathcal{L}[A_{\text{Lie}}]$ is $W - W$. Thus W has inner points in $\mathfrak{a}$ and S has inner points in A_{Lie}. (See II.5.5.) Clearly $S_N := S/N$ does not contain any nontrivial normal subgroups of A_{Lie}/N and S_N has inner points in A_{Lie}/N. Hence S_N is reduced in A_{Lie}/N. By 4.2(iii) the semigroup S_N is weakly exponential.

The assertions (ii), (iii) and (iv) follow from 4.2(i), (ii), and (iii), respectively. □

5. Porcupine Varieties and Lie Semialgebras

Let $\mathfrak{g}$ denote a finite dimensional real Lie algebra. We let $\mathfrak{z}(x)$ denote the centralizer in $\mathfrak{g}$ of an element $x \in \mathfrak{g}$. Recall from III.5.2 that for any subset $\mathfrak{e}$ of $\mathfrak{g}$ we write $\mathfrak{z}(\mathfrak{e})$ for the union of all centralizers $\mathfrak{z}(y)$ with $0 \neq y \in \mathfrak{e}$. In III.5.5(iii) we proved that $\mathfrak{z}(\mathfrak{e})$ is an algebraic variety.

With the aid of the sets $\mathfrak{z}(\mathfrak{e})$ we can now formulate another sufficient condition for the Lie wedge W of a subsemigroup of a Lie group to be a Lie semialgebra. Recall the algebraic boundary $\partial_{alg} W$ of W (see 1.3(iv)).

5.1. COROLLARY. *Let W denote the tangent wedge of a weakly exponential semigroup in a Lie group and suppose that the set*

$$\mathfrak{z}\left(\overline{\text{comp}_G(\mathfrak{g})} \cap H\left(L(\exp^{-1}(\exp W) \setminus W)\right)\right) \cap \partial_{alg} W$$

is nowhere dense in $\partial_{alg} W$. Then W is a Lie semialgebra.

Proof. From Proposition 2.5 we know that W is a semialgebra if

$$L(\exp^{-1}(\exp W) \setminus W) \cap \partial_{alg} W$$

is nowhere dense in $\partial_{alg} W$. By Proposition 3.3 and Lemma 3.2(i) we have

$$L(\exp^{-1}(\exp W) \setminus W) \subseteq \mathfrak{z}\left(\overline{\text{comp}_G(\mathfrak{g})} \cap H\left(L(\exp^{-1}(\exp W) \setminus W)\right)\right).$$

Taken together, these two pieces of information yield the assertion. □

Recall that a subset $\mathfrak{e}$ is *compactly embedded* in $\mathfrak{g}$ if $\overline{e^{\operatorname{ad} \mathbb{R} \cdot x}}$ is compact in $\mathfrak{gl}(\mathfrak{g})$ for all $x \in \mathfrak{e}$. If $\mathfrak{e}$ is a subalgebra of $\mathfrak{g}$, then this is the case iff $\overline{\langle e^{\operatorname{ad} \mathfrak{e}} \rangle}$ is compact in $\mathfrak{gl}(\mathfrak{g})$. (See [27], p. 612.)

5.2. DEFINITION. A subset $\mathfrak{e}$ of a Lie algebra $\mathfrak{g}$ is said to be *lean in* $\mathfrak{g}$ if $\mathfrak{z}(\mathfrak{e})$ is contained in a Zariski closed subset of dimension at most $\dim \mathfrak{g} - 2$.

Note that if $\mathfrak{e} \subseteq \{0\}$ then $\mathfrak{z}(\mathfrak{e}) = \emptyset$, and so $\mathfrak{e}$ is lean in $\mathfrak{g}$. □

By the basic result of Chapter II, Theorem III.7.15, a compactly embedded subalgebra $\mathfrak{e}$ of $\mathfrak{g}$ is lean in $\mathfrak{g}$ if it contains no non-zero ideal of $\mathfrak{g}$,
The significance of lean subsets is the following:

5.3. PROPOSITION. *Let $\mathfrak{g}$ be a Lie algebra and let $\mathfrak{e}$ be a lean subset of $\mathfrak{g}$. Suppose that M is a topological submanifold of $\mathfrak{g}$ with $\dim M = \dim \mathfrak{g} - 1$. Then the following statements hold:*
  (i) $\mathfrak{z}(\mathfrak{e}) \cap M$ *is nowhere dense in* M.
  (ii*) *Let W be the Lie wedge of a weakly exponential subsemigroup of a Lie group G with Lie algebra $\mathfrak{g}$ and suppose that $\mathrm{int}\, W \neq \emptyset$. If*

$$\overline{\mathrm{comp}_G(\mathfrak{g})} \cap H\big(\mathcal{L}(\exp^{-1}(\exp W) \setminus W)\big)$$

  is lean, then W is a semialgebra.
  (ii) *Let W be the Lie wedge of a weakly exponential subsemigroup of a Lie group G with Lie algebra $\mathfrak{g}$ and suppose that $\mathrm{int}\, W \neq \emptyset$. If*

$$\overline{\mathrm{comp}_G(\mathfrak{g})} \cap H(W)$$

  is lean, then W is a semialgebra.

Proof. (i) Suppose that V be a Zariski closed subset of $\mathfrak{g}$ such that $\mathfrak{e} \subseteq V$ and $\dim V \leq \dim \mathfrak{g} - 2$. Suppose now that the claim is false. Then $\overline{V \cap M} \supset \overline{\mathfrak{z}(\mathfrak{e}) \cap M}$ would contain an inner point of the topological $(n-1)$-manifold M, $n = \dim \mathfrak{g}$, and thus there would be a compact topological $(n-1)$-cell C contained in $\overline{V \cap M} \subseteq \overline{V} = V$, in view of the fact that V is Zariski closed and hence is closed in $\mathfrak{g}$. But then $n - 1 \leq \dim V \leq n - 2$, a contradiction.

(ii*) If $\mathrm{int}\, W \neq \emptyset$ then the subspace $M := \partial W$ is a topological submanifold of $\mathfrak{g}$ of dimension $\dim \mathfrak{g} - 1$. We set $\mathfrak{e} = \overline{\mathrm{comp}_G(\mathfrak{g})} \cap H\big(L(\exp^{-1}(\exp W) \setminus W)\big)$ and apply (i) and Proposition 5.1 in order to obtain the assertion.

(ii) is an immediate consequence of (ii*) since $H\big(L(\exp^{-1}(\exp W) \setminus W)\big) \subseteq H(W)$. □

5.4. LEMMA. *Let $p: \mathfrak{g} \to \mathfrak{g}_1$ be a surjective morphism between Lie algebras with kernel $\ker p = \mathfrak{a}$, and let $\mathfrak{e}$ be any subset of $\mathfrak{g}$. Then the following assertions hold:*
  (i) $\mathfrak{z}(\mathfrak{e}, \mathfrak{g}) \subseteq \mathfrak{z}(\mathfrak{e} \cap \mathfrak{a}, \mathfrak{g}) \cup p^{-1}(\mathfrak{z}(p(\mathfrak{e}), \mathfrak{g}_1))$
  (ii) *Suppose that, in addition, $\mathfrak{g} = \mathfrak{a} + \mathfrak{z}(\mathfrak{a}, \mathfrak{g})$. Then $\mathfrak{z}(\mathfrak{e} \cap \mathfrak{a}, \mathfrak{g}) = \mathfrak{z}(\mathfrak{e} \cap \mathfrak{a}, \mathfrak{a}) + \mathfrak{z}(\mathfrak{a}, \mathfrak{g})$.*

Proof. (i) Pick $x \in \mathfrak{z}(\mathfrak{e}, \mathfrak{g})$. Then there exists a non-zero $y \in \mathfrak{e}$ with $[x, y] = 0$. If $y \in \ker p$ then $x \in \mathfrak{z}(\mathfrak{e} \cap \ker p)$; if $p(y) \neq 0$ then $p(x) \in \mathfrak{z}(p(\mathfrak{e}))$, hence $x \in p^{-1}(\mathfrak{z}(p(\mathfrak{e})))$.

(ii) Suppose first that $x \in \mathfrak{z}(\mathfrak{e} \cap \mathfrak{a}, \mathfrak{g})$, and let y be a non-zero element in $\mathfrak{e} \cap \mathfrak{a}$ with $[x, y] = 0$. Write $x = x_\mathfrak{a} + x_{\mathfrak{z}}$ with $x_\mathfrak{a} \in \mathfrak{a}, x_{\mathfrak{z}} \in \mathfrak{z}(\mathfrak{a}, \mathfrak{g})$. Then we conclude from $0 = [x, y] = [x_\mathfrak{a}, y] + [x_{\mathfrak{z}}, y] = [x_\mathfrak{a}, y]$ that $x_\mathfrak{a} \in \mathfrak{z}(\mathfrak{e} \cap \mathfrak{a}, \mathfrak{a})$. Thus $\mathfrak{z}(\mathfrak{e} \cap \mathfrak{a}, \mathfrak{g}) \subseteq \mathfrak{z}(\mathfrak{e} \cap \mathfrak{a}, \mathfrak{a}) + \mathfrak{z}(\mathfrak{a}, \mathfrak{g})$.

Conversely, if $x = x_\mathfrak{a} + x_{\mathfrak{z}}$ with $x_\mathfrak{a} \in \mathfrak{z}(\mathfrak{e} \cap \mathfrak{a}, \mathfrak{a}), x_{\mathfrak{z}} \in \mathfrak{z}(\mathfrak{a}, \mathfrak{g})$ then there exists a non-zero $y \in \mathfrak{e} \cap \mathfrak{a}$ such that $[x_\mathfrak{a}, y] = 0$. But then $[x, y] = [x_\mathfrak{a}, y] + [x_{\mathfrak{z}}, y] = 0$ and therefore $x \in \mathfrak{z}(\mathfrak{e} \cap \mathfrak{a}, \mathfrak{g})$. □

5.5. LEMMA. *Suppose that $p: \mathfrak{g} \to \mathfrak{g}_1$ is a surjective Lie algebra homomorphism with kernel $\mathfrak{a}$. Then a subset $\mathfrak{e}$ of $\mathfrak{g}$ is lean in $\mathfrak{g}$ whenever $\mathfrak{a} \cap \mathfrak{e}$ is lean in $\mathfrak{g}$ and $p(\mathfrak{e})$ is lean in $\mathfrak{g}_1$.*

Proof. If $p(\mathfrak{e})$ is lean in $\mathfrak{b}$ then $p(\mathfrak{e}) \subseteq V$, where V is a subvariety in $\mathfrak{g}_1$ of dimension at most $\dim \mathfrak{g}_1 - 2$. Then $p^{-1}(V)$ is a subvariety in $\mathfrak{g}$ of dimension at most $\dim \mathfrak{g} - 2$ and $p^{-1}\bigl(\mathfrak{z}(p(\mathfrak{e}), \mathfrak{g}_1)\bigr) \subseteq p^{-1}(V)$. Now it follows at once from 5.4(i) that $\mathfrak{e}$ is lean in $\mathfrak{g}$ if also $\mathfrak{e} \cap \mathfrak{a}$ is lean in $\mathfrak{g}$. □

5.6. PROPOSITION. *Suppose that the Lie algebra $\mathfrak{g}$ is the direct ideal sum $\mathfrak{g} = \mathfrak{a} \oplus \mathfrak{b}$ and write $p: \mathfrak{g} \to \mathfrak{b}$ for the projection onto the ideal $\mathfrak{b}$. Then a subset $\mathfrak{e}$ of $\mathfrak{g}$ is lean in $\mathfrak{g}$ whenever $\mathfrak{a} \cap \mathfrak{e}$ is lean in $\mathfrak{a}$ and $p(\mathfrak{e})$ is lean in $\mathfrak{b}$.*

Proof. By Lemma 5.5 it suffices to show that $\mathfrak{e} \cap \mathfrak{a}$ is lean in $\mathfrak{g}$ if it is lean in $\mathfrak{a}$. By Proposition 5.4(ii) we have $\mathfrak{z}(\mathfrak{e} \cap \mathfrak{a}, \mathfrak{g}) = \mathfrak{z}(\mathfrak{e} \cap \mathfrak{a}, \mathfrak{a}) \oplus \mathfrak{b}$. If $\mathfrak{z}(\mathfrak{e} \cap \mathfrak{a}, \mathfrak{a}) \subseteq V$, with V a variety in $\mathfrak{a}$ of codimension at least 2, then $V \oplus \mathfrak{b}$ is a variety in $\mathfrak{g}$ of codimension at least 2, and the assertion follows. □

5.7. PROPOSITION. *Let S be a weakly exponential subsemigroup of a Lie group G such that the wedge $W = \mathcal{L}(S)$ has inner points.*

Suppose further that $\mathfrak{e}_1, \ldots, \mathfrak{e}_j$ are subalgebras of $\mathfrak{g}$ satisfying the following hypotheses:

(a) $\bigcup_{i=1}^{j} \mathfrak{e}_i$ *contains* $\overline{\mathrm{comp}_G(\mathfrak{g})} \cap H\bigl(L(\exp^{-1}(\exp W) \setminus W)\bigr)$.
(b) *All $\mathfrak{e}_i$, $i = 1, \ldots, j$ are lean.*

Then W is a semialgebra.

Condition (b) is satisfied if the following conditions hold:

(c) *Each $\mathfrak{e}_i$ is compactly embedded in $\mathfrak{g}$.*
(d) *None of the algebras $\mathfrak{e}_i$ contains nonzero ideals of $\mathfrak{g}$.*

Proof. Since finite unions of Zariski closed sets are Zariski closed and $\bigcup_{i=1}^{j} \mathfrak{z}(\mathfrak{e}) = \mathfrak{z}(\bigcup_{i=1}^{j} \mathfrak{e}_j)$, then (a) and (b) shows that $\overline{\mathrm{comp}_G(\mathfrak{g})} \cap H\bigl(L(\exp^{-1}(\exp W) \setminus W)\bigr)$ is lean. Then Proposition 5.3(ii) shows that W is a Lie semialgebra. If (c) and (d) hold, then by Theorem III.7.15 all algebras $\mathfrak{e}_i$ are lean. □

We draw attention to the fact that in the last step of this proof we used the culminating result of Chapter II.

5.8. LEMMA. *Let G be a Lie group and $\mathfrak{k}$ a compactly embedded subalgebra of $\mathfrak{g}$ containing $\operatorname{comp}_G(\mathfrak{g})$. Suppose that A is an analytic subgroup and A_{Lie} the group A with its intrinsic Lie group topology. The Lie algebra $\mathfrak{a}$ of A_{Lie} is identified with the subalgebra $\mathfrak{L}[A] \subseteq \mathfrak{g}$. Then $\operatorname{comp}_{A_{\text{Lie}}}(\mathfrak{a}) \subseteq \mathfrak{a} \cap \mathfrak{k}$, and $\mathfrak{a} \cap \mathfrak{k}$ is compactly embedded in $\mathfrak{a}$.*

Proof. If $x \in \operatorname{comp}_{A_{\text{Lie}}}(\mathfrak{a})$ then $C := \overline{\exp \mathbb{R} \cdot x}$ is a compact subgroup of A_{Lie}, and since the identity map $A_{\text{Lie}} \to A$ is continuous, C is a compact subgroup of G. Hence $x \in \operatorname{comp}_G(\mathfrak{g}) \subseteq \mathfrak{k}$. Thus $\operatorname{comp}_{A_{\text{Lie}}}(\mathfrak{a}) \subseteq \mathfrak{a} \cap \mathfrak{k}$ as asserted. If $\mathfrak{k}$ is compactly embedded in $\mathfrak{g}$ then $\mathfrak{a} \cap \mathfrak{k}$ is compactly embedded in $\mathfrak{a}$. □

5.9. COROLLARY. *Let S be a weakly exponential subsemigroup of a Lie group G and suppose that there is a compactly embedded subalgebra $\mathfrak{e}$ containing $H(W) \cap \operatorname{comp}_G(\mathfrak{g})$. Then W is a semialgebra.*

Proof. By Proposition 4.8 and Lemma 5.8 we may assume that S is reduced. Thus Proposition 5.7 applies with $j = 1$ and shows that W is a semialgebra. □

We conclude the section with a remark of independent interest which will not be used in the sequel.

5.10. EXERCISE. Let $p: \mathfrak{g} \to \mathfrak{g}_1$ be a surjective morphism between Lie algebras with kernel $\ker p = \mathfrak{a}$. Let $\mathfrak{e}$ be any subalgebra of $\mathfrak{g}$ satisfying the following conditions:
  (a) $p(\mathfrak{e})$ is compactly embedded in $\mathfrak{g}_1$,
  (b) $p(\mathfrak{e})$ does not contain any nonzero ideal in $\mathfrak{g}_1$.
Then $\dim \mathfrak{g} - \dim 3(\mathfrak{e}, \mathfrak{g}) \geq \min\{\dim \mathfrak{a} - \dim 3(\mathfrak{e} \cap \mathfrak{a}, \mathfrak{a}), 2\}$. In particular, $\mathfrak{e}$ is lean in $\mathfrak{g}$ if $\mathfrak{e} \cap \mathfrak{a}$ is lean in $\mathfrak{a}$.

Proof. By (a) and (b) we can apply Theorem III.7.15 and obtain

$$\dim \mathfrak{g}_1 - \dim 3(p(\mathfrak{e}), \mathfrak{g}_1) \geq 2.$$

Consequently,

$$\dim \mathfrak{g} - \dim p^{-1}\bigl(3(p(\mathfrak{e}), \mathfrak{g}_1)\bigr) \geq \dim \mathfrak{a} + \dim \mathfrak{g}_1 - (\dim 3(p(\mathfrak{e}), \mathfrak{g}_1) + \dim \mathfrak{a}) \geq 2.$$

We first deduce from 5.4(ii) the relation $\dim 3(\mathfrak{e} \cap \mathfrak{a}, \mathfrak{g}) \leq \dim 3(\mathfrak{e} \cap \mathfrak{a}, \mathfrak{a}) + \dim \mathfrak{g}_1$. Then

$$\dim \mathfrak{g} - \dim 3(\mathfrak{e} \cap \mathfrak{a}, \mathfrak{g}) \geq \dim \mathfrak{a} + \dim \mathfrak{g}_1 - (\dim 3(\mathfrak{e} \cap \mathfrak{a}, \mathfrak{a}) + \dim \mathfrak{g}_1)$$
$$= \dim \mathfrak{a} - \dim 3(\mathfrak{e} \cap \mathfrak{a}, \mathfrak{a}).$$

Now from 5.4(i) it follows that

$$\dim \mathfrak{g} - \dim 3(\mathfrak{e}) \geq \min\bigl\{ \dim \mathfrak{g} - \dim 3(\mathfrak{e} \cap \mathfrak{a}, \mathfrak{a}), \dim \mathfrak{g} - \dim p^{-1}(3(p(\mathfrak{e}), \mathfrak{g}_1)) \bigr\}$$
$$\geq \min\{\dim \mathfrak{a} - \dim 3(\mathfrak{e} \cap \mathfrak{a}, \mathfrak{a}), 2\}.$$

This is what we had to show. □

6. Groups with a Unique Maximal Compact Subgroup

6.1. DEFINITION. Let G be a topological group. We shall call G a *UMCS-group* if G contains a unique maximal compact subgroup $K(G)$.

Clearly $K(G)$ is a fully characteristic subgroup of G. □

We observe right away that all connected Lie groups with a compact Lie algebra, all connected nilpotent Lie groups, such groups as e.g. $\widetilde{\mathrm{Sl}}(2,\mathbb{R})$ are UMCS-groups. The class is much vaster than these examples indicate.

6.2. LEMMA. *Let G denote a UMCS-Lie group. Then $\mathfrak{L}(K(G)) = \mathrm{comp}_G(\mathfrak{g}) = \overline{\mathrm{comp}_G(\mathfrak{g})}$.*

Proof. We abbreviate $\mathfrak{L}(K(G))$ by $\mathfrak{k}$. Since $K(G)$ is compact we have $\mathfrak{k} \subseteq \mathrm{comp}_G(\mathfrak{g})$. Now suppose that $x \in \mathrm{comp}_G(\mathfrak{g})$. Then the subgroup $\overline{\exp \mathbb{R} \cdot x}$ is compact, and so, since $K(G)$ is a unique maximal compact subgroup, it is contained in $K(G)$. Hence $x \in \mathfrak{k}$. Thus $\mathfrak{k} = \mathrm{comp}_G(\mathfrak{g})$, and since $\mathfrak{k}$ is closed as a vector subspace, the remainder follows. □

For dimensional reasons, every connected Lie group contains maximal normal connected UMCS-Lie subgroups, but they need not agree.

The following is crucial for our purposes:

6.3. MAIN LEMMA. *Let S be a weakly exponential subsemigroup of a UMCS-Lie group G. Then W is a semialgebra.*

Proof. Since $\mathrm{comp}_G(\mathfrak{g})$ is invariant under all inner automorphisms $\mathrm{Ad}(g)$, $g \in G$ the vector space $\mathrm{comp}_G(\mathfrak{g})$ is an ideal. We set $\mathfrak{e} = H(W) \cap \mathrm{comp}_G(\mathfrak{g}) = H(W) \cap \overline{\mathrm{comp}_G(\mathfrak{g})}$. Then $\mathfrak{e}$ is compactly embedded and thus 5.9 proves the claim. □

6.4. COROLLARY. *Let S be a weakly exponential subsemigroup of a connected Lie group G and $\mathfrak{m}$ a maximal rank subalgebra of $\mathfrak{g}$ which meets the interior of the Lie wedge $W = \mathfrak{L}(S)$. Suppose that the analytic subgroup $M = \langle \exp \mathfrak{m} \rangle$ has a unique maximal compact subgroup. Then $\mathfrak{m} \cap W$ is a semialgebra.*

Proof. By the Maximal Rank Theorem II.4.3 we know that $S \cap M$ is weakly exponential, so we may apply the Main Lemma 6.3 with $S \cap M$ in place of S and the assertion follows. □

6.5. PROPOSITION. *Let S be a weakly exponential subsemigroup of a connected Lie group G and $\mathfrak{h}$ a Cartan algebra of $\mathfrak{g}$. Suppose that $\mathfrak{h} \cap \operatorname{int} W \neq \emptyset$. Then $\mathfrak{h} \cap W$ is a semialgebra and $[\mathfrak{h}, \mathfrak{h}] \subseteq W$.*

Proof. The analytic subgroup $H = \exp \mathfrak{h}$ is closed nilpotent and is therefore a UMCS-group. By Corollary 6.4 we know now that $\mathfrak{h} \cap W$ is a semialgebra. The inclusion $[\mathfrak{h}, \mathfrak{h}] \subseteq W$ follows from the First Triviality Theorem ([27] II.4.13, p. 131). □

All maximally almost periodic connected Lie groups are UMCS-groups. For this class, the conclusion of our Main Lemma 6.3 can be sharpened. Let us take stock of various characterisations of maximally almost periodic connected Lie groups:

6.6. REMARK. For a connected Lie group G the following assertions are equivalent:
   (i) $\mathfrak{g}$ is a compact Lie algebra, that is, the group of its inner automorphisms is compact;
   (ii) G is a covering group of a compact group (or: $\mathfrak{g}$ is the Lie algebra of a compact Lie group);
   (iii) G is maximally almost periodic, that is, it maps injectively into its Bohr compactification;
   (iv) G is the direct product $G = N \times V$ of its maximal compact subgroup N with a central vector subgroup V. □

6.7. THEOREM. *Let G be a connected maximally almost periodic Lie group and let W be the Lie wedge of a weakly exponential subsemigroup S of G with inner points. Then W is invariant.*

Proof. The hypotheses of our Main Lemma 6.3 are satisfied. So W is a semialgebra. By Theorem II.5.3 the wedge W has inner points. But now every generating semialgebra in a compact Lie algebra is invariant (cf. [27], Corollary II.6.5(ii) on p. 159) □

There is also a reverse statement to Theorem 6.7, which is due to MITTENHUBER and NEEB [61].

6.8. THEOREM. *If G is a Lie group whose Lie algebra $\mathfrak{g}$ is compact and S is a closed Lie subsemigroup whose Lie wedge is invariant then S is exponential, $S = \exp W$.* □

6.9. COROLLARY. *Suppose that G is a maximally almost periodic connected Lie group and S is a weakly exponential subsemigroup with inner points. Then S is exponential.* □

The half-space subsemigroups in the motion group show that there exist weakly exponential semigroups which are not exponential.

7. More about Lean Sets

7.1. LEMMA. *If $\mathfrak{g}$ is a direct ideal sum $\mathfrak{g}_1 \oplus \cdots \oplus \mathfrak{g}_n$ and $x = x_1 + \cdots + x_n \in \mathfrak{g}$. Then $\mathfrak{z}(x,\mathfrak{g}) = \bigoplus_{i=1}^n \mathfrak{z}(x_i, \mathfrak{g}_i)$.*

Proof. Straightforward. □

7.2. LEMMA. *Let $\mathfrak{g}$ be a Lie algebra which can be written as the direct ideal sum*
$$\mathfrak{g} = \mathfrak{s}_1 \oplus \mathfrak{s}_2 \oplus \cdots \oplus \mathfrak{s}_n$$
with $n \geq 2$. Let $\varphi_i: \mathfrak{sl}(2,\mathbb{R}) \to \mathfrak{s}_i$ be an isomorphism for $i = 1, \ldots, n$ and define $\Delta: \mathfrak{sl}(2,\mathbb{R}) \to \mathfrak{g}$ by $\Delta(x) = \sum_{i=1}^n \varphi_i(x)$. Then the set $\mathfrak{e} := \Delta\bigl(\mathfrak{sl}(2,\mathbb{R})\bigr)$ is lean in $\mathfrak{g}$.

Proof. The set $\mathfrak{z}(\mathfrak{e})$ is itself a variety by III.5.5(iii). From Lemma 7.1 we see that $\mathfrak{z}(\mathfrak{e}) = \bigcup\{\mathfrak{z}(x,\mathfrak{g}) \mid 0 \neq x \in \mathfrak{e}\} = \{\sum_{i=1}^n r_i \cdot \varphi_i(x) \mid r_i \in \mathbb{R}, x \in \mathfrak{sl}(2,\mathbb{R})\}$. The map

$$(x, r_2, \ldots, r_n) \mapsto \varphi_1(x) + \sum_{i=2}^n r_i \cdot \varphi_i(x) : \bigl(\mathfrak{sl}(2,\mathbb{R}) \setminus \{0\}\bigr) \times \mathbb{R}^{n-1} \to \mathfrak{z}(\mathfrak{e})$$

is a regular bijection onto the subset of all $x_1 + \cdots + x_n \in \mathfrak{z}(\mathfrak{e})$ with $x_1 \neq 0$ which is an open dense submanifold of $\mathfrak{z}(\mathfrak{e})$. Its dimension is $3 + (n-1) = n + 2$. Now $\dim \mathfrak{g} - \mathfrak{z}(\mathfrak{e}) = 3n - (n+2) = 2n - 2 = 2(n-1) \geq 2$. □

7.3. LEMMA. *Suppose that*
$$\mathfrak{g} = \mathfrak{s}_1 \oplus \mathfrak{s}_2 \oplus \cdots \oplus \mathfrak{s}_n, \quad \mathfrak{s}_i \cong \mathfrak{sl}(2,\mathbb{R}), \quad n = 1, 2, \ldots,$$
and that $\mathfrak{u}$ is a subalgebra of $\mathfrak{g}$ not containing any of the direct summands $\mathfrak{s}_i$. Then $\mathfrak{e} := \mathfrak{u} \cap \text{comp}(\mathfrak{g})$ is lean in $\mathfrak{g}$.

Proof. We prove this statement by induction with respect to n. We assume that $\mathfrak{g}$ is a counterexample to the claim with minimal n and derive a contradiction.

Step 1. Claim: $n > 1$. Indeed, suppose $n = 1$, i.e, $\mathfrak{g} \cong \mathfrak{sl}(2,\mathbb{R})$. If $\mathfrak{e} = \{0\}$ then the claim is trivially true. If not, then $\dim \mathfrak{e} = 1$, and then $0 \neq x \in \mathfrak{e}$ implies $\mathfrak{z}(x,\mathfrak{g}) = \mathfrak{e}$ and thus $\mathfrak{z}(\mathfrak{e}) = \mathfrak{e}$.

Step 2. Let $\mathfrak{r}_\mathfrak{u}$ denote the solvable radical of $\mathfrak{u}$. We claim $\mathfrak{r}_\mathfrak{u} = \{0\}$. Suppose not. Then there is an ideal summand $\mathfrak{b} \cong \mathfrak{sl}(2, \mathbb{R})$ such that with the projection $p \colon \mathfrak{g} \to \mathfrak{b}$ we have $p(\mathfrak{r}_\mathfrak{u}) \neq \{0\}$. Since $\mathfrak{sl}(2, \mathbb{R})$ is simple we cannot have $p(\mathfrak{u}) = \mathfrak{b}$. Thus $p(\mathfrak{u})$ is a proper, hence solvable subalgebra of $\mathfrak{b}$. Now $p(\mathfrak{e}) \subseteq p(\mathfrak{u}) \cap \overline{p(\text{comp}(\mathfrak{g}))} \subseteq p(\mathfrak{u}) \cap \text{comp}(\mathfrak{b})$. By the proof of Step 1, $p(\mathfrak{e})$ is lean in $\mathfrak{b}$. The kernel $\mathfrak{a}$ of p is an algebra of $n - 1$ ideal summands isomorphic to $\mathfrak{sl}(2, \mathbb{R})$ and $[\mathfrak{a}, \mathfrak{b}] = \{0\}$, whence $\mathfrak{u} \cap \mathfrak{a}$ is a subalgebra of $\mathfrak{a}$ not containing any of the summands. Since $\mathfrak{a}$ cannot be a counterexample to the proposition, by the minimality of n, we know that $\mathfrak{e} \cap \mathfrak{a}$ is lean in $\mathfrak{a}$. We also observe that $\mathfrak{g}$ is an ideal direct sum $\mathfrak{a} \oplus \mathfrak{b}$. Now Proposition 5.6 shows that $\mathfrak{e}$ is lean in $\mathfrak{g}$, a contradiction. Thus $\mathfrak{r}_\mathfrak{u} = \{0\}$ and $\mathfrak{u}$ is itself a direct sum of simple ideal direct summands of $\mathfrak{u}$.

Step 3. Let $\mathfrak{s}$ be a simple ideal summand of $\mathfrak{u}$. A projection of $\mathfrak{s}$ into any of the simple ideal summands of $\mathfrak{g}$ is either zero or isomorphic to $\mathfrak{sl}(2, \mathbb{R})$. As $\mathfrak{s}$ is simple we conclude $\mathfrak{s} \cong \mathfrak{sl}(2, \mathbb{R})$. We write the unique simple sum decomposition of $\mathfrak{g}$ in the form $\mathfrak{s}_1 \oplus \cdots \oplus \mathfrak{s}_k \oplus \cdots \oplus \mathfrak{s}_n$ in such a fashion that for the projection $\pi_j \colon \mathfrak{g} \to \mathfrak{g}$ onto $\mathfrak{s}_j \cong \mathfrak{sl}(2, \mathbb{R})$, $j = 1, \ldots, n$, we have $\pi_j(\mathfrak{s}) = \mathfrak{s}_j$ for $j = 1, \ldots, k$ and $\pi_j(\mathfrak{s}) = \{0\}$ for $j = k+1, \ldots, n$. We claim that $k \geq 2$. Otherwise $\mathfrak{s} = \mathfrak{s}_1$; hence $\mathfrak{s} \subseteq \mathfrak{u}$ which is impossible by hypothesis. We set $\mathfrak{b} = \mathfrak{s}_1 \oplus \cdots \oplus \mathfrak{s}_k$ and $\mathfrak{a} = \mathfrak{s}_{k+1} \oplus \cdots \oplus \mathfrak{s}_n$ and let $p \colon \mathfrak{g} \to \mathfrak{b}$ denote the projection with kernel $\mathfrak{a}$. We claim $p(\mathfrak{u}) = p(\mathfrak{s})$. Indeed, if $\mathfrak{t}$ is any other simple summand in $\mathfrak{u}$, then $[\mathfrak{s}, \mathfrak{t}] = \{0\}$ and thus $[p(\mathfrak{s}), p(\mathfrak{t})] = \{0\}$. Thus $p(\mathfrak{t})$ is in the centralizer of the diagonal subalgebra $p(\mathfrak{s})$ of $\mathfrak{s}_1 \oplus \cdots \oplus \mathfrak{s}_k$ which is $\{0\}$. This implies $\mathfrak{u} = \mathfrak{s} \oplus \ker p$. Thus $p(\mathfrak{s}) = p(\mathfrak{u})$. Then by Lemma 7.2 we know that $p(\mathfrak{u})$ is lean in $\mathfrak{b}$. But $\mathfrak{a}$ is a Lie algebra of $n - k < n - 1$ summands $\mathfrak{sl}(2, \mathbb{R})$. Again $[\mathfrak{a}, \mathfrak{b}] = \{0\}$. Hence $\mathfrak{u} \cap \mathfrak{a}$ is a subalgebra of $\mathfrak{a}$ none of whose direct ideal $\mathfrak{sl}(2, \mathbb{R})$ summands can be contained in $\mathfrak{u}$. Since $\mathfrak{a}$ cannot be a counterexample to the proposition by the minimality of n, $\mathfrak{e} \cap \mathfrak{a}$ is lean in $\mathfrak{a}$. By Proposition 5.6 we conclude that $\mathfrak{e}$ is lean in $\mathfrak{g}$ which is impossible.

This contradiction finally proves the Lemma. □

7.4. LEMMA. *Suppose that $\mathfrak{g}$ is a direct ideal sum $\mathfrak{a} \oplus \mathfrak{b}$, further $\mathfrak{j}$ a compactly embedded subalgebra of $\mathfrak{a}$, and $\mathfrak{u}$ a subalgebra of $\mathfrak{g}$ such that the following hypotheses are satisfied:*

(a) *$\mathfrak{b}$ is itself a direct sum of finitely many ideals isomorphic with $\mathfrak{sl}(2, \mathbb{R})$.*

(b) *$\mathfrak{a}$ does not contain any subalgebra isomorphic to $\mathfrak{sl}(2, \mathbb{R})$.*

(c) *$\mathfrak{u}$ is an $\overline{\text{ideal free}}$ subalgebra of $\mathfrak{g}$.*

Then $\mathfrak{e} := \mathfrak{u} \cap (\mathfrak{j} \oplus \text{comp}(\mathfrak{b}))$ is lean in $\mathfrak{g}$.

Proof. Let $p \colon \mathfrak{g} \to \mathfrak{b}$ denote the projection onto $\mathfrak{b}$ along $\mathfrak{a}$. We claim that $p(\mathfrak{u})$ is ideal free in $\mathfrak{b}$: Suppose that $\mathfrak{s}$ is a simple ideal summand of $\mathfrak{b}$ contained in $p(\mathfrak{u})$. Then there is a subalgebra $\mathfrak{s}_u$ of $\mathfrak{u}$ mapping isomorphically onto $\mathfrak{s}$. Its projection into $\mathfrak{a}$ is either $\{0\}$ or is faithful since $\mathfrak{s}_u$ is simple. In the first case, $\mathfrak{s}_u \subseteq \mathfrak{b}$; then $\mathfrak{s}_u = \mathfrak{s} \subseteq \mathfrak{u}$ contradicting (c). In the second case, $\mathfrak{a}$ would contain a copy of $\mathfrak{sl}(2, \mathbb{R})$, contradicting (b). Thus the claim is established. Now $p(\mathfrak{e}) \subseteq p(\mathfrak{u}) \cap \text{comp}(\mathfrak{b})$. Then Lemma 7.3 shows that $p(\mathfrak{e})$ is lean.

Next $\mathfrak{e} \cap \mathfrak{a} = \mathfrak{u} \cap \overline{(\text{comp}(\mathfrak{b}) \oplus \mathfrak{j})} \cap \mathfrak{a} = \mathfrak{u} \cap \mathfrak{j}$. This algebra is compactly embedded in $\mathfrak{a}$. It is ideal free in $\mathfrak{a}$ because of $[\mathfrak{a}, \mathfrak{b}] = \{0\}$ and (c). Then it is lean in $\mathfrak{a}$ by Theorem III.7.15. Now Lemma 5.6 shows that $\mathfrak{e}$ is lean in $\mathfrak{g}$. □

7.5. MAIN LEMMA. *Let G be a Lie group whose Lie algebra $\mathfrak{g}$ is the direct sum $\mathfrak{g} = \mathfrak{a} \oplus \mathfrak{b}$ of an ideal $\mathfrak{a}$ in $\mathfrak{g}$ with a direct ideal sum $\mathfrak{b} = \mathfrak{s}_1 \oplus \mathfrak{s}_2 \oplus \ldots \oplus \mathfrak{s}_k$ of isomorphic copies of $\mathfrak{sl}(2, \mathbb{R})$. We assume that the following hypotheses are satisfied:*

(a) *There is a compactly embedded subalgebra $\mathfrak{c}$ of $\mathfrak{a}$ such that*

$$\overline{\text{comp}_G(\mathfrak{g})} \subseteq \mathfrak{c} \oplus \overline{\text{comp}(\mathfrak{b})}.$$

(b) *$\mathfrak{a}$ does not contain any subalgebra isomorphic to $\mathfrak{sl}(2, \mathbb{R})$.*
Then the Lie wedge of any reduced weakly exponential closed subsemigroup S of G is a semialgebra.

Proof. Set $W = \mathfrak{L}(S)$ and $\mathfrak{u} = H(W)$. As S is reduced (see Definition 4.7) and by the Interior Point Theorem II.5.3(ii) we know that int $W \neq \emptyset$ and that $\mathfrak{u}$ is ideal free. We apply Proposition 7.4 and obtain from (a) that $\mathfrak{u} \cap (\mathfrak{c} \oplus \overline{\text{comp}(\mathfrak{g})})$ is lean in $\mathfrak{g}$. Then $H(W) \cap \overline{\text{comp}_G(\mathfrak{g})}$, which is contained in $\mathfrak{u} \cap (\mathfrak{c} \oplus \overline{\text{comp}(\mathfrak{g})})$, is lean. Now 5.3(ii) shows that W is a semialgebra. □

CHAPTER 5

MORE EXAMPLES

In the foregoing chapters we have pushed the *general* theory to a first limit, to go beyond we now need detailed and explicit information about a a few very *special* situations. With this information at our hands we then shall be able to unravel the exact structure of Lie groups containing a reduced weakly exponential subsemigroup. The results are of a negative nature: we show that the Lie wedge of the weakly exponential subsemigroup cannot contain compact elements in its interior which lie in a copy of one of the algebras $\mathfrak{sl}(2,\mathbb{R})$, mot, osc or certain extensions of them.

On the other hand the required explicit calculations also provide a good testing ground as well as a pool of interesting examples for the theory introduced in the previous chapters. Not surprisingly, the Master Examples introduced right in the beginning again play a significant role in our considerations.

We collect the main information involving low dimensional 'test examples' in the present chapter and draw the conlusions in the next one.

1. Examples Connected with the Special Linear Group

1.1. NOTATION. In the following we consider $\mathfrak{g} = \mathfrak{sl}(2,\mathbb{R})$ as the set of all 2×2 matrices X over $\mathbb{R}$ with $\operatorname{tr} X = 0$. We write k for the Cartan-Killing form $(X,Y) \mapsto \operatorname{tr}(\operatorname{ad} X \operatorname{ad} Y)$ of $\mathfrak{sl}(2,\mathbb{R})$ and put

$$\mathcal{D}^+ = \{X \in \mathfrak{g} \mid k(X,X) > 0\}, \qquad \mathcal{D}^- = \{X \in \mathfrak{g} \mid k(X,X) < 0\},$$
$$\mathcal{D}^0 = \{X \in \mathfrak{g} \mid k(X,X) = 0\}.$$

The set $\mathcal{K} = \overline{\mathcal{D}^-} = \mathcal{D}^- \cup \mathcal{D}^0$ is called the *Standard Double Cone* in $\mathfrak{sl}(2,\mathbb{R})$. □

1.2. REMARK. Note that $k(X,X) = -8 \det X$, hence $\mathcal{D}^+$ is exactly the set of all real 2×2-matrices with two real nonzero eigenvalues r and $-r$; similarly, $\mathcal{D}^-$ is the set of all 2×2-matrices with two nonzero purely imaginary eigenvalues ir and $-ir$. □

1.3. PROPOSITION. *Suppose that G is a Lie group with Lie algebra $\mathfrak{g} = \mathfrak{sl}(2,\mathbb{R})$ and write $Z(G)$ for the center of G. Then the following assertions hold:*
  (i) *The partition $\mathfrak{g} = \mathcal{D}^+ \cup \mathcal{D}^0 \cup \mathcal{D}^-$ of $\mathfrak{g}$ induces a partition*
   $\exp \mathfrak{g} = \exp(\mathcal{D}^+) \cup \exp(\mathcal{D}^0) \cup \exp(\mathcal{D}^-)$.
  (ii) *The set $\exp \mathcal{D}^+$ is open in G and*

$$G \setminus \exp \mathfrak{g} = \bigcup_{1 \neq z \in Z(G)} z \exp\bigl((\mathcal{D}^+ \cup \mathcal{D}^0) \setminus \{1\}\bigr).$$

In particular, $\bigl(z \exp \overline{\mathcal{D}^+} \setminus \{0\}\bigr) \cap \exp \mathfrak{g} = \emptyset$ whenever $1 \neq z \in Z(G)$.
  (iii) *The restriction of the exponential function to the set $\overline{\mathcal{D}^+} = \{X \in \mathfrak{g} \mid k(X,X) \geq 0\} = \mathfrak{g} \setminus \mathcal{D}^-$ is a homeomorphic embedding; similarily, the restriction of $\exp$ to every connected component of the open set $\mathcal{D}^- \cap \mathrm{reg}\exp$ is a homeomorphic embedding.*
  (iv) *A nonzero element $X \in \mathfrak{g}$ is compact if and only if it lies in $\mathcal{D}^-$, i.e., $\mathrm{comp}\bigl(\mathfrak{sl}(2,\mathbb{R})\bigr) \setminus \{0\} = \mathcal{D}^-$. Thus the set $\overline{\mathrm{comp}\,\mathfrak{g}}$ is just the standard double cone. Moreover, $Z(G) \subseteq \exp \mathbb{R}{\cdot}X$ for every $X \in \mathcal{D}^-$, and $\exp \mathcal{D}^- \setminus Z(G) = \exp(\mathcal{D}^- \cap \mathrm{reg}\exp)$.*
  (v) *Every $X \in \mathcal{D}^+$ satisfies the following condition $(*)$:*
   $(*)$ *If $\exp X = \lim z_n \exp X_n$, with $z_n \in Z(G)$ and $X_n \in \mathfrak{g}$, then $X = \lim X_n$.*
  (vi) $\mathrm{Irr}(G) = Z(G) \exp \mathcal{D}^0$.

Proof. We leave the proof of assertions (i)–(iv) to the reader (hint: use the preceding remark or consult [27], section V.4).

(v) Suppose that $0 \neq X \in \overline{\mathcal{D}^+}$ such that $\exp X = \lim z_n \exp X_n$ with $z_n \in Z(G)$, $X_n \in \mathfrak{g}$. By the above assertion (ii) the set $\exp \mathcal{D}^+$ is open in G, so for all sufficiently large indexes n we have $z_n \exp X_n \in \exp \mathcal{D}^+ \subset \exp \mathfrak{g}$. But, also by (ii), $z_n \exp \mathcal{D}^+ \cap \exp \mathfrak{g} = \emptyset$ whenever $z_n \neq 1$, thus eventually $z_n = 1$ and $\exp X_n \in \exp \mathcal{D}^+$, which means that $X_n \in \mathcal{D}^+$, by (i). Since by (iii) the restriction of $\exp$ to $\overline{\mathcal{D}^+}$ is a homeomorphism this implies $\lim X_n = X$.

(vi) $\mathcal{D}^0$ is precisely the set $\mathfrak{g} \setminus \mathrm{reg}(\mathfrak{g})$. Thus $Z(G) \exp \mathcal{D}^0 \subseteq \mathrm{Irr}(G)$. Likewise $\exp(\mathfrak{g} \setminus \mathcal{D}^0) \subseteq \mathrm{Reg}(G)$. Hence $Z(G) \exp(\mathfrak{g} \setminus \mathcal{D}^0) \subseteq \mathrm{Reg}(G)$. Further $\exp(\mathfrak{g} \setminus \mathcal{D}^0) = \exp \mathfrak{g} \setminus \exp \mathcal{D}^0$ by (i). This implies the assertion. □

1.4. REMARK. Note that the above assertion 1.3(ii) explicitly describes the 'maximal open hole' in the exponential image of $\mathfrak{sl}(2,\mathbb{R})$ as the union of the sets $z \exp \mathcal{D}^+$ with $z \in Z(G) \setminus \{1\}$. Assertion (v) is essentially a paraphrase of this fact. □

1.5. EXERCISE. Show that for every $X \in \mathcal{D}^-$ with $\exp X \notin Z(G)$ the following condition (∗∗) holds:
(∗∗) If $\exp X = \lim z_n \exp X_n$ with $z_n \in Z(G)$, $X_n \in \mathfrak{g}$, then there is a sequence $\sigma_n \in \mathbb{R}$ with $X = \lim \sigma_n \cdot X_n$.
(Note that this condition is slightly weaker than condition (∗) above.) Hint: If $X \in \mathcal{D}^-$ such that $\exp X \notin Z(G)$ then it lies in the interior of $\exp \mathcal{D}^-$. Hence eventually $X_n \in \mathcal{D}^- \cap \mathrm{reg}\,\exp$, by (i) and (ii). Thus for sufficiently large n there exist real numbers σ_n such that $z_n = \exp(\sigma_n - 1) \cdot X_n$. Choose the numbers σ_n so that all vectors $\sigma_n \cdot X_n$ lie in the same connected component $\mathcal{D}_*^-$ of $\mathcal{D}^- \cap \mathrm{reg}\,\exp$. Then $\exp X = \lim z_n \exp X_n = \lim \exp \sigma_n \cdot X_n$, and an application of (iii) shows $X = \lim \sigma_n \cdot X_n$. □

We also recall that every semialgebra with inner points in $\mathfrak{sl}(2, \mathbb{R})$ is the intersection of half-spaces bounded by tangent planes of the standard double cone. The following observation slightly sharpens results in [27] (p. 110, and p. 418, Proposition V.4.24).

1.6. PROPOSITION. *If W is a semialgebra with nonvoid interior and contained in $\overline{\mathcal{D}^+}$ then W is the intersection of conjugates of the set*

$$\mathfrak{sl}(2, \mathbb{R})^+ := \left\{ \begin{pmatrix} a & b \\ c & -a \end{pmatrix} \Big| a \in \mathbb{R},\, b,\, c \geq 0 \right\}$$

of all matrices m in $\mathfrak{sl}(2, \mathbb{R})$ such that $\exp r \cdot m$ has nonnegative entries for every $r \geq 0$, and hence $\exp W$ is a closed subsemigroup of G. □

Note that the above two propositions are uniformly valid for all Lie groups G with $L(G) = \mathfrak{g}$. Our next proposition characterizes the Lie wedges of weakly exponential subsemigroups containing inner points.

1.7. PROPOSITION. *Assume that G is a Lie group with Lie algebra $\mathfrak{g} = \mathfrak{sl}(2, \mathbb{R})$ and let S be a closed proper subsemigroup of G. We assume that the Lie wedge W of S has inner points in $\mathfrak{g}$. Then the following assertions are equivalent:*
(i) *S is weakly exponential.*
(ii) *W is a semialgebra and contained in $\overline{\mathcal{D}^+}$.*
(iii) *S is exponential.*

Proof. (i)⇒(ii) Assume that S is weakly exponential and suppose first that W is not contained in $\overline{\mathcal{D}^+}$. Since $\mathcal{D}^-$ is open in $\mathfrak{g}$ this means that the interior of W contains a nonzero element $U \in \mathrm{comp}\,\mathfrak{g}$. Thus G is simply connected (for otherwise $S = G$) and the exponential image z of a suitable positive multiple of U lies in the center of G.

Since U is an inner point of W we conclude that every point $\exp \lambda \cdot U$ with $\lambda > 0$ is an inner point of S. (The easy proof of this statement is postponed to the following Lemma, which we shall use again when discussing weakly exponential subsemigroups of Mot and Osc.) In particular, z is an inner point of S.

Now $z \neq \mathbf{1}$, hence $z\mathcal{D}^+ \subseteq G \setminus \exp \mathfrak{g}$. Since the intersection $z\mathcal{D}^+ \cap \operatorname{int} S$ is nonvoid this means that S cannot be weakly exponential, a contradiction. Thus $W \subseteq \overline{\mathcal{D}^+}$.

Suppose now that W is not a semialgebra. Then by Proposition IV.5.1(ii) the edge $H(W)$ of W contains a nonzero element $X \in \operatorname{comp} \mathfrak{g} = \overline{\mathcal{D}^-}$. Since W has interior points in $\mathfrak{g}$ there exists an element $Y \in \operatorname{int} W$ with $k(X, Y) \neq 0$. But then for any $t < -k(Y,Y)/k(X,Y)$ we have $k(Y + t \cdot X, Y + t \cdot X) = k(Y,Y) + 2tk(X,Y) + t^2 \cdot k(X,X) < 0$, so $0 \neq Y + t \cdot X \in W \cap \operatorname{comp} \mathfrak{g} = W \cap \mathcal{D}^- = \emptyset$, a contradicition.

(ii)$\Rightarrow$(iii) Applying the above observation 1.6 we see that $\exp W$ is a closed subsemigroup of G, hence agrees with S.

The implication (iii)$\Rightarrow$(i) is obvious. □

1.8. LEMMA. *Let $U \in \mathfrak{g}$ be an interior point in the Lie wedge W of a closed subsemigroup S of a Lie group G. Then for every $\lambda > 0$ the exponential image $\exp \lambda \cdot U$ is an inner point of S.*

Proof. Choose a positive real $\varepsilon < 1$ such that $\varepsilon \lambda \cdot U$ lies in an open Campbell Hausdorff neighborhood B in $\mathfrak{g}$. Then $\exp(B \cap W)$ is a neighborhood of $\exp \varepsilon \lambda \cdot U$ and $\exp(1 - \varepsilon)\lambda \cdot U \exp(B \cap W)$ is a neighborhood of $\exp U$ and contained in S. □

2. Some Results on Injective Images of Lie Subsemigroups

Let S be a Lie subsemigroup of a Lie group G and suppose that we are given a continuous homomorphism $f: G \to G_1$ which is injective on S. It is of general interest to know conditions which guarantee that $f(S)$ is closed in G_1. Our first instance for such a situation concerns Lie groups whose Lie algebra is isomorphic to a direct product $\mathfrak{sl}(2, \mathbb{R})^k$. This result will provide an example of a weakly exponential subsemigroup generating a non-closed analytic subgroup, we shall use it for other purposes later.

2.1. PROPOSITION. *Suppose that G is a connected Lie group whose Lie algebra $\mathfrak{g}$ contains a Lie semialgebra W such that the Lie algebra $W - W$ is isomorphic to $\mathfrak{sl}(2, \mathbb{R})^k$. We assume that W corresponds to a product $W_1 \times \cdots \times W_k \subseteq \mathfrak{sl}(2, \mathbb{R})^k$ with $W_j \subseteq \overline{\mathcal{D}^+}$. Then*
  (i) *$\exp W$ is a closed exponential subsemigroup,*
  (ii) *$\exp W$ is the (intrinsic) direct product $\exp W_1 \times \cdots \times \exp W_k$.*
  (iii) *Let G_1 denote a Lie group with a closed Lie subsemigroup T_1 and $f: G_1 \to G$ a morphism such that $\mathfrak{L}(f): \mathfrak{g}_1 \to \mathfrak{g}$ maps $\mathfrak{g}_1$ isomorphically onto $W - W$ and $\mathfrak{L}(T_1)$ isomorphically onto W. Then f maps T_1 isomorphically onto $\exp W$.*

Proof. Step 1. We consider the exponential functions $\exp_1: \mathfrak{sl}(2, \mathbb{R}) \to \widetilde{\mathrm{Sl}}(2, \mathbb{R})$ and $\exp_2: \mathfrak{sl}(2, \mathbb{R}) \to \mathrm{PSl}(2, \mathbb{R})$. The semigroup $\widetilde{S}_j = \exp_1 W_j \subseteq \widetilde{\mathrm{Sl}}(2, \mathbb{R})$ is closed and exponential (see 1.7). Likewise, $\mathcal{S}_j = \exp_2 W_j \subseteq \mathrm{PSl}(2,\mathbb{R})$ is closed and exponential in view of 1.3(iii). If $\pi\colon \widetilde{\mathrm{Sl}}(2,\mathbb{R}) \to \mathrm{PSl}(2,\mathbb{R})$ denotes the universal covering then $\pi|\widetilde{S}_j : \widetilde{S}_j \to \mathcal{S}_j$ is an algebraic and topological isomorphism. Indeed, all maps in the following diagram are homeomorphisms:

$$\begin{array}{ccc} W_j & \xrightarrow{\exp_1} & \widetilde{S}_j \\ \mathrm{id}_{W_j} \downarrow & & \downarrow \pi \\ W_j & \xrightarrow{\exp_2} & \mathcal{S}_j \end{array}$$

Set $\widetilde{S} = \widetilde{S}_1 \times \cdots \times \widetilde{S}_k$, and $\mathcal{S} = \mathcal{S}_1 \times \cdots \times \mathcal{S}_k$. Then the covering morphism π^k maps $\widetilde{S}$ isomorphically onto $\mathcal{S}$.

Step 2. We set $\mathfrak{a} = W - W$, $A = \langle \exp \mathfrak{a} \rangle$, and assume that $G = \overline{A}$; it obviously suffices to show the proposition under this additional assumption. Then $\mathfrak{g}$ is reductive and $\mathfrak{g}' = [\mathfrak{g}, \mathfrak{g}] = \mathfrak{a}$, so $\mathfrak{g}$ is the ideal direct sum $\mathfrak{g} = \mathfrak{a} \oplus \mathfrak{z}$ where $\mathfrak{z}$ is the center. We let $Z = \exp \mathfrak{z}$. Then Z is the identity component of the center of G and is therefore closed. We denote with q the quotient morphism $G \to G/Z$. Furthermore we define A_{Lie} to be the group A endowed with its intrinsic Lie group topology, and write $j\colon A_{\mathrm{Lie}} \to G$ for the natural injection.

Now there are covering homomorphisms

$$\widetilde{\mathrm{Sl}}(2,\mathbb{R})^k \xrightarrow{\kappa_1} A_{\mathrm{Lie}} \xrightarrow{\kappa_2} G/Z \xrightarrow{\kappa_3} \mathrm{PSl}(2,\mathbb{R})^k,$$

where $\kappa_2 = qj$ and the composition $\kappa_3\kappa_2\kappa_1$ is the universal covering. Note that $j\kappa_1\colon \widetilde{\mathrm{Sl}}(2,\mathbb{R})^k \to G$ maps $\widetilde{S}$ surjectively to $S := \exp_G W$ and $\kappa_3 q\colon G \to \mathrm{PSl}(2,\mathbb{R})^k$ maps S surjectively to $\mathcal{S}$. It follows from Step 1 that the induced maps in the top row of the following diagram are isomorphisms:

(*) $$\begin{array}{ccccc} \widetilde{S} & \xrightarrow{\varphi} & S & \xrightarrow{\psi} & \mathcal{S} \\ \mathrm{incl} \downarrow & & \mathrm{incl} \downarrow & & \downarrow \mathrm{incl} \\ \widetilde{\mathrm{Sl}}(2,\mathbb{R})^k & \xrightarrow{j\kappa_1} & G & \xrightarrow{\kappa_3 q} & \mathrm{PSl}(2,\mathbb{R})^k. \end{array}$$

Step 3. We know that $\mathcal{S}$ is closed in $\mathrm{PSl}(2,\mathbb{R})^k$. Hence $\mathcal{S} \subseteq \kappa_3 q(\overline{S}) \subseteq \overline{\mathcal{S}} = \mathcal{S}$ and thus the map $\alpha\colon \overline{S} \to S$, $\alpha(x) = \psi^{-1}\kappa_3 q(x)$ is a well-defined continuous map which by the commutativity of (*) is a retraction. Since retracts of Hausdorff spaces are always closed it follows that S is closed in $\overline{S}$, and this means that $\overline{S} = S$. This proves (i). Since $S \cong \widetilde{S} = \widetilde{S}_1 \times \cdots \times \widetilde{S}_k$ and since $j\kappa_1(\widetilde{S}_j) = j\kappa_1(\exp_1 W) = \exp_G W_j$ we conclude (ii). Finally, in order to verify (iii) we note that there is a covering morphism $F\colon \widetilde{\mathrm{Sl}}(2,\mathbb{R})^k \to G_1$ such that $fF = j\kappa_1\colon \widetilde{\mathrm{Sl}}(2,\mathbb{R})^k \to G$. Then F maps $\widetilde{S}$ surjectively onto T_1 and f maps T_1 onto S. By Step 2, the morphism $fF = j\kappa_1$ maps $\widetilde{S}$ isomorphically onto S. It follows that f maps T_1 isomorphically onto S. □

2.2. COROLLARY. *The semigroup* $\mathrm{Sl}(2)^+$ *is absolutely closed.*

Proof. Straightforward from the definition of $\mathrm{Sl}(2)^+$ and the above Proposition. □

In the situation we are dealing with next the map p is a covering homomorphism. Recall that a Lie subsemigroup of a Lie group is by definition the closure of the subsemigroup generated by the exponential image of its Lie wedge.

2.3. LEMMA. *Let $p: G \to G^*$ be a covering homomorphism of connected Lie groups and suppose that T is a Lie subsemigroup of G with inner points. If p maps $\operatorname{int} T$ injectively into G^* then $p(T)$ is closed in G^*.*

Proof. Write $\mathfrak{g}$ for the Lie algebra of G and identify the Lie algebra of G^* with $\mathfrak{g}$, so that $\mathfrak{L}(p) = \operatorname{id}_\mathfrak{g}$. We write W for the Lie wedge of T and S for the semigroup $\langle \exp_G W \rangle$ generated by its exponential image in G. Then $T = \overline{S}$ by definition and thus $p(T) \subseteq \overline{p(S)}$. By [27], p. 377, Theorem V.1.10 we know that $\operatorname{int} S = \operatorname{int} T$ is a dense ideal of T. Similarily we know from [27], p. 370, Proposition V.0.15 that $\operatorname{int} p(S) = \operatorname{int} \overline{p(S)} = \operatorname{int} \overline{p(T)}$.

We now assume that the covering map p is injective on $\operatorname{int} T = \operatorname{int} S$. Then p induces a homeomorphism $p_*: \operatorname{int} S \to \operatorname{int} p(S)$, since covering maps are open. Let $x \in \overline{p(T)} = \overline{p(S)}$. We note that $p(S) = \langle p(\exp_G W) \rangle$. By [43], p. 177, Corollary 2.7(iii) there is a smooth arc $\psi: [0,1] \to \overline{p(S)}$ such that $\psi(1) = x$ and $\psi([0,1[) \subseteq \operatorname{int} p(S)$. Since $\operatorname{int} p(S) = p(\operatorname{int} S)$ we have $\psi(0) = p(s)$ for some point $s \in \operatorname{int} S$, let $\Psi: [0,1] \to G$ be the unique lifting with $\Psi(1) = s$, $\psi = p \circ \Psi$. Then the two curves $\Psi | [0,1[: [0,1[\to G$ and $r \mapsto p_*^{-1}(r): [0,1[\to G$ are both liftings of $\psi | [0,1[$ and take the same value s at $r = 0$, thus they must coincide. Now we have $\Psi([0,1[) = p_*^{-1}\psi([0,1[) \subseteq p_*^{-1}(\operatorname{int} p(S)) = \operatorname{int} S$ and therefore $\Psi(1) \in \overline{\Psi([0,1[)} \subseteq \overline{\operatorname{int} S} = \overline{S} = T$. Thus $x = \psi(1) = p(\Psi(1)) \in p(T)$. □

We illustrate the above principle by applying it to locally direct products, this result will be useful later.

2.4. PROPOSITION. *Suppose that $p: G_1 \times G_2 \to G$ is a covering homomorphism of connected Lie groups with Lie algebra $\mathfrak{g} = \mathfrak{g}_1 \oplus \mathfrak{g}_2$. Let T_j be closed subsemigroups of G_j, $j = 1,2$, and write $i_j: T_j \to G_1 \times G_2$, $i_1(t) = (t,1)$ and $i_2(t) = (1,t)$. We assume that the following hypotheses are satisfied:*
 (a) $E_j := p(i_j(T_j))$ *is closed in G for $j = 1,2$.*
 (b) $p i_1: T_1 \to G$ *is injective.*
Then we have the following conclusions:
 (i) $\mathfrak{L}(p i_j): T_j \to G$ *induces isomorphisms $\mathfrak{L}(T_j) \cong \mathfrak{L}(E_j)$ for $j = 1,2$.*
 (ii) *The map $m: E_1 \times E_2 \to E := p(T_1 \times T_2)$, $m(e_1, e_2) = e_1 e_2$, is a bijective continuous homomorphism onto a subsemigroup of G.*
Suppose that, in addition, the following hypothesis holds for $j = 1, 2$:

(c) T_j is a Lie subsemigroup of G_j with interior points.

Then we also have that

(iii) E is closed in G.

Proof. (i) The covering map $p\colon G_1 \times G_2 \to G$ induces the identity map $\mathfrak{L}(p)\colon \mathfrak{g}_1 \oplus \mathfrak{g}_2 \to \mathfrak{g}$ by our convention. Since pi_j maps T_j onto E_j, then $\mathfrak{L}(pi_j)$ maps $\mathfrak{L}(T_j)$ to $\mathfrak{L}(E_j)$. The assertion follows.

(ii) Since $E = p(T_1 \times T_2) = p(T_1 \times \{1\})p(\{1\} \times T_2) = E_1 E_2$, clearly m is surjective. Next we show that m is injective. Suppose $m(e_1, e_2) = m(e_1', e_2')$. There are elements $t_j, t_j' \in T_j$ such that $e_j = pi_j(t_j)$, $e_j' = pi_j(t_j')$. Then $p(t_1, t_2) = p(t_1, 1)p(1, t_2) = e_1 e_2 = m(e_1, e_2)$. Likewise $p(t_1', t_2') = m(e_1', e_2')$. Thus there is an element $(z_1, z_2) \in \ker p$ such that $(t_1 z_1, t_2 z_2) = (t_1', t_2')$. The relation $t_1 z_1 = t_1'$ means $pi_1(t_1) = p(t_1, 1) = p(t_1', 1) = pi_1(t_1')$ and by (b) this implies $t_1 = t_1'$, i.e., $z_1 = 1$. Thus $(1, z_2) \in \ker p$ and therefore $e_2 = p(1, t_2) = p(1, t_2') = e_2'$. This shows that m is injective and therefore also bijective.

(iii) We set $A = p(G_1 \times \{1\})$ and $B = p(\{1\} \times G_2)$. Then A and B are analytic subgroups of G such that $AB = G$. As usual we let A_{Lie} and B_{Lie} denote these groups with their intrinsic Lie group topology. Let E_1^* denote E_1 with the topology induced from A_{Lie}. Then E_1^* is a closed subsemigroup of A_{Lie} such that $\mathfrak{L}(E_1^*) = \mathfrak{L}(E_1) \cong \mathfrak{L}(T_1)$ in view of (i). Corresponding statements hold for E_2 in place of E_1. Let $\mu\colon A_{\text{Lie}} \times B_{\text{Lie}} \to G$ denote the covering homomorphism given by $\mu(a, b) = ab$. On the level of the Lie algebras, $\mathfrak{L}(\mu)\colon \mathfrak{a} \oplus \mathfrak{b} \to \mathfrak{g}$ is the identity. The closed subsemigroup $E_1^* \times E_2^*$ of $A_{\text{Lie}} \times B_{\text{Lie}}$ has the tangent wedge $\mathfrak{L}(T_1) \oplus \mathfrak{L}(T_2)$.

Since we now assume hypothesis (c), in view of (ii), we can apply Lemma 2.3 with $A_{\text{Lie}} \times B_{\text{Lie}}$ in place of G, $E_1^* \times E_2^*$ in place of T and μ in place of φ. We conclude that $E = m(E_1 \times E_2)$ is closed. □

2.5. COROLLARY. *Suppose that $p\colon G_1 \times G_2 \to G$ is a covering homomorphism of connected Lie groups. We write $\mathfrak{g} = \mathfrak{g}_1 \oplus \mathfrak{g}_2$ and may write $\mathfrak{L}(p) = \text{id}_{\mathfrak{g}}$. Let T_j be closed subsemigroups of G_j, $j = 1, 2$, and write $i_j\colon T_j \to G_1 \times G_2$, $i_1(t) = (t, 1)$ and $i_2(t) = (1, t)$. We assume that the following hypotheses are satisfied:*

(a) $\mathfrak{g}_1 \cong \mathfrak{sl}(2, \mathbb{R})^k$ and $\mathfrak{L}(T_1) \cong W_1 \times \cdots \times W_k \subseteq \mathfrak{sl}(2, \mathbb{R})^k$ with $W_j \subseteq \overline{D^+}$.

(b) $E_2 := p(i_2(T_2))$ is closed in G.

(c) T_2 is a Lie subsemigroup of G_2 with interior points.

Then $E_1 E_2$ is closed in G.

Proof. We verify the hypotheses of Proposition 2.4: The present hypotheses (a) and (b) together with Proposition 2.1(i) secures hypothesis (a) of Proposition 2.4. The present hypothesis (a) yields (b) of 2.4. Finally, hypotheses (a) and (c) above guarantee 2.4(c). Thus 2.4 applies and proves the assertion. □

3. Examples Connected with the Motion Algebra

3.1. NOTATION. In the following we consider the *Motion Algebra* $\mathfrak{mot}$ as the real Lie algebra, which is defined on the product space $\mathbb{C} \times \mathbb{R}$, with Lie brackets
$$[(\beta,\gamma),(\beta',\gamma')] = (2i(\beta\gamma' - \beta'\gamma), 0).$$
As a model for the simply connected Lie group $\widetilde{\mathrm{Mot}}$ with Lie algebra $\mathfrak{mot}$ we take the product $\mathbb{C} \times \mathbb{R}$, with multiplication
$$(z,r)(z',r') = (ze^{-ir'} + z'e^{ir}, 0).$$
(Note that in the usual geometric interpretation of $\mathfrak{mot}$ as the Lie algebra of the group of all Euclidean motions in the plane the parameter r corresponds to *half* of the rotation angle.) With respect to this parametrization the exponential function $\widetilde{\exp}$ of $\widetilde{\mathrm{Mot}}$ is given by the formula
$$\widetilde{\exp}(\beta,\gamma) = \begin{cases} (\beta, 0) & \text{if } \gamma = 0; \\ (\frac{\beta}{\gamma}\sin\gamma, \gamma) & \text{if } \gamma \neq 0. \end{cases}$$
As in the case of $\mathfrak{sl}(2,\mathbb{R})$ we write k for the Cartan-Killing form of $\mathfrak{g} = \mathfrak{mot}$ and note that for $X = (\beta,\gamma)$ we have $k(X,X) = -8\gamma^2$. Thus $\mathfrak{g} = \mathcal{D}^- \cup \mathcal{D}^0 = \overline{\mathcal{D}^-}$, where $\mathcal{D}^- = \{X \in \mathfrak{g} \mid k(X,X) < 0\}$ and $\mathcal{D}^0 = \{X \in \mathfrak{g} \mid k(X,X) = 0\} = \mathbb{C}$. □

These formulas and the elementary properties of covering groups imply the following proposition:

3.2. PROPOSITION. *Let G be a connected Lie group with Lie algebra $\mathfrak{g} = \mathfrak{mot}$. We write $\mathfrak{n}$ for the unique proper ideal of $\mathfrak{g}$, that is, $\mathfrak{n}$ is the vector subspace $\mathbb{C} \times \{0\} = \mathcal{D}^0$ of the product space $\mathfrak{g} = \mathbb{C} \times \mathbb{R}$. Then the following assertions hold:*
(i) *The center $Z(G)$ of G is the discrete subgroup $\{\exp(0,k\pi) \mid k \in \mathbb{Z}\}$; it is contained in every one-parameter subgroup $\exp \mathbb{R} \cdot X$ with $X \in \mathfrak{g} \setminus \mathfrak{n}$.*
(ii) *$G = \exp \mathfrak{g} \cup Z(G) \exp \mathfrak{n}$ and $\exp \mathfrak{g} \cap Z(G)\exp \mathfrak{n} = Z(G) \cup \exp \mathfrak{n}$. Furthermore, the set $\exp \mathfrak{g} \setminus Z(G)$ is open in G.*
(iii) *$\mathcal{D}^- = \mathrm{comp}\,\mathfrak{g} \setminus \{0\} = \mathfrak{g} \setminus \mathfrak{n}$*
(iv) *Every $X \in \mathfrak{g}$ with $\exp X \notin Z(G)$ satisfies condition $(**)$ of Proposition 1.3(vi):*
$(**)$ *If $\exp X = \lim z_n \exp X_n$ with $z_n \in Z(G)$, $X_n \in \mathfrak{g}$, then there is a sequence $\sigma_n \in \mathbb{R}$ with $X = \lim \sigma_n \cdot X_n$.*

Proof. In view of the preceding formulas we only have to verify assertion (iv). Suppose that $\exp X = \lim z_n \exp X_n \notin Z(G)$, where $z_n \in Z(G)$, $X_n \in \mathfrak{g}$ and $X \in \mathfrak{g}$. Since $\exp \mathfrak{g} \setminus Z(G)$ is open in G we may suppose that $z_n \exp X_n \in \exp \mathfrak{g} \setminus Z(G)$ for all indices n. Then by (ii) we must have $z_n = 1$ whenever $X_n \in \mathfrak{n}$.

If $X_n \in \mathfrak{g} \setminus \mathfrak{n}$ then $Z(G) \subset \exp \mathbb{R} \cdot X_n$, and therefore $z_n \exp X_n = \exp \sigma_n \cdot X_n$ for some $\sigma_n \in \mathbb{R}$. For $X_n \in \mathfrak{n}$ we have $z_n = 1$, so we put $\sigma_n = 1$. The so defined numbers σ_n satisfy $z_n \exp X_n = \exp \sigma_n \cdot X_n$ and the assertion follows. □

3.3. REMARK. If in the above proposition $G = \widetilde{\text{Mot}}$ then the map

$$\{(\beta,\gamma) \mid \gamma \notin (\mathbb{Z} \setminus \{0\})\pi\} \to \exp\mathfrak{g} \setminus (Z(G) \setminus \{1\}), \quad X \mapsto \exp X = \widetilde{\exp}X$$

is a homeomorphism. □

3.4. PROPOSITION. *Suppose that W is the Lie wedge of a closed subsemigroup of a Lie group G with Lie algebra $L(G) = \text{mot}$. Then at least one of the following assertions holds:*
  (i) *S is a group;*
  (ii) *S is a half-space subsemigroup bounded by $N = \exp\mathfrak{n}$;*
  (iii) *$S \subseteq N$;*
  (iv) *$H(W) = \{0\}$.*

Proof. Suppose first that $H(W)$ contains a nonzero element $X \in \text{comp}\,\mathfrak{g}$. Then for every $Y \in W \cap \mathcal{D}^-$ we have by 3.2(ii) $\exp\mathbb{R}^+\cdot Y \cap \exp\mathbb{R}\cdot X \neq \{1\}$ and hence $Y \in H(W)$. Thus $W \cap \mathcal{D}^- \subseteq H(W)$, so either $W = \mathfrak{g}$ or $W = \mathbb{R}\cdot X$, and therefore S is a group.

Suppose now that $H(W) \cap \text{comp}\,\mathfrak{g} = \{0\}$, that is, $H(W) \subseteq \mathcal{D}^0 = \mathfrak{n}$, and assume that neither (iii) nor (iv) is satisfied. Then there exist elements $Y = (\beta,\gamma) \in W$ and $X = (\beta_0, 0) \in H(W)$ with $\gamma \neq 0$ and $\beta_0 \neq 0$. But a straightforward calculation shows that

$$\lim \frac{1}{n}\cdot e^{\pm n\cdot \text{ad}\,X}Y = \lim(\frac{1}{n}\cdot Y \pm (2i\beta_0\gamma, 0)) = \pm(2i\beta_0, 0) \in W,$$

so $H(W)$ contains both $(\beta_0, 0)$ and $(i\beta_0, 0)$, hence $\mathfrak{n} \subseteq H(W)$. Thus in this case either (i) or (ii) holds. □

3.5. PROPOSITION. *Assume that G is a Lie group with Lie algebra $\mathfrak{g} = \text{mot}$ and let S be a closed proper subsemigroup of G. We assume that the Lie wedge W of S has inner points in $\mathfrak{g}$. Then G is simply connected (hence $\cong \widetilde{\text{Mot}}$), and the following assertions are equivalent:*
  (i) *S is weakly exponential.*
  (ii) *W is a semialgebra.*
  (iii) *S is one of the two half-space semigroups bounded by $\exp\mathfrak{n}$.*

Proof. By Proposition 3.2(iii) the interior of W contains a nonzero element $U \in \text{comp}\,\mathfrak{g} \setminus \{0\} = \mathfrak{g} \setminus \mathfrak{n}$, and by 3.2(i) $\exp\mathbb{R}\cdot U$ contains the center of G, hence the exponential image z of a suitable positive multiple of U lies in the center of G. Since U is an inner point of W, Lemma 1.8 shows that every point $\exp\lambda\cdot U$ with $\lambda > 0$ is an inner point of S. In particular, z is an inner point of S.

If G were not simply connected then the center of G were finite and hence $z^n = 1$ for some $n \in \mathbb{N}$, so $\mathbf{1}$ were an inner point of S and therefore $S = G$, a contradiction to our assumption that S is a proper subsemigroup of G. Thus G is simply connected.

(i)$\Rightarrow$(ii) Assume that S is weakly exponential and suppose that W is not a semialgebra. Then by Proposition IV.5.1(ii) the edge $H(W)$ of W contains a nonzero element X. Since $S \neq G$ and since S contains inner points in G, Proposition 3.4 implies that S is a halfspace semigroup bounded by N. Thus W is one of the two half-spaces bounded by $\mathfrak{n}$, hence a semialgebra, contrary to our assumption.

The implication (ii)$\Rightarrow$(iii) is well known from the so-called 'First Classification Theorem of Low Dimensional Semialgebras' (cf. [27], II.3.4, p.104). The implication (iii)$\Rightarrow$(i) is obvious. $\square$

4. Examples Connected with the Oscillator Group

For the oscillator algebra $\mathfrak{osc}$ and the simply connected oscillator group Osc we use the following parametrisation from DÖRR's paper [12] (see also [27] for a slightly different version):

We define the oscillator algebra $\mathfrak{osc}$ on the space $\mathbb{R} \oplus \mathbb{C} \oplus \mathbb{R}$, with Lie brackets

$$(1) \qquad [(x,c,r),(x',c',r')] = \left(\operatorname{Im}(\bar{c}c'), 2i\cdot \det \begin{pmatrix} r & r' \\ c & c' \end{pmatrix}, 0\right).$$

The associated simply connected Lie group Osc is then defined on the space $\mathbb{R} \times \mathbb{C} \times \mathbb{R}$ with multiplication

$$(2) \qquad (x,c,r)\cdot(x',c',r') = \left(x + x' + \frac{1}{2}\operatorname{Im}(e^{i(r+r')}\bar{c}c'), e^{ir}c' + e^{-ir'}c, r + r'\right).$$

The exponential function going with this parametrization is given by

$$(3) \qquad \exp(x,c,r) = \begin{cases} \left(x + \frac{|c|^2}{4r} - \frac{|c|^2 \sin 2r}{8r^2}, \frac{c \sin r}{r}, r\right) & \text{for } r \neq 0 \\ (x,c,0) & \text{for } r = 0. \end{cases}$$

4.1. LEMMA. *The oscillator algebra has the following subalgebras $\mathfrak{a}$:*
(i) $\dim \mathfrak{a} = 1$: *All one dimensional subspaces. In particular, $\mathfrak{z} = \mathfrak{g}''$, the center and second commutator algebra, is one-dimensional.*
(ii) $\dim \mathfrak{a} = 2$: $\mathfrak{a} = \mathfrak{z} + \mathbb{R}\cdot a$ *with any element $a \notin \mathfrak{z}$. Such a subalgebra $\mathfrak{a}$ is compactly embedded if and only if it is a Cartan subalgebra, or equivalently, $a \notin \mathfrak{g}'$.*
(iii) $\dim \mathfrak{a} = 3$: $\mathfrak{a} = \mathfrak{g}'$, *the Heisenberg subalgebra.*

Proof. The proof is a straightforward exercise. $\square$

4.2. DEFINITION. The *standard Lorentzian cone* in $\mathfrak{osc}$ is defined as the set $W_{\text{Lor}} := \{(x,c,r) \in \mathfrak{osc} \mid |c|^2 + 4rx \leq 0, r \geq 0, x \leq 0\}$. The pair $(\mathfrak{osc}, W_{\text{Lor}})$ is sometimes also called the *standard pair*. $\square$

4.3. PROPOSITION.

(i) *The Standard Lorentzian Cone W_{Lor} is invariant in osc and for every invariant cone W in osc there is a Lie algebra automorphism of osc mapping W onto W_{Lor}.*

(ii) *Let W be a proper semialgebra in osc with inner points. Then there are two cases:*

Case (A): $\mathfrak{g}'' \subseteq W$. *Then W is a half-space semialgebra bounded by $\mathfrak{g}'$.*
Case (B): $\mathfrak{g}'' \not\subseteq W$. *Then the pair (osc, W) is isomorphic with the Standard Pair $(\text{osc}, W_{\text{Lor}})$.*

Proof. Assertion (i) is known in much more general form from [27], p. 111–120.

(ii) In Case (A) $W/\mathfrak{g}''$ is a semialgebra in $\text{osc}/\mathfrak{g}'' \cong \text{mot}$ with inner points. Since it is proper it is a half-space semialgebra, so W is a half-space semialgebra too. For Case (B) the assertion is shown in [27], p.120, Theorem II.3.20. □

Note that Osc is simply connected, so every automorphism of osc naturally induces an automorphism of Osc. Thus if S is a subsemigroup of Osc with Lie wedge W such that (osc, W) is isomorphic with the Standard Pair then upon reparametrization we may assume that $W = W_{\text{Lor}}$.

4.4. LEMMA. (DÖRR, [12])
Let S be a closed subsemigroup of Osc with Lie wedge W_{Lor}. Then S contains the half-space $H = \{(u, v, w) \in \text{Osc} : w \geq \pi\}$. In particular, $(1, 0, \pi) \in S$.

Proof. Consider $(u, v, w) \in \text{Osc}$ with $w > \pi$. Now

$$(u, v, w) = (u, e^{(w-w_0)i}v, w_0)(0, 0, w - w_0) \quad \text{for all} \quad w_0 \in \mathbb{R}.$$

Since $\lim_{w \to \pi - 0} = -\infty$ we find a $w_0 \in]\frac{\pi}{2}, \pi[$ such that $u \leq -\frac{|v|^2}{4} \cot w_0$. Hence $(u, v, w_0) \in \exp W_{\text{Lor}}$ by Example II.1.7(ii)(4). Obviously, $w - w_0 \geq 0$ and thus $(0, 0, w - w_0) \in \exp W_{\text{Lor}}$. Hence $(u, v, w) \in (\exp W_{\text{Lor}})^2 \subseteq S$. Since S is closed, we have $(u, v, w) \in S$ even if $w = \pi$. □

4.5. PROPOSITION.
Let S be a weakly exponential subsemigroup of the Lie group $G = \text{Osc}$ with inner points and $S \neq \text{Osc}$. Then S is a half-space semigroup bounded by Osc'.

Proof. Since Osc does not contain any compact subgroups except $\{1\}$ we see that $\text{comp}_G(\mathfrak{g}) = \{0\}$. Thus by Corollary IV.3.4 the Lie wedge of any weakly exponential subsemigroup in Osc must be a semialgebra. Now it follows from 4.3(ii) that there are two cases:

Case (A): W is one of the two half-space semialgebras bounded by osc'. Then the assertion follows.

Case (B): The pair (osc, W) is isomorphic with $(\text{osc}, W_{\text{Lor}})$. In this case we may assume that $W = W_{\text{Lor}}$. Then, in the notation of Lemma 4.4, $\mathfrak{h} := \mathbb{R} \cdot v \oplus \mathbb{R} \cdot w$ is a Cartan subalgebra of osc and meets the interior of W. Thus by Theorem

II.4.2(ii) we know that $\exp \mathfrak{h} \cap S = \exp(W \cap \mathfrak{h})$. The restriction of exp to $\mathfrak{h}$ is a diffeomorphism onto $\exp \mathfrak{h}$, and by Lemma 4.4 $\exp \mathfrak{h} \cap S$ contains $\exp v \exp \mathbb{R} \cdot w = \exp(v + \mathbb{R} \cdot w)$. Hence $v + \mathbb{R} \cdot w \subseteq W$ and therefore $\mathbb{R} \cdot w = \mathfrak{osc}'' \subseteq W$, by the usual limit argument. This contradiction shows that Case (B) cannot occur, and the proof is finished. □

We close this section with the construction of two examples in central extensions of Mot and Osc.

4.6. EXAMPLE. Let B_1 be a connected Lie group with a central element $z_1 \neq 1$ and with Lie algebra $\mathfrak{b}_1$. Define $\mathfrak{g} = \mathfrak{b}_1 \times \mathbb{R}$, $\mathfrak{a} = \{0\} \times \mathbb{R}$, $\mathfrak{b} = \mathfrak{b}_1 \times \{0\}$. Let $G_1 = B_1 \times \mathbb{R}$ and $\Delta = \{(z_1^n, -n) \mid n \in \mathbb{Z}\}$. Set $G = G_1/\Delta$, $A = (\{1\} \times \mathbb{R})\Delta/\Delta$, $B = B_1\Delta/\Delta \cong B_1$, and $z = (z_1, 0)\Delta$. We may write $\exp: \mathfrak{g} \to G$, $\exp(x, r) = (\exp_{B_1} x, r)\Delta$.

Then for any $(x, r) \in \mathfrak{g}$ we have the equivalences

$$\exp(x, r) \in A \iff (\exp_{B_1} x, r) \in (\{1\} \times \mathbb{R})\Delta = \langle z \rangle \times \mathbb{R}$$
$$\iff \exp_{B_1} x \in \langle z \rangle \iff (\exists n \in \mathbb{Z}) \, z^n = \exp_{B_1} x.$$

Similarly we get

$$\exp(x, r) \in B \iff (\exp_{B_1} x, r) \in (B_1 \times \{0\})\Delta = \mathfrak{b}_1 \times \mathbb{Z} \iff r \in \mathbb{Z}.$$

Thus it follows

(1) $$\exp^{-1} A = \exp_{B_1}^{-1}\langle z \rangle \times \mathbb{R},$$
(2) $$\exp^{-1} B = \mathfrak{b}_1 \times \mathbb{Z} = \mathfrak{b} \oplus \mathbb{Z} \cdot (0, 1),$$
(3) $$\exp^{-1}(A \cap B) = \exp_{B_1}^{-1}\langle z \rangle \times \mathbb{Z}.$$

From (2) we deduce

$$\exp(\exp^{-1} B) = \big(\exp_{B_1}(\mathfrak{b}_1) \times \mathbb{Z}\big)\Delta/\Delta = \big(\langle z \rangle \exp_{B_1} \mathfrak{b}_1\big) \times \mathbb{Z})/\Delta \cong \langle z \rangle(\exp_{B_1} \mathfrak{b}_1).$$

Define a homomorphism $q: B_1 \to G$ by $q(b_1) = (b_1, 0)\Delta$. Then the restriction and corestriction of q,

(4) $\quad q^*: \langle z \rangle(\exp_{B_1}(\mathfrak{b}_1)) \to \exp\big(\exp^{-1}(B)\big) \quad$ is a homeomorphism.

In other words: *In the central extension G of B_1 with $\dim G = \dim B_1 + 1$ the elements of $\langle z \rangle(\exp_{B_1}(\mathfrak{b}_1))$ are accessible by one-parameter subgroups in G.*

Now suppose that W_1 is a half-space in $\mathfrak{b}_1$ containing $\mathfrak{b}'_1 = [\mathfrak{b}_1, \mathfrak{b}_1]$. We assume that B_1 is simply connected. Then B'_1 is closed and B_1/B'_1 is a vector group isomorphic to $\mathfrak{b}_1/\mathfrak{b}'_1$ under the exponential map. Then $W_1/\mathfrak{b}_1$ has an isomorphic copy in B_1/B'_1 which is necessarily of the form S_1/B'_1. Then S_1 is a closed half-space semigroup in B_1 with $\mathcal{L}(S_1) = W_1$. Examples are:

(a) $B_1 = \text{Mot}$ as in 3.1 with $W_1 = \mathbb{C} \times \mathbb{R}^+$ and $z = (0, \pi)$.

(b) $B_1 = \text{Osc}$ with $W_1 = \mathbb{R} \times \mathbb{C} \times \mathbb{R}^+$ and $z = (0, 0, \pi)$.

Define $W^* = W_1 \times \mathbb{R}^+$ and let W denote any wedge surrounding W^* with $H(W) = H(W_1) \times \{0\}$ (see [27], p.304, Definition IV.2.10), i.e., W contains $W^* \setminus H(W^*)$ in its interior. $(B_1 \times \mathbb{R})/(B_1' \times \{0\}) \cong (\mathfrak{b}_1 \times \mathbb{R})/(\mathfrak{b}_1' \times \{0\})$ is a vector group. Then $W/(\mathfrak{b}_1' \times \{0\})$ has an isomorphic copy in $(B_1 \times \mathbb{R})/(B_1' \times \{0\})$ which is necessarily of the form $T/(B_1' \times \{0\})$. Then T is a closed semigroup in $B_1 \times \mathbb{R}$ with Lie wedge W. Note that $(B_1 \times \mathbb{R})/(H(T) \times \{0\}) \cong \mathfrak{g}/(H(W_1) \times \{0\})$ is a two dimensional vector group containing a subgroup $\Delta^* = \Delta(H(T) \times \{0\})/(H(T) \times \{0\})$ isomorphic to $\mathbb{Z}$. We may choose W so small that $(T/(H(T) \times \{0\})) \cap \Delta^*$ is singleton. Then $T \cap \Delta = \{(1,0)\}$. Since $T\Delta/(H(T) \times \{0\}) = (T/(H(T) \times \{0\}))\Delta^*$ and this latter subsemigroup of the plane is closed (see Figure 2 on p. 401 of [27]), we know that $T\Delta$ is closed in $B_1 \times \mathbb{R}$. Thus $S := T\Delta/\Delta$ is a closed subsemigroup of G locally isomorphic to T and with Lie wedge W.

If $(x, r) \in W$ and $(\exp_{B_1} x, r)\Delta = \exp(x, r) \in B$ then for some $b \in B_1$ and some $n \in \mathbb{Z}$ we have $(z^n \exp_{B_1} x, r - n) = (\exp_{B_1} x, r)(z^n, -n) = (b, 0)$, i.e, $r \in \mathbb{Z}$ and $\exp(x, r) = (z^r \exp_{B_1} x, 0)\Delta$. But the choice of W implies that $(x, r) \in W$ entails $x \in W_1$ and $r \geq 0$. Therefore,

(5) $$B \cap \exp W = [(z^{\mathbb{N}_0} \exp_{B_1}(W_1)) \times \{0\}]\Delta/\Delta.$$

In the examples $\mathfrak{b}_1 = \mathfrak{mot}$ and $\mathfrak{b}_1 = \mathfrak{osc}$ we have $z^{\mathbb{N}_0} \exp_{B_1}(W_1) = S_1$. □

As a consequence of (5) we may state:

(mot) *There is a 4-dimensional Lie group G homeomorphic to $\mathbb{R}^3 \times \mathbb{S}^1$ and Lie algebra $\mathfrak{mot} \times \mathbb{R}$ such that G contains a proper closed exponential semigroup S containing G' and $\exp U$ for a compact element U in the interior of W.*

(osc) *There is a 5-dimensional Lie group G homeomorphic to $\mathbb{R}^4 \times \mathbb{S}^1$ and Lie algebra $\mathfrak{osc} \times \mathbb{R}$ such that G contains a proper closed exponential semigroup S containing G' and $\exp U$ for a compact element U in the interior of W.*

5. Examples Connected with Compact Lie Algebras

The invariant wedges W in a compact Lie algebra are classified by Theorem III.2.1 in [27] (p.190). If $\mathfrak{g}$ is not semisimple, they exist in abundance, among them many which are not vector spaces.

5.1. PROPOSITION. *Let G be a noncompact connected Lie group whose Lie algebra $\mathfrak{g}$ is compact. Then there is a surjective morphism $p: G \to \mathbb{R}$. For any such morphism let $B = \ker p$ and $\mathfrak{a}$ a central one-dimensional subspace such that $p(\exp \mathfrak{a}) = \mathbb{R}$.*

Let C be a compact convex $e^{\operatorname{ad} \mathfrak{g}}$-invariant 0-neighborhood in $\mathfrak{b} = \mathfrak{L}(B)$, further $0 \neq a \in \mathfrak{a}$ and set $W = \mathbb{R}^+ \cdot (a + C)$. Then $S = \exp W$ is a closed exponential reduced subsemigroup of G and $\mathfrak{L}(S) = W$.

Proof. Since the Lie group G is not compact and has a compact Lie algebra, there is a unique maximal compact subgroup N and a closed central vector subgroup $V \cong \mathbb{R}^n$, $n \geq 1$ such that $(n, v) \mapsto nv : N \times V \to G$ is an isomorphism. The commutator group G' of G is closed, semisimple, and contained in N. This implies the existence of p, B, $\mathfrak{a}$. Since $e^{\text{ad}\,\mathfrak{g}} = e^{\text{ad}\,\mathfrak{g}'}$ is a compact group of automorphisms of $\mathfrak{g}$ leaving $\mathfrak{b}$ invariant the existence of C is guaranteed. Now W is a pointed, hence reduced invariant cone such that $W \cap \mathfrak{g}' = \{0\}$. Proposition VI.5.8 on p. 526 of [27] shows that the closed subsemigroup $S = \langle \exp W \rangle$ satisfies $\mathfrak{L}(S) = W$. Clearly, S is invariant under inner automorphisms. The Theorem of MITTENHUBER and NEEB (Corollary 11 of [61]) now proves $S = \exp W$. □

Cf. also Theorem III.2.1 in [27] (p.190).

5.2. COROLLARY. *Every simply connected noncompact Lie group with compact Lie algebra contains a closed exponential semigroup with trivial group of units and inner points.* □

CHAPTER 6

TEST ALGEBRAS AND GROUPS

1. Extensions of the Motion Algebra and of the Oscillator Algebra

The following Lemma will provide a unifying background for the 'Test objects' we shall define in the next section.

1.1. LEMMA. *Assume that* $\mathfrak{b}$ *be a real Lie algebra which is spanned by vectors* u, v, x, y, z, *subject to the following relations:*

$$[u, v] = 0, \quad [u, x] = x, \quad [u, y] = y,$$
$$[v, x] = y, \quad [v, y] = -x,$$
$$[x, y] = z, \quad [x, z] = 0$$
$$[y, z] = 0.$$

We write $\mathfrak{n}$ *for the ideal* $\mathbb{R} \cdot x + \mathbb{R} \cdot y + \mathbb{R} \cdot z$ *and* $\mathfrak{m}$ *for the ideal* $\mathbb{R} \cdot v + \mathfrak{n}$.
Then we have the following assertions:
  (i) $[u, z] = 2 \cdot z$, *and* $[v, z] = 0$.
  (ii) *If the vectors* u, v, x, y *are linearly dependent then* $x = y = z = 0$ *and hence* $\mathfrak{b} = \mathbb{R} \cdot u + \mathbb{R} \cdot v$ *is abelian.*
  (iii) *If* x, y, z *are linearly independent then* $\mathfrak{n}$ *is isomorphic with the three-dimensional Heisenberg algebra.*
  (iv) *If* v, x, y, z *are linearly independent then* $\mathfrak{m}$ *is isomorphic with the Oscillator algebra* osc.
  (v) *Each Cartan subalgebra of* $\mathfrak{b}$ *is conjugate to* $\mathfrak{h} := \mathbb{R} \cdot u + \mathbb{R} \cdot v$. *Also,*

$$\mathrm{reg}\,\mathfrak{b} = \begin{cases} \mathfrak{b} & \text{if } u = 0, \\ \mathfrak{b} \setminus \mathfrak{n} & \text{if } u \neq 0 = z, \text{ and} \\ \mathfrak{b} \setminus \mathfrak{m} & \text{if } u \neq 0 \neq z. \end{cases}$$

  (vi) *If* $\mathfrak{b}$ *is not abelian then its proper ideals are the subspaces* $\{0\}$, $\mathbb{R} \cdot z$, *all vector subspaces containing* $\mathfrak{n}$. *Every ideal of* $\mathfrak{m}$ *is also an ideal of* $\mathfrak{b}$.
  (vii) *For any* $\lambda \in \mathbb{R}$ *the proper ideals of* $\mathbb{R} \cdot (\lambda \cdot u + v) \oplus \mathfrak{n}$ *are* $\{0\}$, $\mathbb{R} \cdot z$, $\mathfrak{n}$.
  (viii) *If* $u \neq 0$ *then* $\mathrm{comp}\,\mathfrak{b} \subseteq \mathfrak{m}$. *More explicitly,*

$$\mathrm{comp}\,\mathfrak{b} = \{\lambda \cdot e^{\mathrm{ad}\,b} v \mid b \in \mathfrak{b}, \lambda \in \mathbb{R}\}$$
$$= \mathbb{R} \cdot \{v + \lambda \cdot x + \mu \cdot y - (\lambda^2 + \mu^2) \cdot z \mid \lambda, \mu \in \mathbb{R}\}.$$

$$\overline{\mathrm{comp}\,\mathfrak{b}} = \begin{cases} \mathrm{comp}\,\mathfrak{b} \cup \mathbb{R} \cdot z, & \text{if } z \neq 0, \\ \mathrm{comp}\,\mathfrak{b} \cup \mathfrak{n} = \mathbb{R} \cdot v + \mathfrak{n}, & \text{if } z = 0. \end{cases}$$

Proof. Assertion (i) follows from the Jacobi identity.

To see (ii), suppose that for some non-zero $(\alpha, \beta, \gamma, \delta) \in \mathbb{R}^4$ the linear combination $\mathfrak{b} = \alpha \cdot u + \beta \cdot v + \gamma \cdot x + \delta \cdot y$ vanishes. Then

$$0 = [u, \mathfrak{b}] = \gamma \cdot x + \delta \cdot y \text{ and}$$
$$0 = [v, \mathfrak{b}] = -\delta \cdot x + \gamma \cdot y,$$

If $0 \neq \gamma^2 + \delta^2 = \begin{vmatrix} \gamma & \delta \\ -\delta & \gamma \end{vmatrix}$ then $x = y = 0$ and then also $z = 0$. Then the assertion follows. Otherwise $\gamma = \delta = 0$. Then u, v are linearly dependent, hence the identities above imply that $x = y = z = 0$. It follows again that $\mathfrak{b} = \mathbb{R} \cdot u + \mathbb{R} \cdot v$.

Assertions (iii) and (iv) follow from the definitions and our given relations.

(v) Our defining relations imply that ad u is diagonalizable. Also, ad $u(\mathfrak{b}) = [\mathfrak{b}, \mathfrak{b}] = \mathfrak{n}$. Thus the nilspace $\mathfrak{h}$ of ad u has minimal dimension among the maps ad $\mathfrak{b}$, $\mathfrak{b} \in \mathfrak{b}$, and therefore u is regular and $\mathfrak{h}$ is a Cartan subalgebra. Since all Cartan subalgebras of a solvable Lie algebra are conjugate this proves the first part of the assertion. For the second part we note that for $u \neq 0$ the conjugates of v are regular in $\mathfrak{b}$ if and only if $z = 0$. If $u = 0$ then $\mathfrak{b}$ is abelian by (ii) and thus reg $\mathfrak{b} = \mathfrak{b}$.

Assertions (vi), (vii) are straightforward from the definitions.

(viii) Suppose that $u \neq 0$. Then the given relations imply that all compact elements of $\mathfrak{b}$ are conjugate to a scalar multiple of v. Since v commutes with u and z a straightforward calculation shows that

$$\text{comp } \mathfrak{b} = \mathbb{R} \cdot \langle e^{\text{ad } \mathfrak{b}} \rangle v = \mathbb{R} \cdot e^{\text{ad } \mathfrak{n}} v = \mathbb{R} \cdot e^{\text{ad } (\mathbb{R} \cdot x + \mathbb{R} \cdot y)} v$$
$$= \mathbb{R} \cdot \{ e^{\text{ad } (-\mu \cdot x + \lambda \cdot y)} v \mid \lambda, \mu \in \mathbb{R} \}$$
$$= \mathbb{R} \cdot \{ (v + \lambda \cdot x + \mu \cdot y - (\lambda^2 + \mu^2) \cdot z) \mid \lambda, \mu \in \mathbb{R} \}.$$

The formula for $\overline{\text{comp } \mathfrak{b}}$ follows by passing to limits. □

1.2. DEFINITION. (i) The algebra spanned by five linearly independent elements u, v, x, y, z as in 1.1 will be denoted by $\mathfrak{extosc}$. For $\lambda \in \mathbb{R}$ we write $\mathfrak{osc}(\lambda)$ for the subalgebra $\mathbb{R} \cdot (\lambda \cdot u + v) + \mathfrak{n}$ of $\mathfrak{extosc}$.

(ii) The algebra spanned by four linearly independent elements u, v, x, y, and $z = 0$ as in 1.1 will be denoted by $\mathfrak{extmot}$. For $\lambda \in \mathbb{R}$ we write $\mathfrak{mot}(\lambda)$ for the subalgebra $\mathbb{R} \cdot (\lambda \cdot u + v) + \mathfrak{n}$ of $\mathfrak{extosc}$.

(iii) We define $\mathcal{B}$ to be the class of all Lie algebras which are isomorphic to one of the following:

$\{0\}$, $\mathfrak{so}(3)$, $\mathfrak{sl}(2, \mathbb{R})$, $\mathfrak{extosc}$, $\mathfrak{osc}(\lambda)$, $\mathfrak{extmot}$, $\mathfrak{mot}(\lambda)$, where $\lambda \in \mathbb{R}$. □

1.3. REMARKS. (i) We have the isomorphisms $\mathfrak{osc}(0) \cong \mathfrak{osc}$ and $\mathfrak{mot}(0) \cong \mathfrak{mot}$. Each member of the class $\mathcal{B}$ is isomorphic to an ideal of either $\mathfrak{extosc}$ or $\mathfrak{extmot}$.

(ii) For any proper nonzero ideal $\mathfrak{j}$ of $\mathfrak{extosc}$ the quotient $\mathfrak{extosc}/\mathfrak{j}$ is either abelian or isomorphic to $\mathfrak{extmot}$. If $\mathfrak{j}$ is a proper nonzero ideal of $\mathfrak{extmot}$ then $\mathfrak{extmot}/\mathfrak{j}$ is abelian.

(iii) Let $\lambda \in \mathbb{R}$. For any proper nonzero ideal $\mathfrak{j}$ of $\mathfrak{osc}(\lambda)$ the quotient $\mathfrak{osc}(\lambda)/\mathfrak{j}$ is either abelian or isomorphic to $\mathfrak{mot}(\lambda)$). If $\mathfrak{j}$ is a proper nonzero ideal of $\mathfrak{mot}(\lambda)$ then $\mathfrak{mot}(\lambda)/\mathfrak{j}$ is abelian.

(iv) By (ii), (iii) above every homomorphic image of an algebra in $\mathcal{B}$ either also belongs to $\mathcal{B}$ or is abelian.

(v) Every connected Lie group with Lie algebra $\mathfrak{mot}(\lambda)$ or $\mathfrak{osc}(\lambda)$ for some $\lambda \neq 0$ is a UMCS-group because it has no nondegenerate compact subgroups. □

2. Test Objects

2.1. DEFINITION. A *test algebra* is a Lie algebra which is an ideal direct sum $\mathfrak{a} \oplus \mathfrak{b}$ where $\mathfrak{a}$ is abelian and $\mathfrak{b}$ belongs to the class $\mathcal{B}$ (see 1.2).
A *test group* is a connected Lie group G whose Lie algebra is a test algebra. We shall denote the analytic groups $\langle \exp \mathfrak{a} \rangle$ and $\langle \exp \mathfrak{b} \rangle$ by A and B, respectively.
□

We observe that a test algebra is reductive if it is either abelian or contains a copy of $\mathfrak{sl}(2, \mathbb{R})$ or a copy of $\mathfrak{so}(3)$. In a reductive test algebra the summand $\mathfrak{a}$ is uniquely determined as the center of $\mathfrak{g}$, and the summand $\mathfrak{b}$ is uniquely determined as the Levi complement. All non-reductive test algebras are solvable and the following Lemma shows that in such algebras the summand $\mathfrak{b}$ can be choosen so as to contain any given regular element.

2.2. LEMMA. *Let* $\mathfrak{g} = \mathfrak{a} \oplus \mathfrak{b}$ *with a central ideal* $\mathfrak{a}$ *and such that* $\mathfrak{b}' = [\mathfrak{b}, \mathfrak{b}] \neq \mathfrak{b}$. *Suppose that* $Y \in \operatorname{reg} \mathfrak{g}$. *Then there is an isomorphic copy* $\mathfrak{b}_1$ *of* $\mathfrak{b}$ *in* $\mathfrak{g}$ *with* $Y \in \mathfrak{b}_1$, $\mathfrak{b}' \subseteq \mathfrak{b}_1$ *and* $\mathfrak{g} = \mathfrak{a} \oplus \mathfrak{b}_1$.

Proof. Write $Y = Y_\mathfrak{a} + Y_\mathfrak{b}$ with $Y_\mathfrak{a} \in \mathfrak{a}$ and $Y_\mathfrak{b} \in \mathfrak{b}$. Since Y is regular in $\mathfrak{g}$ the element $Y_\mathfrak{b}$ is regular in $\mathfrak{g}$ and thus $Y_\mathfrak{b} \notin \mathfrak{b}'$. Let $\mathfrak{n}$ denote a hyperplane ideal of $\mathfrak{b}$ containing $\mathfrak{b}'$ and not containing $Y_\mathfrak{b}$. Now we define a Lie algebra morphism $\varphi \colon \mathfrak{b} \to \mathfrak{a}$ by $\ker \varphi = \mathfrak{n}$ and $\varphi(Y_\mathfrak{b}) = Y_\mathfrak{a}$.

Set $\mathfrak{b}_1 = \mathbb{R} \cdot Y + \mathfrak{n}$. Then $\mathfrak{b}_1$ is an ideal since it contains $\mathfrak{g}' = \mathfrak{b}'$. Now define $\Phi \colon \mathfrak{b} \to \mathfrak{g}$ by $\Phi(X) = X + \varphi(X)$. Since $\varphi(X) \in \mathfrak{a}$ and $\mathfrak{a}$ is central, Φ is a morphism of Lie algebras. If $\Phi(X) = 0$, then $X = -\varphi(X) \in \mathfrak{a} \cap \mathfrak{b} = \{0\}$. It follows that Φ is injective. Further, $\Phi(Y_\mathfrak{b}) = Y_\mathfrak{a} + Y_\mathfrak{b} = Y$, and $\Phi(\mathfrak{n}) = \mathfrak{n}$. Thus $\mathfrak{b}_1 = \mathbb{R} \cdot Y + \mathfrak{n} \subseteq \operatorname{im} \Phi$. But $Y \notin \mathfrak{n}$ implies $\dim \mathfrak{b}_1 = 1 + \dim \mathfrak{n} = \dim \mathfrak{b}$. Thus $\operatorname{im} \Phi = \mathfrak{b}_1$ follows and the Lemma is proved. □

2.3. DEFINITION. A weakly exponential subsemigroup S of a connected Lie group is called *trivial* if it contains the commutator subgroup G' of the analytic subgroup $G = \langle S, S^{-1} \rangle$ it generates. □

We note that a weakly exponential semigroup with inner points in G is trivial if and only if its Lie wedge contains the commutator algebra $\mathfrak{g}'$ of $\mathfrak{g} = \mathcal{L}(G)$, i.e, is a trivial semialgebra in the sense of [27], p. 130, II.4.11.

2.4. DEFINITION. A Lie wedge W is called a *test wedge* if
(a) $W - W$ is a test algebra $\mathfrak{g}$, and
(b) either $\mathfrak{g}$ is solvable or there is a nonzero compact element of the Levi complement $\mathfrak{b}$ which lies in $\mathrm{int}_{\mathfrak{b}}(\mathfrak{b} \cap W)$.

A *test semigroup* is a weakly exponential semigroup whose Lie wedge is a test wedge. □

The main result of this chapter will be the following theorem:

2.5. MAIN LEMMA. ('The Testing Theorem') *All test semigroups are trivial.*
□

The proof of this theorem requires considerable effort. Our first aim is showing that the Lie wedge of a weakly exponential semigroup with interior points in a test group is a semialgebra. We start with a natural reduction Lemma.

2.6. LEMMA. (i) *If $\mathfrak{g}$ is a test algebra and $\mathfrak{n}$ an ideal of $\mathfrak{g}$ then the quotient $\mathfrak{g}/\mathfrak{n}$ is a test algebra.*
(ii) *If G is a test group and N a closed normal subgroup, then G/N is a test group.*

Proof. Clearly (i)⇒(ii). We show (i). It is clear that $\mathfrak{g}/(\mathfrak{a} \cap \mathfrak{n}) \cong \frac{\mathfrak{a}}{\mathfrak{a} \cap \mathfrak{n}} \times \mathfrak{b}$ is a test algebra. So we assume that $\mathfrak{a} \cap \mathfrak{n} = \{0\}$. If we set $\mathfrak{a}_n = (\mathfrak{n} + \mathfrak{b}) \cap \mathfrak{a}$, and write $\mathfrak{a} = \mathfrak{a}_n \oplus \mathfrak{a}_0$ with a suitable complement $\mathfrak{a}_0$, then it suffices to show that $\frac{\mathfrak{a}_n \oplus \mathfrak{b}}{\mathfrak{n}}$ is a test algebra. Thus we may assume that $\mathfrak{g} = \mathfrak{n} + \mathfrak{b}$.

Under this assumption $\mathfrak{g}/\mathfrak{n} \cong \mathfrak{b}/(\mathfrak{b} \cap \mathfrak{n})$ is a homomorphic image of $\mathfrak{b}$. But by remark 1.3(iv) the homomorphic image of a member of $\mathcal{B}$ either also belongs to $\mathcal{B}$ or is abelian. Thus $\mathfrak{b}/(\mathfrak{b} \cap \mathfrak{n})$ is a test algebra. □

2.7. PROPOSITION. *Let W be a generating wedge in a Lie algebra $\mathfrak{g}$. Then for every $x \in \mathfrak{g}$ with $e^{\mathbb{R} \cdot \mathrm{ad}\, x} W \subseteq W$ the following assertions hold:*
(i) *If $(\mathrm{ad}\, x)^2 = 0$ then $[x, \mathfrak{g}] \subseteq H(W)$.*
(ii) *If W is a Lie wedge and x lies in some commutative ideal then $H(W)$ contains the ideal generated by $[x, \mathfrak{g}]$.*

(iii) *If W is a Lie wedge with $H(W)$ ideal free and x lies in a commutative ideal of $\mathfrak{g}$ then x is central in $\mathfrak{g}$.*

Proof. (i) Suppose that $(\operatorname{ad} x)^2 = 0$ and pick an element $w \in W$. Then $[x, [x, w]] = 0$, and therefore $e^{t \operatorname{ad} x} w = w + t[x, w]$ for all $t \in \mathbb{R}$. It follows that

$$\pm[x, w] = \lim_{t \to \pm\infty} \frac{1}{|t|}(w + [x, w]) = \lim_{t \to \pm\infty} \frac{1}{|t|} e^{\operatorname{ad} tx} w \in W$$

Thus $[x, w] \in H(W)$. Since W is generating in $\mathfrak{g}$ we conclude $[x, \mathfrak{g}] = [x, W] - [x, W] \subseteq H(W)$.

(ii) If W is a Lie wedge then $e^{\operatorname{ad} H(W)} W \subseteq W$, so (ii) follows by applying (i) to the elements of $[x, \mathfrak{g}]$ and iterating. (Note that $(\operatorname{ad} x)^2 = 0$ whenever x lies in a commutative ideal of $\mathfrak{g}$.)

(iii) If W is an ideal free Lie wedge and x lies in some commutative ideal of $\mathfrak{g}$ then (ii) implies that $[x, \mathfrak{g}] = \{0\}$. Thus x is central in $\mathfrak{g}$. □

2.8. LEMMA. *Let $\mathfrak{g}$ be the ideal direct sum $\mathfrak{g} = \mathfrak{a} \oplus \mathfrak{b}$, where $\mathfrak{a}$ is abelian. We write p for the projection $\mathfrak{g} \to \mathfrak{b}$ onto $\mathfrak{b}$ with kernel $\mathfrak{a}$. Suppose that W is a generating Lie wedge in $\mathfrak{g}$ whose edge $H(W)$ is ideal free. Then $p(H(W))$ is ideal free.*

Proof. Suppose that $\mathfrak{j}$ is an ideal of $\mathfrak{b}$ with $\mathfrak{j} \subseteq p(H(W))$. The commutator $[\mathfrak{j}, \mathfrak{j}]$ is a characteristic ideal of $\mathfrak{j}$, hence an ideal of $\mathfrak{g}$. But $[\mathfrak{j}, \mathfrak{j}] \subseteq [\mathfrak{a} + H(W), \mathfrak{a} + H(W)] \subseteq H(W)$ and $H(W)$ is ideal free, so $[\mathfrak{j}, \mathfrak{j}] = \{0\}$. Also, $e^{\operatorname{ad}(\mathfrak{j})} W = e^{\operatorname{ad}(\mathfrak{j}+\mathfrak{a})} W \subseteq e^{\operatorname{ad}(H(W))} W \subseteq W$ since W is a Lie wedge. Applying Proposition 2.7(iii) we conclude that $\mathfrak{j}$ lies in the center of $\mathfrak{g}$. Now $\mathfrak{j} \subseteq p(H(W)) \cap \mathfrak{z}(\mathfrak{b}) = p(H(W) \cap \mathfrak{z}(\mathfrak{g})) = p(\{0\}) = \{0\}$, so $\mathfrak{j} = \{0\}$. □

2.9. REMARK. Recall from Definition IV.5.3 that a subset $\mathfrak{e}$ of a Lie algebra $\mathfrak{g}$ is called *lean in* $\mathfrak{g}$ if the set $\mathfrak{z}(\mathfrak{e}) = \bigcup \{\mathfrak{z}(x, \mathfrak{g}) \mid 0 \neq x \in \mathfrak{e}\}$ is contained in a Zariski closed subset of dimension at most $\dim \mathfrak{g} - 2$.

The following observations are straightforward, we only sketch the proofs.

(i) In the algebra $\mathfrak{g} = \mathfrak{mot}$ every Cartan subalgebra $\mathfrak{h}$ is lean (since $\mathfrak{z}(\mathfrak{h}) = \mathfrak{z}(\mathfrak{h}) = \mathfrak{h}$), and every proper subalgebra which is not a Cartan subalgebra lies in the commutator algebra (all elements outside the commutator algebra are regular).

(ii) In the algebra $\mathfrak{g}$ described in 1.1 (namely, $\mathfrak{g} = \mathfrak{extosc}$ or $\mathfrak{g} = \mathfrak{extmot}$) every proper subalgebra $\mathfrak{c}$ of the ideal $\mathfrak{m} = \mathbb{R} \cdot x + \mathbb{R} \cdot y + \mathbb{R} \cdot z$ with $\mathbb{R} \cdot z \cap \mathfrak{c} = \{0\}$ is one-dimensional and lean. (If $\mathfrak{c}$ contains an element c outside of $\mathfrak{m}'$ then $\mathfrak{c} = \mathbb{R} \cdot c$, for otherwise $0 \neq z$ and $z \in \mathfrak{c}$. Each such subalgebra satisfies $\mathfrak{z}(\mathfrak{c}) = \mathfrak{z}(\mathfrak{c}) = \mathfrak{c} + \mathbb{R} \cdot z$. The only nonzero subalgebras of $\mathfrak{m}' = \mathbb{R} \cdot x + \mathbb{R} \cdot y + \mathbb{R} \cdot z$ with $\mathfrak{c} \cap \mathbb{R} \cdot z = \{0\}$ are the lines $\mathbb{R} \cdot l$ with $l \notin \mathbb{R} \cdot z$, and these satisfy $\mathfrak{z}(\mathfrak{c}) = \mathfrak{z}(\mathfrak{c}) = \mathfrak{c} + \mathbb{R} \cdot z$ as well.) □

Our next Lemma extends the above observations to ideal free subalgebras of test algebras with $\mathfrak{b} \cong \mathfrak{mot}$ or $\mathfrak{b} \cong \mathfrak{osc}$.

2.10. LEMMA. *Let $\mathfrak{g} = \mathfrak{a} \oplus \mathfrak{b}$ be a solvable test algebra with $\mathfrak{b}$ isomorphic to one of the algebras* mot, osc, extmot, extosc. *We denote with p the projection $\mathfrak{g} \to \mathfrak{b}$ onto $\mathfrak{b}$ with kernel $\mathfrak{a}$. Suppose that $\mathfrak{e}$ is a subalgebra of $\mathfrak{g}$ such that both $\mathfrak{e}$ and $p(\mathfrak{e})$ are ideal free, and assume that $\mathfrak{e}$ lies in the linear span of* comp $\mathfrak{g}$.

  (i) *If $\mathfrak{b} \cong$ mot then either $\mathfrak{e}$ is lean in $\mathfrak{g}$, or $p(\mathfrak{e})$ is nonzero and contained in the commutator algebra $[\mathfrak{b}, \mathfrak{b}] = [\mathfrak{g}, \mathfrak{g}]$, which is a commutative ideal.*
  (ii) *In all other cases $\mathfrak{e}$ is lean in $\mathfrak{g}$.*

Proof. Note first that the center $\mathfrak{z}(\mathfrak{g})$ of $\mathfrak{g}$ is the direct sum $\mathfrak{z}(\mathfrak{g}) = \mathfrak{a} \oplus \mathfrak{z}(\mathfrak{b})$. We know from Lemma IV.5.8 that $\mathfrak{e}$ is lean in $\mathfrak{g}$ if $\mathfrak{a} \cap \mathfrak{e}$ is lean in $\mathfrak{a}$ and $p(\mathfrak{e})$ is lean in $\mathfrak{b}$. Since $\mathfrak{e}$ is ideal free we have $(\mathfrak{a} \oplus \mathfrak{b}) \cap \mathfrak{e} = \mathfrak{z}(\mathfrak{g}) \cap \mathfrak{e} = \{0\}$, hence $\mathfrak{a} \cap \mathfrak{e}$ is lean in $\mathfrak{g}$. Since $p(\mathfrak{e})$ is ideal free, by assumption, assertions (i), (ii) follow from the corresponding assertions (i),(ii) of Remark 2.9. □

Now we are ready for the first major step towards the proof of the Testing Theorem.

2.11. PROPOSITION. *Let S be a weakly exponential semigroup with inner points in a test group G. Then the following assertions hold:*

  (i) $W = \mathfrak{L}(S)$ *is a semialgebra.*
  (ii) *If $\mathfrak{b} \cong$ mot then $\mathfrak{L}(S)$ is a trivial semialgebra, i.e., contains $[\mathfrak{g}, \mathfrak{g}]$.*
  (iii) *If $\mathfrak{b} \cong$ osc then W is invariant.*

Proof. Throughout the following proof we shall denote by $p \colon \mathfrak{g} \to \mathfrak{b}$ the projection onto $\mathfrak{b}$ with kernel $\mathfrak{a}$. Also, we shall abbreviate $H(W)$ by $\mathfrak{e}$ and write $\mathfrak{p} = p(\mathfrak{e})$.

(i) Recall from the Main Lemma IV.6.3 that the Lie wedge of a weakly exponential subsemigroup must be a semialgebra if G is a UMCS-group. Now all connected Lie groups with Lie algebra mot(λ) or osc(λ), $\lambda \neq 0$, are UMCS-groups, and so are the test groups with $\mathfrak{b} \cong \mathfrak{so}(3)$. If $\mathfrak{b} \cong \mathfrak{sl}(2, \mathbb{R})$ then W is a semialgebra by our Main Lemma IV.7.5 (with $\mathfrak{c} = \mathfrak{a}$). Thus we are left with the cases where $\mathfrak{b}$ is isomorphic with one of the algebras mot, extmot, osc, extosc.

By Proposition IV.4.8 and since quotients of test algebras are test algebras (2.6) we may factor the largest ideal $\mathfrak{j}$ of $\mathfrak{g}$ contained in $\mathfrak{e}$, by Theorem II.5.8 we may suppose that W has interior points in $\mathfrak{g}$.

Thus we henceforth assume that S is reduced.

Since S is reduced it follows by Lemma 2.8 that $p(H(W))$ is ideal free in $\mathfrak{b}$. Thus the assumptions of Lemma 2.10 are satisfied and therefore either $\mathfrak{e}$ is lean or $\mathfrak{b} \cong$ mot and $p(\mathfrak{e})$ is contained in the commutative ideal $\mathfrak{g}' = [\mathfrak{b}, \mathfrak{b}]$. But if $p(\mathfrak{e}) \subseteq \mathfrak{g}' = [\mathfrak{b}, \mathfrak{b}]$ then by Lemma 2.7 it follows that $p(\mathfrak{e})$ is central in $\mathfrak{g}$, which is impossible since $p(\mathfrak{e})$ is ideal free.

Thus $\mathfrak{e}$ is lean and therefore W is a semialgebra.

By Lemma 2.2 we may assume that $\mathfrak{b}$ meets the interior of W. Now (ii) follows from [27], Theorem II.3.4.(i), p. 104, and (iii) follows from [27], Theorem 3.20, p. 120. □

3. Locally Direct Products of Lie Groups

Throughout this section we consider a Lie group G whose Lie algebra $\mathfrak{g}$ is the direct sum $\mathfrak{g} = \mathfrak{a} \oplus \mathfrak{b}$ of two *ideals* $\mathfrak{a}$ and $\mathfrak{b}$. We write A, B for the corresponding analytic normal subgroups. We want to have information about the extent to which the direct sum decomposition of $\mathfrak{g}$ is reflected by a 'nearly direct' product structure of G. If the analytic subgroups A, B are closed then there is a straightforward answer to this question.

3.1. NOTATION. In the following we denote the center of a topological group G with $Z(G)$ and the identity component of $Z(G)$ with $Z_0(G)$. □

3.2. LEMMA. *Let G be a connected Lie group whose Lie algebra $\mathfrak{g}$ is the direct sum $\mathfrak{g} = \mathfrak{a} \oplus \mathfrak{b}$ of two ideals $\mathfrak{a}$ and $\mathfrak{b}$. We write for the normal analytic subgroups $A = \langle \exp \mathfrak{a} \rangle$ and $B = \langle \exp \mathfrak{b} \rangle$. Then the following assertions hold:*
   (i) *The intersection $A \cap B$ is a finitely generated central subgroup of G.*
   (ii) *The map $q: A_{\mathrm{Lie}} \times B_{\mathrm{Lie}} \to G$, $(a,b) \mapsto ab$ is a covering morphism with kernel $\ker q = \{(x, x^{-1}) \mid x \in A \cap B\}$.*
   (iii) *If $A \cap B = \{1\}$ then A and B are closed and G is algebraically and topologically the direct product of A and B.*
   (iv) *If G contains no nontrivial central torus subgroup then neither A nor B contains a nontrivial central torus and $\exp \mathfrak{z}(\mathfrak{a}) \cap \exp \mathfrak{z}(\mathfrak{b}) = \{1\}$.*

Proof. (i) Since $A \cap B$ is closed this is obvious from the general correspondence between closed subgroups of G and subalgebras of $\mathfrak{g}$.

(ii) Since every element of A commutes with every element of B the map q is a morphism. If $(x, y) \in \ker q$ then $xy = 1$ and therefore $x = y^{-1} \in A \cap B$; conversely, if $x \in A \cap B$ then $(x, x^{-1}) \in \ker q$. Thus $\ker q = \{(x, x^{-1}) \mid x \in A \cap B\}$. Since q is discrete, the map q is a covering morphism.

(iii) If $A \cap B = \{1\}$ then (ii) shows that $q: A_{\mathrm{Lie}} \times B_{\mathrm{Lie}} \to G$ is an isomorphism and thus $A \cong A_{\mathrm{Lie}}$ and $B \cong B_{\mathrm{Lie}}$. Thus A and B are locally compact and therefore closed.

(iv) Assume that G does not contain any nontrivial central torus. Since every central subgroup of A or B is central also in G it is obvious that neither A nor B contains a nontrivial central torus. Let $x \in \exp \mathfrak{z}(\mathfrak{a}) \cap \exp \mathfrak{z}(\mathfrak{b})$. Then there are central elements $X \in \mathfrak{a}$ and $Y \in \mathfrak{b}$ with $x = \exp X = \exp Y$. Since both X and Y are central in $\mathfrak{g}$, so is their difference $X - Y$, and $\exp(X - Y) = 1$. If $X - Y \neq 0$ then $\exp \mathbb{R} \cdot (X - Y)$ is a central torus group, contradicting our assumption. Thus $X = Y \in \mathfrak{a} \cap \mathfrak{b} = \{0\}$ and hence $x = \exp X = 1$. □

We next look for conditions ensuring that A is closed in G, under the additional assumption that the center of G is torus-free.

3.3. NOTATION. Let G be a Lie group with Lie algebra $\mathfrak{g}$ and let $A = \langle \exp \mathfrak{a} \rangle$ be the analytic subgroup corresponding to a subalgebra of $\mathfrak{g}$. Then we write A_{Lie} for the group underlying A, but endowed with its intrinsic Lie group topology. □

3.4. DEFINITION. A Lie group G is said to have *slim center* if the torsion-free rank of the quotient $Z(G)/Z_0(G)$ is at most 1. □

3.5. PROPOSITION. *Let A_* be a connected Lie group with Lie algebra $\mathfrak{a}$. Then the following assertions are equivalent:*
  (i) *A_* has slim center.*
  (ii) *Let G be a connected Lie group whose Lie algebra $\mathfrak{g}$ is the direct Lie algebra sum $\mathfrak{g} = \mathfrak{a} \oplus \mathfrak{b}$ of $\mathfrak{a}$ with some Lie algebra $\mathfrak{b}$, and write $A = \langle \exp \mathfrak{a} \rangle$. Suppose that G contains no nontrivial central torus subgroup and that A_{Lie} is canonically isomorphic with A_*. Then A is closed in G.*

Proof. (i)⇒(ii) Let G, $\mathfrak{b}$ as in the hypothesis of (ii), and write $B = \langle \exp \mathfrak{b} \rangle$.

The closure $G_1 := \overline{A}$ is a connected closed Lie subgroup whose Lie algebra contains $\mathfrak{a}$, hence is of the form $\mathfrak{g}_1 = \mathfrak{a} \oplus \mathfrak{b}_1$ with $\mathfrak{b}_1 = \mathfrak{b} \cap \mathfrak{g}_1 \subseteq \mathfrak{b}$. If T is a central torus subgroup of G_1 then it is central in G, since the closure G_1 of A is centralized by $\overline{B}$. Thus we see that our assumptions remain intact if we replace $\mathfrak{g}$ by $\mathfrak{g}_1$ and $\mathfrak{b}$ by $\mathfrak{b}_1$, so we henceforth assume that $\mathfrak{g} = \mathfrak{g}_1$, $\mathfrak{b} = \mathfrak{b}_1$; that is to say, A is dense in G.

Since A is dense in G we have $[\mathfrak{g}, \mathfrak{g}] \subseteq \mathfrak{a}$, so $[\mathfrak{b}, \mathfrak{b}] \subseteq [\mathfrak{g}, \mathfrak{g}] \cap \mathfrak{b} \subseteq \mathfrak{a} \cap \mathfrak{b} = \{0\}$. Hence the algebra $\mathfrak{b}$ is abelian and thus in fact central in $\mathfrak{g}$. The closure $\overline{B}$ is closed and free of nontrivial compact subgroups, otherwise G would contain a nontrivial central torus. Hence $\overline{B}$ is a vector group and therefore $B = \overline{B}$.

We next show that under this assumption $A \cap B$ is discrete. The quotient morphism $\kappa: G \to G/B$ induces a Lie algebra morphism $\mathfrak{g} \to \mathfrak{g}/\mathfrak{b}$ which maps the subalgebra $\mathfrak{a}$ isomorphically onto $\mathfrak{g}/\mathfrak{b} \cong \mathfrak{a}$. From the corresponding commutative diagram

$$\begin{array}{ccccc} \mathfrak{a} & \to & \mathfrak{g} & \to & \mathfrak{g}/\mathfrak{b} \cong \mathfrak{a} \\ \downarrow & & \downarrow & & \downarrow \\ A_{\text{Lie}} & \to & G & \to & G/B \end{array}$$

we conclude that the composition $A_{\text{Lie}} \to G/B$, $a \mapsto aB$ is a a covering morphism whose kernel is the group $A \cap B$ with the discrete topology. Being the kernel of a covering morphism, $A \cap B$ must be a finitely generated abelian subgroup of B, and it must be a free abelian group since B is a vector group.

Suppose that $x \in A \cap B \cap Z_0(A_{\text{Lie}})$. Then $x = \exp a = \exp b$ for some $a \in \mathfrak{a}$, $b \in \mathfrak{b}$. (Note that B is a vector group.) But then $\exp(a - b) = 1$, so $a = b \in \mathfrak{a} \cap \mathfrak{b} = \{0\}$, since G does not contain a nontrivial central torus subgroup. Thus $x = 1$ and we conclude that the restriction of the quotient map $Z(A_{\text{Lie}}) \to Z(A_{\text{Lie}})/Z_0(A_{\text{Lie}})$ to $A \cap B$ is injective. Since A_{Lie} has slim center, the torsion-free rank of $Z(A_{\text{Lie}})/Z_0(A_{\text{Lie}})$ is at most 1, so $A \cap B$ has torsion free rank at most 1, hence is cyclic.

Since every cyclic subgroup of a vector group is discrete, we conclude that $A \cap B$ is discrete in B and therefore a discrete subgroup of G.

Now, applying the Open Mapping Theorem, we see that the induced map $A_{\text{Lie}}/(A \cap B) \to G/B$, $a(A \cap B) \mapsto aB$, is an isomorphism of Lie groups which factors as

$$A_{\text{Lie}}/(A \cap B) \xrightarrow{j} A/(A \cap B) \xrightarrow{q} G/B$$

with bijective morphisms j, q of topological groups. It follows that both j and q are isomorphisms of topological groups. Hence $A/(A \cap B)$ is locally compact and thus A is locally compact and therefore closed in G. This establishes (ii).

(ii)$\Rightarrow$(i) Suppose that, on the contrary, (ii) holds but the center of A_* is not slim. Then there are two elements $z_1, z_2 \in Z(A_*)$ such that $z_1 Z_0(A_*)$ and $z_2 Z_0(A_*)$ generate a free abelian subgroup of rank 2 in $Z(A_*)/Z_0(A_*)$. We define $\widetilde{G} = A_* \times \mathbb{R}$, $D = \{(z_1^m z_2^n, m + n\sqrt{2}) \mid m, n \in \mathbb{Z}\}$, and let $G = \widetilde{G}/D$. Then the Lie algebra of G has the form $\mathfrak{g} = \mathfrak{a} \oplus \mathfrak{b}$, where $\mathfrak{b} = \mathbb{R}$ with trivial brackets, $A = \langle \exp \mathfrak{a} \rangle$ is the image of $A_* \times \{0\}$ under the quotient map $\widetilde{G} \to G$, and $A_{\text{Lie}} \cong A_*$. The group G has no nontrivial central torus subgroup, and A is dense in G. This is a contradiction to (ii) and the proof is finished. □

3.6. EXAMPLES. (i) If $\widetilde{G}$ is a covering group of G and $\widetilde{G}$ has slim center then G has slim center.

(ii) The direct product of a Lie group with slim center and a Lie group with connected center has slim center.

(iii) Any connected simple Lie group has slim center. (Inspecting a table of the simple Lie groups we see that the center of a simply connected simple Lie group is always either finite or the product of a finite group with $\mathbb{Z}$.)

(iv) Any connected linear semisimple Lie group has finite, hence slim center.

(v) If $\mathfrak{g}$ is isomorphic to one of the Lie algebras $\mathfrak{sl}(2, \mathbb{R})$, $\mathfrak{mot}$, $\mathfrak{osc}$ then every connected Lie group with Lie algebra $\mathfrak{g}$ has slim center. □

3.7. PROPOSITION. *Let G be a connected Lie group whose Lie algebra $\mathfrak{g}$ is the direct Lie algebra sum $\mathfrak{g} = \mathfrak{a} \oplus \mathfrak{b}$ of $\mathfrak{a}$ with some Lie algebra $\mathfrak{b}$, and write $A = \langle \exp \mathfrak{a} \rangle$, $B = \langle \exp \mathfrak{b} \rangle$. Suppose that G contains no nontrivial central torus subgroup and that A_{Lie} and B_{Lie} have slim center. Then the following assertions hold:*

(i) *The analytic subgroups A and B are closed in G.*

(ii) *The intersection $A \cap B$ is a closed discrete subgroup of G and $Z_0(A) \cap Z_0(B) = \{1\}$.*

(iii) *The map $q: A \times B \to G$, $(a, b) \mapsto ab$ is a covering morphism with kernel $\ker q = \{(x, x^{-1}) \mid x \in A \cap B\}$.*

Proof. These assertions follow by combining 3.5 with 3.2. □

3.8. COROLLARY. *Suppose that under the assumptions of* Proposition 3.7 *we have, in addition, $A \cap B \subseteq Z_0(A)$. Then $A \cap B$ is isomorphic with a subgroup of $Z(B)/Z_0(B)$.*
In particular, if $\mathfrak{a}$ is abelian then $A \cap B$ is either singleton or isomorphic with $\mathbb{Z}$.

Proof. Consider the quotient map $\kappa: Z(B) \to Z(B)/Z(B)_0$. We know from 3.7(ii) that the restriction of κ to $A \cap B = Z_0(A) \cap B$ is injective, hence induces the required isomorphism. The remainder of the assertion follows by specialization and the well known fact that any subgroup of a cyclic group is cyclic. □

We finally use the information listed above to compute, in some special cases, the maximal compact subgroups of G. Recall that we agreed to write $\text{comp}_G(\mathfrak{g})$ for the set of all vectors $x \in \mathfrak{g}$ such that $\exp \mathbb{R} \cdot x$ is precompact.

3.9. PROPOSITION. *Let G be a connected Lie group containing no nontrivial central torus subgroup. We assume that the Lie algebra $\mathfrak{g}$ of G is a direct ideal sum $\mathfrak{g} = \mathfrak{a} \oplus \mathfrak{b}$ with $\mathfrak{a}$ abelian. We write $A = \langle \exp \mathfrak{a} \rangle$, $B = \langle \exp \mathfrak{b} \rangle$. Suppose that*
  (a) *B contains no nontrivial compact subgroups and the group $Z(B)/Z_0(B)$ is cyclic;*
  (b) *G contains a nontrivial compact subgroup.*
Then the following assertions hold:
  (i) *The analytical subgroups A and B are closed, and A is a vector group.*
  (ii) *$A \cap B$ is infinite cyclic, $A \cap B \cong \mathbb{Z}$.*
  (iii) *The maximal compact subgroups of G are one-dimensional.*
  (iv) *Let $a \in \mathfrak{a}$, $b \in \mathfrak{b}$, such that $z = \exp a = \exp b$ is a generator of $A \cap B$. Then $\text{comp}_G(\mathfrak{g}) = \mathbb{R} \cdot \{a - x \mid x \in (\text{Inn } \mathfrak{b})b\}$.*

Proof. (i), (ii) We know from Proposition 3.7 that the analytic subgroups A, B are closed and that A must be a vector group. Assumption (a) implies that $A \cap B$ must be cyclic, by Corollary 3.8, it cannot be finite since B does not contain nontrivial compact subgroups.

(iii) Let K be a maximal compact subgroup of G and consider the inverse image $\widetilde{K}$ of K under the covering map $q: A \times B \to G$, $(x, y) \mapsto xy$. Since $A \times B$ does not contain any compact subgroup, the group $\widetilde{K}$ must be a vector group and K is isomorphic with the quotient $\widetilde{K}/(D \cap \widetilde{K})$, where D denotes the kernel

$$\{(x, x^{-1}) : x \in A \cap B\} \cong A \cap B \cong \mathbb{Z}$$

of q. Since $D \cap \widetilde{K}$ is cyclic and K is nondegenerate (by (b)) we conclude that K is a one-dimensional torus.

(iv) It is obvious that $\exp(a - b) = 1$. Since $A \cap B \neq \{1\}$, both a and b are nonzero, and hence $a - b \neq 0$, for otherwise $\mathfrak{a} \cap \mathfrak{b} \neq \{0\}$. Thus $\exp \mathbb{R} \cdot (a - b)$ is a maximal compact subgroup of G. But the maximal compact subgroups of G are conjugate and $\mathfrak{a}$ is central, so

$$\exp \mathbb{R} \cdot (a - \text{Inn}(\mathfrak{b})b)) = \exp \mathbb{R} \cdot \text{Inn}(\mathfrak{g})(a - b) = \exp \mathbb{R} \cdot \text{Ad}(G)(a - b))$$
$$= \{g^{-1}(\exp \mathbb{R} \cdot (a - b))g \mid g \in G\},$$

and the assertion follows. □

4. The Proof of the Testing Theorem

We now investigate weakly exponential subsemigroups in test groups (cf. Definition 2.1).

4.1. LEMMA. *Suppose that S is a weakly exponential subsemigroup of G with Lie wedge $W = \mathfrak{L}(S)$ and that K is a compact connected normal subgroup of G. Then*
   (i) $T = KS$ *is a weakly exponential subsemigroup with Lie wedge* $\mathfrak{L}(T) = \mathfrak{k} + \mathfrak{L}(S)$, *and*
   (ii) $T/K = SK/K$ *is a weakly exponential semigroup of G/K.*
If S has inner points then so has T as well as T/K.

Proof. (i) Since S is closed and K is compact normal, their product $SK = KS$ is closed. If $\operatorname{int} S \neq \emptyset$, then $\emptyset \neq K \operatorname{int} S \subseteq \operatorname{int} T$. Clearly, both $\mathfrak{k} = \mathfrak{L}(K)$ and W are contained in $\mathfrak{L}(T)$. As K is compact, $K = \exp \mathfrak{k}$. Then $SK = \overline{\exp W} \exp \mathfrak{k} \subseteq \overline{\exp WK}$. Let $X \in W$. If, firstly, $\exp \mathbb{R} \cdot X$ is relatively compact, then $\overline{\exp \mathbb{R} \cdot X} \subseteq S \cap S^{-1}$, and thus $\overline{\exp \mathbb{R} \cdot X} \subseteq W$. Now $\overline{\exp \mathbb{R} \cdot X} K$ is a compact connected group and thus equals $\exp(\mathfrak{L}(\overline{\exp \mathbb{R} \cdot X}) + \mathfrak{k})$. It is therefore contained in $\overline{\exp(W + \mathfrak{k})}$. If, secondly, $\exp \mathbb{R} \cdot X$ is not relatively compact, then $r \mapsto \exp r \cdot X \colon \mathbb{R} \to \exp \mathbb{R} \cdot X$ is an isomorphism onto a closed subgroup. Then $K \exp \mathbb{R} \cdot X$ is isomorphic to $K \times \mathbb{R}$. In particular, the exponential function is surjective; thus $K \exp \mathbb{R} \cdot X = \exp(\mathfrak{k} + \mathbb{R} \cdot X)$ and $K \exp \mathbb{R}^+ \cdot X = \exp(\mathfrak{k} + \mathbb{R}^+ \cdot X) \subseteq \exp(\mathfrak{k} + W)$. We conclude that $K \exp W \subseteq \exp(\mathfrak{k} + W)$. Hence $T = SK \subseteq \overline{\exp(\mathfrak{k} + W)} \subseteq \overline{\exp \mathfrak{L}(T)}$. This proves the assertion.

The proof also shows that $\mathfrak{L}(T) = \mathfrak{k} + W$.

(ii) Since $(K \exp_G W)/K \subseteq \exp_{G/K} \mathfrak{L}(SK/K)$ then SK/K is weakly exponential. If T has inner points, then T/K has inner points since the quotient morphism is open. □

4.2. LEMMA. *Let G be a connected Lie group and suppose that its Lie algebra $\mathfrak{g}$ is the direct Lie algebra sum $\mathfrak{g} = \mathfrak{a} \oplus \mathfrak{b}$, where $\mathfrak{a}$ is central in $\mathfrak{g}$. We write A, B for the analytic subgroups associated with $\mathfrak{a}$, $\mathfrak{b}$ and assume that the intersection $A \cap B$ is isomorphic with $\mathbb{Z}$. We write K for the maximal compact normal subgroup of G and $\mathfrak{k} = \mathfrak{L}(K)$.*

Further, let S be a weakly exponential subsemigroup with Lie wedge W, and let $X \in \mathfrak{b}$, $Y \in \mathfrak{a}$ such that $z = \exp Y \in A \cap B$ and $\exp(X + Y) = z \exp X \in S$. We assume that the following condition (∗) is satisfied:

(∗) *If $\exp X = \lim z_n \exp X_n$ with $z_n \in A \cap B$ and $X_n \in \mathfrak{b}$ then $\lim X_n = X$.*

Then $X + Y \in \mathfrak{k} + W$.

Proof. First a reduction: Denote with κ the quotient map $\mathfrak{g} \to \mathfrak{g}/\mathfrak{k}$. We know from Lemma 4.1 that the product KS is a closed weakly exponential semigroup with Lie wedge $\mathfrak{k} + W$. Thus it suffices to show $X + Y \in \mathfrak{k} + W$ under the additional assumption that $K \subseteq S$, so that the claim becomes $X + Y \in W$. We assume $K \subseteq S$ and thus $\mathfrak{k} \subseteq W$. Then $X + Y \in W$ if and only if $\kappa(X) \in \kappa(W)$. So we can factor K and assume that $K = \{1\}$.

Now Proposition 3.7 applies and shows that A and B are closed and that A is a vector group.

By assumption we have $z \exp X \in S$, and since S is weakly exponential there is a sequence of elements P_n in W such that $z \exp X = \lim \exp P_n$. Write $P_n = P_n^\mathfrak{a} + P_n^\mathfrak{b}$, with $P_n^\mathfrak{a} \in \mathfrak{a}$, $P_n^\mathfrak{b} \in \mathfrak{b}$, and note that $\exp P_n = \exp P_n^\mathfrak{a} \exp P_n^\mathfrak{b}$. Consider the covering map $q\colon A \times B \to G$, $q(x,y) = xy$. Note that $\ker q = \{(x, x^{-1}) \mid x \in A \cap B\}$. Every convergent sequence in G is the q-image of a convergent sequence in $A \times B$, thus it follows from $\lim \exp P_n = z \exp X$ that there must exist elements $d_n \in A \cap B$ such that the limit

(†) $\quad (a,b) = \lim(\exp P_n^\mathfrak{a}, \exp P_n^\mathfrak{b})(d_n, d_n^{-1}) = \lim(d_n \exp P_n^\mathfrak{a}, d_n^{-1} \exp P_n^\mathfrak{b})$

exists and $q(a,b) = ab = z \exp X$. Since $ab = z \exp X \in B$ we see that $a \in A \cap B$. Also, looking at the first components in (†) we observe that $\lim a d_n^{-1} \exp(-P_n^\mathfrak{a}) = 1$ and therefore

$$z \exp X = ab = \lim(\exp P_n^\mathfrak{a})(\exp(P_n^\mathfrak{b}))$$
$$= \lim a d_n^{-1} (\exp -P_n^\mathfrak{a}) \lim(\exp P_n^\mathfrak{a}) \exp(P_n^\mathfrak{b})$$
$$= a \lim d_n^{-1} \exp P_n^\mathfrak{b}.$$

If we put $z_n = z^{-1} a d_n^{-1}$, $X_n = P_n^\mathfrak{b}$ we have $\exp X = \lim z_n \exp X_n$ and thus condition (∗) implies $X = \lim P_n^\mathfrak{b}$. Now

$$\exp Y \exp X = \exp(X + Y) = z \exp X$$
$$= \lim \exp P_n = \lim \exp P_n^\mathfrak{a} \lim \exp P_n^\mathfrak{b}$$
$$= \lim \exp P_n^\mathfrak{a} \exp X,$$

and thus $\exp Y = \lim \exp P_n^\mathfrak{a}$. But A is a vector group and thus $\exp |\mathfrak{a}$ is a homeomorphism. So $\lim P_n^\mathfrak{a} = Y$. Therefore,

$$X + Y = \lim P_n^\mathfrak{b} + \lim P_n^\mathfrak{a} = \lim(P_n^\mathfrak{a} + P_n^\mathfrak{b}) = \lim P_n \in W,$$

which establishes our assertion. $\square$

The case $\mathfrak{b} \cong \mathfrak{sl}(2,\mathbb{R})$

4.3. LEMMA. *Let G be a connected Lie group. We assume that the Lie algebra $\mathfrak{g}$ of G is a direct ideal sum $\mathfrak{g} = \mathfrak{a} \oplus \mathfrak{b}$. We write $A = \langle \exp \mathfrak{a} \rangle$, $B = \langle \exp \mathfrak{b} \rangle$ and let S be a closed subsemigroup of G. We suppose that the following conditions are satisfied:*
 (a) *$\mathfrak{a}$ is abelian.*
 (b) *$A \cap B = \{1\}$.*
 (c) *There is a $b \in (\operatorname{int}_\mathfrak{b}(\mathfrak{b} \cap W)) \setminus \overline{\exp_B(\mathfrak{b})}$.*

Then S is not weakly exponential.

Proof. From (b) we conclude that $G = A \times B$. We write $\mathfrak{g} = \mathfrak{a} \times \mathfrak{b}$ and $\exp_G(X_\mathfrak{a}, X_\mathfrak{b}) = (\exp_A X_\mathfrak{a}, \exp_B X_\mathfrak{b})$. Since by (a) $\mathfrak{a}$ is abelian, $\exp_G(\mathfrak{g}) = A \times \exp_B \mathfrak{b}$. In particular, $U := A \times \exp_B \mathfrak{b}$ is a neighorhood of $(1_A, b)$ in $G = A \times B$. By our choice of b in (c) the set $S \cap U$ is a neighborhood of b in G which is contained in S and satisfies $S \cap U \subseteq S \setminus \overline{\exp \mathfrak{g}} \subseteq S \setminus \overline{\exp W}$, where $W = \mathfrak{L}(S)$. Thus S is not weakly exponential. □

4.4. PROPOSITION. *Let G be a connected Lie group whose Lie algebra $\mathfrak{g} = \mathfrak{L}(G)$ is the ideal direct sum $\mathfrak{g} = \mathfrak{a} \oplus \mathfrak{b}$. We write $A = \langle \exp \mathfrak{a} \rangle$, $B = \langle \exp \mathfrak{b} \rangle$ for the associated analytic subgroups.*

Further, we assume that S is a subsemigroup of G with Lie wedge $W = \mathfrak{L}(S)$ and that the following hypotheses are satisfied:
  (a) *$\mathfrak{a}$ is central in $\mathfrak{g}$.*
  (b) *$\mathfrak{b} \cong \mathfrak{sl}(2, \mathbb{R})$ and $B \cong \widetilde{\mathrm{Sl}}(2, \mathbb{R})$.*
  (c) *$S = \overline{\exp W}$, i.e. S is closed and weakly exponential.*
  (d) *There is a compact element $U \in \mathrm{int}_\mathfrak{b}(\mathfrak{b} \cap W)$.*

Then $G' \subseteq S$.

If also
  (d*) *the interior of W contains a compact element $U \in \mathfrak{b}$*

then $S = G$.

Proof. (A) By Proposition 2.11 we know that W is a semialgebra. The radical $A = \exp \mathfrak{a}$ of G is closed, and B is closed by hypothesis (b).

(B) Let K denote the maximal compact normal subgroup of G and $\mathfrak{k} = \mathfrak{L}(K)$. Now $\mathfrak{k}$, as a compact ideal of $\mathfrak{g}$ is necessarily contained in $\mathfrak{a}$. Let us momentarily assume that $\mathfrak{b} \subseteq \mathfrak{k} + W$ is established. Then $\mathfrak{b} \subseteq \mathfrak{k} + H(W)$ and thus $\mathfrak{k} + H(W) = \mathfrak{a}_1 \oplus \mathfrak{b}$ with $\mathfrak{a}_1 = \mathfrak{a} \cap (\mathfrak{k} + H(W))$. Since $\mathfrak{k} \subseteq \mathfrak{a}_1$, the projection of $H(W)$ into $\mathfrak{b}$ is surjective. Therefore, and by (b), the Lie algebra $H(W)$ contains a subalgebra isomorphic to $\mathfrak{sl}(2, \mathbb{R})$, but by (a) and (b) there is only one such in $\mathfrak{g}$, namely, $\mathfrak{b}$. It follows that $\mathfrak{b} \subseteq H(W)$ which implies $G' \subseteq G$. If also (d*) holds then $U \in (\mathrm{int}\, W) \cap H(W)$; hence $H(W)$ has inner points and therefore $\mathfrak{g} = H(W)$, which implies $G = S$. It will therefore be our task to show that $\mathfrak{b} \subseteq \mathfrak{k} + W$.

(C) Replacing S by S^{-1} and U by a positive scalar multiple, if necessary, we may assume that U is conjugate (under $\mathrm{Ad}(G)$) to $\begin{pmatrix} 0 & 1 \\ -1 & 0 \end{pmatrix}$. It is therefore no loss of generality to assume

$$U = \begin{pmatrix} 0 & 1 \\ -1 & 0 \end{pmatrix}.$$

Note that $\mathbb{Z}\pi \cdot U$ is the center of $\mathfrak{b}$. (Figure 5 in II.1.7(i) may help to visualize the situation.) Also, $A \cap B$ must be a cyclic subgroup of the center of B. Hence either $A \cap B = \{1\}$ or there is a nonzero integer m such that $A \cap B = \exp m\pi \mathbb{Z} \cdot U \in A \cap B$. By Lemma 4.3 $A \cap B = \{1\}$ does not occur.

(D) Assume $A \cap B = \exp m\pi \mathbb{Z} \cdot U$. Then there is a unique $Y \in \mathfrak{a}$ such that $z := \exp m\pi \cdot U = \exp Y$. Since W and therefore also $W \cap \mathfrak{b}$ is a Lie semialgebra and since $U \in \text{int } W$ we know that S contains the Lie semigroup Σ^+ (see [27], p.433) generated by the wedge $\mathcal{K}$ containing U and satisfying $\mathcal{D}^- \cup \mathcal{D}^0 = \mathcal{K} \cup -\mathcal{K}$. This follows e.g. from [27], p. 432, Theorem V.4.45. (Figure 5 in II.1.7(i) is helpful.) In particular, because of $z = \exp m\pi \cdot U$, for each $X \in \mathcal{D}^+$ we have $z \exp \mathbb{R} \cdot X \subseteq \Sigma^+ \subseteq S$. By V.1.3(v) condition $(*)$ of Lemma 4.2 is satisfied for every element in $\mathcal{D}^+$, e.g. for $n \cdot X$ for $n \in \mathbb{N}$. So we conclude $n \cdot X + Y \in \mathfrak{k} + W$, i.e., $X \in \mathfrak{k} + W - \frac{1}{n} \cdot Y$ for any $n \in \mathbb{N}$. Thus $X \in \mathfrak{k} + W$. Since $X \in \mathcal{D}^+$ was arbitrary, we see that $\mathcal{D}^+ \subseteq \mathfrak{k} + W$, and hence $\mathfrak{b} = \mathcal{D}^+ + \mathcal{D}^+ \subseteq \mathfrak{k} + W$. The proposition is proved. $\square$

The case $\mathfrak{b} \cong \mathfrak{mot}$

4.5. PROPOSITION. *Let G be a connected Lie group whose Lie algebra $\mathfrak{g} = \mathcal{L}(G)$ is the ideal direct sum $\mathfrak{g} = \mathfrak{a} \oplus \mathfrak{b}$ and let S be a subsemigroup of G with Lie wedge $W = L(S)$. Write A, B for the analytic subgroups associated with $\mathfrak{a}$, $\mathfrak{b}$ and suppose the following hypotheses:*
  (a) *$\mathfrak{a}$ is abelian.*
  (b) *$\mathfrak{b} \cong \mathfrak{mot}$ and $B \cong \widetilde{\text{Mot}}$.*
  (c) *$S = \overline{\exp W}$, i.e., S is closed and weakly exponential.*
  (d) *The interior of W contains a compact nonzero element U of $\mathfrak{b}$.*

Then $G' \subseteq S$. In particular, the semigroup in G_1 corresponding to the half-space semigroup of $\widetilde{\text{Mot}}$ containing the element corresponding to U is contained in S.

Proof. By Proposition 2.11 the wedge W is a trivial Lie semialgebra. Thus $\mathfrak{g}' = [\mathfrak{g}, \mathfrak{g}] \subseteq W$. Hence $G' \subseteq S$.

We identify $\mathfrak{b}$ with $\mathfrak{mot}$ and B with $\widetilde{\text{Mot}}$. Then up to conjugation, $U = (0, r) \in \mathbb{C} \times \mathbb{R}$ with $0 \neq r \in \mathbb{R}$ in the notation of V.2.1. Now $W \cap \mathfrak{b}$ contains $\mathfrak{b}' = \mathbb{C} \times \{0\}$ and $(0, r)$ and thus the half-space $\mathbb{C} \times r\mathbb{R}^+$. $\square$

Example V.3.4 shows that 4-dimensional examples exist where $W \neq \mathfrak{g}$ and S is exponential.

The case $\mathfrak{b} \cong \mathfrak{osc}$

4.6. PROPOSITION. *Let G be a connected Lie group and let S be a weakly exponential subsemigroup of G with Lie wedge $W = \mathcal{L}(S)$. Suppose that the following conditions are satisfied:*
  (a) *The Lie algebra $\mathfrak{g} = L(G)$ may be identified with the ideal direct sum $\mathfrak{g} = \mathfrak{a} \oplus \mathfrak{b}$ with $\mathfrak{a}$ abelian and $\mathfrak{b} \cong \mathfrak{osc}$. We write $A = \exp \mathfrak{a}$, $B = \langle \exp \mathfrak{b} \rangle$ for the corresponding analytic subgroups.*
  (b) *G does not contain a non-degenerate central torus subgroup.*
  (c) *The interior of W is nonempty.*

Then W is a trivial semialgebra, i.e. $[\mathfrak{g}, \mathfrak{g}] \subseteq W$.

Proof. Since the regular elements of $\mathfrak{g}$ are dense in $\mathfrak{g}$ hypothesis (c) implies that we can find a regular element $Y \in \operatorname{int} W$. If $Y \in \operatorname{comp}_G(\mathfrak{g})$ then $Y \in H(W) \cap \operatorname{int} W$ and therefore $\mathfrak{g} = W$, so that the assertion becomes trivial. Thus we henceforth assume that $Y \notin \operatorname{comp}_G(\mathfrak{g})$. By Weil's Lemma this means that $\exp \mathbb{R} \cdot Y$ is closed and topologically isomorphic with $\mathbb{R}$.

Now using Lemma 2.2 we construct an ideal $\mathfrak{b}_1$ of $\mathfrak{g}$, isomorphic with $\mathfrak{b}$ and containing both Y and $\mathfrak{b}'$, such that $\mathfrak{g} = \mathfrak{a} \oplus \mathfrak{b}_1$. Our assumptions remain intact with $\mathfrak{b}_1$ in place of $\mathfrak{b}$, therefore we henceforth suppose that $\mathfrak{b} = \mathfrak{b}_1$, that is, $Y \in \mathfrak{b} \cap \operatorname{int} W$. The Lie group B_{Lie} has slim center (since $\mathfrak{b} \cong \mathfrak{osc}$, as observed in 3.6), and G has no nondegenerate central torus groups by (b). Thus the hypotheses of Proposition 3.7 are satisfied and we conclude that the analytic subgroup $B = \langle \exp \mathfrak{b} \rangle$ is closed in G. Now we observe that B has no nondegenerate central torus subgroups (since G has none) and that Y is a regular element in $\mathfrak{b}$ such that $\exp \mathbb{R} \cdot Y$ is topologically isomorphic with $\mathbb{R}$. Hence $B \cong \operatorname{Osc}$.

From Proposition 2.11 we know that W is an invariant wedge. Write $\mathfrak{h}_\mathfrak{b}$ for the Cartan subalgebra of $\mathfrak{b}$ generated by Y. Then $\mathfrak{h} = \mathfrak{a} \oplus \mathfrak{h}_\mathfrak{b}$ is the Cartan subalgebra of $\mathfrak{g}$ generated by Y. Since $\mathfrak{h}$ is a maximal rank subalgebra meeting the interior of W we deduce from Theorem II.4.2(ii) that

$$(*) \qquad \overline{\exp(\mathfrak{h} \cap W)} = (\exp \mathfrak{h}) \cap S.$$

It is well known that every pair $(\mathfrak{osc}, W_i)$, with W_i an invariant cone in $\mathfrak{osc}$, is canonically isomorphic with the standard Lorentzian pair of dimension 4 (cf.[27], II.3.20, p.120). Thus there is an isomorphism $\varphi: \mathfrak{osc} \to \mathfrak{b}$ such that $\mathfrak{b} \cap W$ meets $\mathfrak{h}_\mathfrak{b}$ in the cone $\varphi(-\mathbb{R}^+ \times \{0\} \times \mathbb{R}^+)$.

The universal covering group of G can be written $\mathbb{R}^n \times \operatorname{Osc}$ so that the covering homomorphism $\kappa: \mathbb{R}^n \times \operatorname{Osc} \to G$ maps $\mathbb{R}^n \times \{0\}$ to A and $\{0\} \times \operatorname{Osc}$ to B. We define $\Delta = \ker \kappa \cong A \cap B$. There are two cases: (A) $A \cap B \neq 1$, and (B) $A \cap B = \{1\}$. Then

$$\ker \kappa = \begin{cases} \{(-nd, 0, 0, \pi k n) \in \mathbb{R}^n \times \operatorname{Osc} : n \in \mathbb{Z}\} & \text{in Case (A)}, \\ \{(0, 0, 0, 0)\} & \text{in Case (B)}, \end{cases}$$

for a suitable $d \in \mathbb{R}^n$. Now we may write $G = \frac{\mathbb{R}^n \times \operatorname{Osc}}{\Delta}$. Then

$$(*) \qquad (\exp \mathfrak{h}) \cap S = \overline{\exp(\mathfrak{h} \cap W)} = \overline{(\mathbb{R}^n \times (-\mathbb{R}^+ \times \{0\} \times \mathbb{R}^+))\Delta/\Delta}$$

From Lemma V.4.4 we know that $S \cap \exp \mathfrak{h}$ contains the element $(0, 1, 0, \pi)\Delta \in G$. But this element is not contained in $\overline{(\mathbb{R}^n \times (-\mathbb{R}^+ \times \{0\} \times \mathbb{R}^+))\Delta/\Delta}$. This contradicts $(*)$ which completes the proof. $\square$

Example V.3.4 shows that 5-dimensional examples exist where $W \neq \mathfrak{g}$ and S is exponential.

The case $\mathfrak{b} \cong \mathfrak{extmot}, \mathfrak{extosc}$

The following lemma deals with the case where $\mathfrak{b}$ is isomorphic with either $\mathfrak{extmot}$ or $\mathfrak{extosc}$ or the subalgebras $\mathfrak{mot}(\lambda)$ of $\mathfrak{extmot}$ and $\mathfrak{osc}(\lambda)$ of $\mathfrak{extosc}$ which we recall from 1.2:

(†) $\mathrm{span}\{\lambda \cdot u + v, x, y, z\}$ with $\mu \neq 0$.

4.7. LEMMA. *Let G be a Lie group whose Lie algebra $\mathfrak{g}$ can be decomposed into an ideal direct sum $\mathfrak{g} = \mathfrak{a} \oplus \mathfrak{b}$, where $\mathfrak{a}$ is a central subalgebra of $\mathfrak{g}$ and $\mathfrak{b}$ is a subalgebra isomorphic to $\mathfrak{extmot}, \mathfrak{extosc}, \mathfrak{mot}(\lambda),$ or $\mathfrak{osc}(\lambda)$ with $\lambda \neq 0$. Suppose that S is a weakly exponential subsemigroup of G such that the tangent wedge $W = L(S)$ has interior points in $\mathfrak{g}$.*

Then the tangent wedge W of S is a trivial semialgebra, i.e., it contains the ideal $\mathfrak{g}' = \mathfrak{n}$ spanned by x, y and z.

Proof. We note first that $\mathfrak{n} = \mathfrak{g}'$ unless $\mathfrak{g}$ is abelian. We assume now that $\mathfrak{g}$ is not abelian.

From Propositon 2.11 we know that W is a semialgebra.

It remains to show here that $\mathfrak{n} \subseteq W$. Let $\mathfrak{j}$ denote the largest ideal of $\mathfrak{g}$ contained in W. Then $W/\mathfrak{j}$ is a reduced semialgebra in $\mathfrak{g}/\mathfrak{j}$. Let $w \in \mathrm{int}\, W$. Then w is of the form $a + \alpha \cdot u + \beta \cdot v + \xi \cdot x + \eta \cdot y + \zeta \cdot z$ with $a \in \mathfrak{a}$, and $\alpha, \beta, \xi, \eta, \zeta \in \mathbb{R}$ so that in the case of an algebra of type (†) we have $\alpha = t\lambda$ and $\beta = t$ with $t \in \mathbb{R}$. Since w is taken in the interior of W and $\lambda \neq 0$ we may assume that $\alpha \neq 0 \neq \beta$ in all cases. Thus $\alpha + i\beta \notin \mathbb{R} \cup i\mathbb{R}$. By Proposition II.5.28 of [27] (p.148) we have $\mathrm{Spec}_{\mathfrak{g}/\mathfrak{j}}(\mathrm{ad}(w+\mathfrak{j})) \subseteq \mathbb{R} \cup i\mathbb{R}$. If $\mathfrak{j}$ does not contain $\mathfrak{n}$, then from 1.1(vi) we know that $\mathrm{ad}(w+\mathfrak{j})\big|\frac{\mathfrak{n}+\mathfrak{j}}{\mathfrak{j}} : \frac{\mathfrak{n}+\mathfrak{j}}{\mathfrak{j}} \to \frac{\mathfrak{n}+\mathfrak{j}}{\mathfrak{j}}$ induces on $\mathfrak{n}/\mathbb{R} \cdot z \cong \mathbb{R}^2$ a linear map φ which on the basis vectors $x' = x + \mathbb{R} \cdot z$ and $y' = y + \mathbb{R} \cdot z$ operates as follows:

$$\varphi(x') = \alpha \cdot x' + \beta \cdot y',$$
$$\varphi(y') = -\beta \cdot x' + \alpha \cdot y'.$$

Therefore $\alpha + i\beta \in \mathrm{Spec}\,\varphi$. But $\mathrm{Spec}\,\varphi \subseteq \mathrm{Spec}_{\mathfrak{b}/\mathfrak{j}}(\mathrm{ad}(w+\mathfrak{j}))$. Thus $\alpha + i\beta \in \mathbb{R} \cup i\mathbb{R}$. This contradiction proves the claim. □

Proof of the Testing Theorem 2.5. Because of Lemma 4.7 we are left with the cases where $\mathfrak{b}$ is isomorphic with one of the algebras $\mathfrak{so}(3), \mathfrak{sl}(2, \mathbb{R}), \mathfrak{mot}, \mathfrak{osc}$.

Step 1. By the Intrinsic Embedding Theorem II.5.8 we may assume that $\mathrm{int}(W) \neq \emptyset$.

Step 2. We may assume that $H(W)$ is ideal free: For let N be the largest normal subgroup of G contained in S. Then we consider G/N and S/N. By Proposition IV.4.2 and Lemma 2.6 these quotient objects satisfy again the hypotheses. If we can show that $(\mathfrak{g}/\mathfrak{n})' \subseteq W/\mathfrak{n}$, then $\mathfrak{g}' \subseteq W$ will follow.

By Definition 2.4(b) and Lemma 2.2 we may assume that in addition the following hypothesis is satisfied:

(‡) $\mathrm{int}_{\mathfrak{b}}(\mathfrak{b} \cap W)$ contains a regular compact element $U \neq 0$ of $\mathfrak{b}$.

Step 3. If $\overline{\exp \mathbb{R}\cdot U}$ is compact, then $U \in H(W)$. By (‡), $\in \text{int}_{\mathfrak{b}}(\mathfrak{b} \cap W)$; hence $\mathfrak{g}' \subseteq \mathfrak{b} \subseteq H(W)$, establishing the assertion. In particular, this takes care of the case that $\mathfrak{b} \cong \mathfrak{so}(3)$.

So we now assume that $\exp \mathbb{R}\cdot U$ is algebraically and topologically isomorphic with $\mathbb{R}$.

Step 4. Let C be the maximal torus of the center of G. We have $SC = \overline{\exp(W + \mathcal{L}(C))}$ by Proposition 4.1. Then G/C does not contain a nondegenerate central torus. Then by Proposition 3.7, AC/C and BC/C are closed where $A = \exp \mathfrak{a}$ and $B = \langle \exp \mathfrak{b} \rangle$. Since $(\exp \mathbb{R}\cdot U)C/C \cong \exp \mathbb{R}\cdot U \cong \mathbb{R}$ we know that BC/C is simply connected. Now $\mathcal{L}(BC/C) \cong \mathfrak{sl}(2,\mathbb{R})$, $\mathfrak{mot}$ or $\mathfrak{osc}$. Then Proposition 4.4, 4.5, or 4.6 applies, respectively, and shows that $\mathfrak{g}' \subseteq H(W) + \mathcal{L}(C)$. Since $\mathcal{L}(C)$ is central, $\mathfrak{g}'' = [\mathfrak{g}', \mathfrak{g}'] \subseteq [H(W) + \mathcal{L}(C), H(W) + \mathcal{L}(C)] \subseteq H(W)' \subseteq H(W)$. But $H(W)$ is ideal free. Hence $\mathfrak{g}'' = \{0\}$. This rules out the possibilities that $\mathfrak{b} \cong \mathfrak{sl}(2,\mathbb{R})$, $\mathfrak{osc}$. Thus $\mathfrak{b} \cong \mathfrak{mot}$ and $C \subseteq A$. Since $\exp \mathbb{R}\cdot U \cong \mathbb{R}$ then $B \cong \widetilde{\text{Mot}}$ and B is closed. Then 4.5 proves the claim.

Now we may assume that G does not contain any nondegenerate central torus. Then B is closed by 3.7 and is simply connected as $\exp \mathbb{R}\cdot U \cong \mathbb{R}$.

Step 5. If $\mathfrak{b} \cong \mathfrak{sl}(2,\mathbb{R})$, $\mathfrak{mot}$, respectively, $\mathfrak{osc}$ then 4.4, 4.5, respectively, 4.6 proves the assertion. □

4.8. LEMMA. *If, under the general hypotheses of our Main Lemma 2.5, the algebra $\mathfrak{g}$ is reductive and a compact element of $\mathfrak{g}' = \mathfrak{b}$ is contained in* int W, *then $S = G$.*

Proof. Under the additional hypothesis $H(W)$ contains inner points of $\mathfrak{g}$ and thus agrees with $\mathfrak{g}$. □

In Definition 2.4 we formulated the following condition for test wedges:
(b) either $\mathfrak{g}$ is solvable or there is a nonzero compact element of the Levi complement $\mathfrak{b}$ which lies in $\text{int}_{\mathfrak{b}}(\mathfrak{b} \cap W)$.

This condition is in fact indispensable for the conclusion of Theorem 2.5. This can be seen for compact Lie algebras $\mathfrak{g}$ from Proposition V.5.1 and its Corollary V.5.2. In the same vein V.1.6 shows that every connected Lie group with Lie algebra $\mathfrak{sl}(2,\mathbb{R})$ contains proper exponential closed subsemigroups, among them also semigroups with trivial groups of units. In these cases the wedge W does not contain any nonzero compact element whatsoever, so condition (b) is not satisfied.

CHAPTER 7

GROUPS SUPPORTING REDUCED WEAKLY EXPONENTIAL SEMIGROUPS

1. 'Occam's Razor' and Reduced Weakly Exponential Subsemigroups

This section deals with an effective reduction technique, needed in subsequent sections.

1.1. LEMMA. *Let W be a wedge in a Lie algebra $\mathfrak{g}$ and let $\mathfrak{i}$ denote the largest ideal of $\mathfrak{g}$ contained in W. Then for any $x \in \operatorname{int} W$ there is a neighborhood U_x of 0 in $\mathfrak{g}$ such that for every symmetric neighborhood U of 0 contained in U_x the wedge*
$$W_U := \bigcap_{u \in U} e^{\operatorname{ad} u} W \subseteq W$$
has the following properties:
  (i) $x \in \operatorname{int} W_U$, *and*
  (ii) $H(W_U) = \mathfrak{i}$.

Proof. The proof of Lemma II.4.5 in [27] shows the existence of U_x such that for every symmetric 0-neighborhood $U \subseteq U_x$ the wedge W_U has an ideal as its edge $H(W_U)$, so that $H(W_U) \subseteq \mathfrak{i}$. However, $e^{\operatorname{ad} -u}\mathfrak{i} \subseteq \mathfrak{i}$ for all $u \in U = -U$ and thus $\mathfrak{i} \subseteq W_U$. Hence $\mathfrak{i} \subseteq H(W_U)$. □

Note that we do not know in general whether the closed semigroup
$$S_U = \bigcap_{u \in U} (\exp u) S (\exp -u)$$
must be exponential [weakly exponential] if S is exponential [weakly exponential].

Let us now introduce 'Occam's Razor.' It is, properly speaking, not a proposition on wedges and Lie algebras but a proposition on propositions about wedges and Lie algebras, an assertion of the form: if something can be proved for *all* subobjects of a given type then actually more is true for these subobjects.

1.2. PROPOSITION. *Let G be a connected Lie group, $\mathfrak{g}$ its Lie algebra. Suppose that $\mathcal{C}$ is a family of closed subsemigroups of G which is closed under the application of inner automorphisms of G. Then for every given point $h \in \mathfrak{g}$ and any given vector subspace $V \subseteq \mathfrak{g}$ the following conditions (a), (b) are equivalent:*
  (a) *For every subsemigroup $S \in \mathcal{C}$ the associated Lie wedge $W = \mathfrak{L}(S)$ contains V whenever it contains h in its interior.*
  (b) *If $S \in \mathcal{C}$ and $\mathrm{i}(W)$ is the maximal $\mathfrak{g}$-ideal contained in the Lie wedge $W := \mathfrak{L}(S)$, then $\mathrm{i}(W)$ contains V whenever $\mathfrak{L}(S)$ contains h in its interior.*

Proof. The implication (b)⇒(a) is trivial. Suppose now that (a) holds. Then we apply the above Lemma 1.1 with $x = h$ and $\mathrm{i} = \mathrm{i}(W)$, finding a neighborhood U_h of 0 in $\mathfrak{g}$ such that for every symmetric 0-neighborhood U with $U \subseteq U_h$ the wedge

$$W_U = \bigcap_{u \in U} e^{\mathrm{ad}\, u} W \subseteq W$$

contains h in its interior and has $\mathrm{i}(W)$ as its edge, $h \in \mathrm{int}\, W_U$ and $H(W_U) = \mathrm{i}(W)$.

Obviously, for every $u \in U$ the wedge $e^{\mathrm{ad}\, u} W$ is the Lie wedge of the subsemigroup $(\exp u) S (\exp -u) \in \mathcal{C}$. Thus condition (a) implies that $V \subseteq e^{\mathrm{ad}\, u} W$, for every $u \in U$. Since V is a vector subspace of $\mathfrak{g}$ this means that $V \subseteq H(e^{\mathrm{ad}\, u} W)$, for all u, so $V \subseteq \mathrm{i}$. □

It is readily seen that the family of all weakly exponential closed subsemigroups and the family of all exponential closed subsemigroups of G are closed under the application of inner automorphisms.

1.3. THEOREM. ('Occam's Razor') *Let G be a connected Lie group, $\mathfrak{g}$ its Lie algebra and let V be a vector subspace of $\mathfrak{g}$. Suppose that $h \in \mathrm{int}(\mathfrak{L}(S))$ for a reduced weakly exponential subsemigroup S of G.*

Suppose that for every $g \in G$ one has

(Occ) $\qquad \mathrm{Ad}(g)h \in \mathrm{int}\, \mathfrak{L}(S) \Rightarrow \mathrm{Ad}(g)V \subseteq \mathfrak{L}(S).$

Then $V = 0$.

Proof. By (Occ) we have $V \subseteq \mathfrak{L}(S)$. Let i be the largest ideal of $\mathfrak{g}$ contained in $\mathfrak{L}(S)$. Then by Proposition 1.2, applied to the class $\mathcal{C}$ of conjugates of S in G, we see that $V \subseteq \mathrm{i}$. By the definition of reduced weakly exponential subsemigroups (IV.4.7), $\mathrm{i} = \{0\}$. Hence $V = \{0\}$. □

Our first application of Occam's Razor is to prove that the Cartan subalgebras of a Lie group containing a reduced weakly exponential semigroup must be abelian.

1.4. MAIN LEMMA. *Suppose that a Lie group G contains a reduced weakly exponential subsemigroup S. Then all Cartan subalgebras of $\mathfrak{g}$ are abelian.*

Proof. Let h be a regular element of $\mathfrak{g}$ in the interior of $\mathfrak{L}(S)$ and let $\mathfrak{h}$ denote the Cartan subalgebra generated by h. Define V to be the commutator subalgebra $[\mathfrak{h}, \mathfrak{h}]$ of $\mathfrak{h}$. Suppose now that $g \in G$ and that $\mathrm{Ad}(g)h \in \mathrm{int}\,\mathfrak{L}(S)$. By Proposition IV.6.5, $\mathrm{Ad}(g)V$ must be contained in $\mathfrak{L}(S)$, so condition (Occ) of Theorem 1.3 is satisfied. Then $[\mathfrak{h}, \mathfrak{h}] = V = \{0\}$, that is, $\mathfrak{h}$ is abelian. Thus all Cartan subalgebras of $\mathfrak{g}$ are abelian (since their complexifications are the Cartan subalgebras of the complexification of $\mathfrak{g}$, which are all conjugate). □

2. The Spectrum of ad$W$ is Contained in $\mathbb{R} \cup i\mathbb{R}$

One of the most important features of reduced semialgebras is that for every point x in a reduced semialgebra the spectrum of $\mathrm{ad}\,x$ is contained in $\mathbb{R} \cup i\mathbb{R}$ (cf. [27], II.5.28, p.148). We shall show in this section that the Lie wedges of reduced weakly exponential subsemigroups also have this property. This requires a device allowing the characterisation of roots with respect to $\mathfrak{h}$ by their values at a single point.

2.1. DEFINITION. Let h be a regular element in a Lie algebra $\mathfrak{g}$, and denote with $\mathfrak{h}$ the Cartan algebra generated by h. Then h is called *strictly regular* if different roots of the Cartan algebra $\mathfrak{h}$ take on different values at h. □

(For this definition cf. SAN MARTIN [78] and EGGERT [18]. A forerunner of this concept was used by JURDEVIC and KUPKA in [50]; see pp. 166 ff.)

2.2. PROPOSITION. (EGGERT, [18])
  (i) *The set* $\mathrm{sreg}(\mathfrak{g})$ *of strictly regular elements is open and dense in* $\mathfrak{g}$.
  (ii) $\mathrm{sreg}(\mathfrak{g})$ *is invariant under all automorphisms of the Lie algebra $\mathfrak{g}$. It is also invariant under non-zero scalar multiplication.*
  (iii) *If h is a regular element and $\mathfrak{h}$ the Cartan algebra generated by h then each eigenvalue μ of $\mathrm{ad}\,h$ can be written as $\mu = \lambda(h)$, where λ is a root of $\mathfrak{h}$. If, in addition, h is strictly regular then λ is uniquely determined. Thus if h is strictly regular then every weight space with respect to $\mathrm{ad}\,h$ is a root space with respect to $\mathfrak{h}$.*

Proof. See Theorem 3.18 of EGGERT's thesis [18], p.52. □

The proof of this proposition is not complicated and the reader could fill in without too much effort the necessary details.

2.3. NOTATION. *In the following we shall view every real Lie algebra $\mathfrak{g}$ as a subalgebra of its complexification $\mathfrak{g}_\mathbb{C}$. Thus for any $z = x + i \cdot y \in \mathfrak{g}_\mathbb{C}$ with $x, y \in \mathfrak{g}$ the conjugate $\bar{z} = x - i \cdot y$ is well defined.*

If $\mathfrak{h}$ is a Cartan subalgebra of $\mathfrak{g}$ then the roots of $\mathfrak{g}_\mathbb{C}$ with respect to $\mathfrak{h}_\mathbb{C}$ are completely determined by their values on $\mathfrak{h}$, so we may define the complex conjugate $\bar{\lambda}$ of a root λ by the formula

$$\bar{\lambda}(x) = \overline{\lambda(x)} \quad \text{for all } x \in \mathfrak{h}.$$

Obviously, if $z = x + i \cdot y \in \mathfrak{h}_\mathbb{C}$ for some $x, y \in \mathfrak{h}$ then $\bar{\lambda}(z) = \overline{\lambda(\bar{z})} = \overline{\lambda(x - i \cdot y)}$. □

2.4. LEMMA. *Let h be a strictly regular element in a real Lie algebra $\mathfrak{g}$, and write $\mathfrak{h}$ for the Cartan algebra generated by h. Consider the associated root space decomposition of the complexification $\mathfrak{g}_\mathbb{C}$ with respect to $\mathfrak{h}_\mathbb{C}$, and let λ be a root of $\mathfrak{h}_\mathbb{C}$.*

Then $\lambda(\mathfrak{h}) \subseteq \mathbb{R}$ if and only if $\lambda(h) \in \mathbb{R}$.

Proof. For any eigenvalue c of $\operatorname{ad} h$ let $\mathfrak{g}_\mathbb{C}^c$ denote the associated weight space. Then $\mathfrak{g}_\mathbb{C}^\lambda = \mathfrak{g}_\mathbb{C}^{\lambda(h)}$ and $\mathfrak{g}_\mathbb{C}^{\bar{\lambda}} = \mathfrak{g}_\mathbb{C}^{\bar{\lambda}(h)}$. If $\lambda(h)$ is real then $\mathfrak{g}_\mathbb{C}^{\bar{\lambda}} = \mathfrak{g}_\mathbb{C}^{\bar{\lambda}(h)} = \mathfrak{g}_\mathbb{C}^{\lambda(h)} = \mathfrak{g}_\mathbb{C}^\lambda$, and hence $\lambda = \bar{\lambda}$. The converse is obvious. □

2.5. EXERCISE. *(i) Use 2.4(i) to show that if $\operatorname{reg} \mathfrak{g} \cap W$ lies in a single conjugacy class (e.g. if $\mathfrak{g}$ is solvable) and h is a strictly regular element in W such that $\operatorname{spec}(\operatorname{ad} h) \subseteq \mathbb{R}$ then $\operatorname{spec}(\operatorname{ad} w) \subseteq \mathbb{R}$ for all $w \in W$.*

(ii) Show that if $\lambda(h) \in i\mathbb{R}$, then either the root spaces $\mathfrak{g}_\mathbb{C}^\lambda$ and $\mathfrak{g}_\mathbb{C}^{\bar{\lambda}}$ commute, $[\mathfrak{g}_\mathbb{C}^\lambda, \mathfrak{g}_\mathbb{C}^{\bar{\lambda}}] = 0$, or all values of λ on $\mathfrak{h}$ are purely imaginary.

(Hint: The assumption implies that $[\mathfrak{g}_\mathbb{C}^\lambda, \mathfrak{g}_\mathbb{C}^{\bar{\lambda}}] = [\mathfrak{g}_\mathbb{C}^{\lambda(h)}, \mathfrak{g}_\mathbb{C}^{\overline{\lambda(h)}}] \subseteq \mathfrak{g}_\mathbb{C}^{\lambda(h) + \overline{\lambda(h)}} = \mathfrak{g}_\mathbb{C}^0 = \mathfrak{h}$. Now the assertion follows from $[\mathfrak{g}_\mathbb{C}^\lambda, \mathfrak{g}_\mathbb{C}^{\bar{\lambda}}] \subseteq \mathfrak{g}_\mathbb{C}^{\lambda + \bar{\lambda}} \cap \mathfrak{h}$.) □

In the following proposition we will encounter the algebras $\mathfrak{extmot}$ and $\mathfrak{extosc}$ which we defined in VI.1.2. These algebras are prototypes of Lie algebras with roots taking on values in $\mathbb{C} \setminus (\mathbb{R} \cup i\mathbb{R})$: Every Lie algebra with this property must contain a copy of one of them.

Recall from IV.4.7 that the Lie wedge of a reduced weakly exponential semigroup has inner points.

2.6. PROPOSITION. *Suppose that S is a reduced weakly exponential subsemigroup of a Lie group G and let W be its Lie wedge. Then for all points $w \in W$ we have*

$$\operatorname{Spec} \operatorname{ad} w \subseteq \mathbb{R} \cup i\mathbb{R}.$$

Proof. We fix an element $w \in W$. Since $\operatorname{Spec} \operatorname{ad} w$ depends continuously on w it suffices to show the above inclusion under the additional assumption that w is a strictly regular point of $\mathfrak{g}$ and lies in the interior of W.

Let $\mathfrak{h}$ be the Cartan algebra generated by w and suppose that Spec ad $w =$ $\{\mu(w) \mid \mu$ is a root w.r.t. $\mathfrak{h}\}$ is not contained in $\mathbb{R} \cup i\mathbb{R}$. Choose a root λ of $\mathfrak{h}$, say, $\lambda\colon \mathfrak{h} \to \mathbb{C},\ h \mapsto \alpha(h) + i\beta(h)$, where α and β are real functionals, such that $\lambda(w) \notin \mathbb{R} \cup i\mathbb{R}$ and such that the map $\mathfrak{h} \to \mathbb{C}, h \mapsto 3\alpha(h) + i\beta(h)$, is not a root of $\mathfrak{h}$.

Then there exist non-zero vectors $x, y \in \mathfrak{g}$ with

$$\operatorname{ad} h(x) = \alpha(h)\cdot x + \beta(h)\cdot y$$
$$\operatorname{ad} h(y) = -\beta(h)\cdot x + \alpha(h)\cdot y$$

for all $h \in \mathfrak{h}$. Note that $z := [x, y] \in \mathfrak{g}_{\mathbb{C}}^{2\alpha} \cap \mathfrak{g}$. By our choice of λ we know that $\mathfrak{g}_{\mathbb{C}}^{3\alpha+i\beta} = \{0\}$, so $[x, [x, y]] = [y, [y, x]] = 0$. Thus the elements x, y and z span a nonzero nilpotent subalgebra $\mathfrak{n}$ of $\mathfrak{g}$ which is normalized by $\mathfrak{h}$ by the choice of x and y. It follows that $\mathfrak{m} := \mathfrak{h} \oplus \mathfrak{n}$ is a maximal rank subalgebra of $\mathfrak{g}$ which meets the interior of W. Write M for the analytic subgroup of G corresponding to $\mathfrak{m}$, and recall that M is the normalizer of $\mathfrak{m}$, hence is closed. The center $\mathfrak{a}$ of $\mathfrak{m}$ is exactly the kernel of λ. Set $\alpha_0 = \alpha(w)$, $\beta_0 = \beta(w)$. Then $\lambda(w) \notin \mathbb{R} \cup i\mathbb{R}$ implies $\alpha_0 \neq 0 \neq \beta$. Define $\mathfrak{b} = \mathbb{R}\cdot w + \mathfrak{n}$. Then $\mathfrak{b} \in \mathcal{A}$ as defined in VI.1.2, and $\mathfrak{m} = \mathfrak{a} \oplus \mathfrak{b}$. Thus $\mathfrak{m}$ is a test algebra (VI.2.1) and M is a test group. By the Maximal Rank Theorem II.4.3 $S \cap M$ is weakly exponential, and $\mathcal{L}(S \cap M) = W \cap \mathfrak{m}$. Then by the Testing Theorem VI.2.5 we have $\mathfrak{n} \subseteq H(W)$.

Now we apply Occam's Razor 1.3 with $h = w$ and $V = \mathfrak{n}$ and conclude that $\mathfrak{n} = \{0\}$. This contradiction proves the proposition. $\square$

2.7. MAIN LEMMA. *Let S be a reduced weakly exponential subsemigroup of a Lie group G and suppose that $\mathfrak{h}$ is a Cartan subalgebra of $\mathfrak{g} = \mathcal{L}(G)$ which meets the interior of W. Then for every root λ with respect to $\mathfrak{h}_{\mathbb{C}}$ we either have $\lambda(\mathfrak{h}) \subseteq \mathbb{R}$ or $\lambda(\mathfrak{h}) \subseteq i\mathbb{R}$.*

Proof. The set $\{x \in \mathfrak{h} \mid \lambda(x) \not\subseteq \mathbb{R} \cup i\mathbb{R}\}$ is Zariski open, hence either void or dense in $\mathfrak{h}$. But by Proposition 2.6 it misses the nonvoid open subset $\mathfrak{h} \cap \operatorname{int} W$, so it cannot be dense in $\mathfrak{h}$, and therefore $\lambda(\mathfrak{h}) \subseteq \mathbb{R} \cup i\mathbb{R}$.

Suppose now that $\lambda(\mathfrak{h}) \not\subseteq i\mathbb{R}$, that is, $0 \neq \lambda(x) \in \mathbb{R}$ for some $x \in \mathfrak{h}$. Then $0 \neq \lambda(y) \in \mathbb{R}$ for all y in a neighborhood of x, and thus $0 \neq \lambda(h) \in \mathbb{R}$ for some strictly regular element $h \in \mathfrak{h}$. By Lemma 2.4 this implies that $\lambda(\mathfrak{h}) \subseteq \mathbb{R}$ and the proof is completed. $\square$

2.8. DEFINITION. *Let $\mathfrak{h}$ be a Cartan subalgebra of a Lie algebra $\mathfrak{g}$, and let λ be a root with respect $\mathfrak{h}_{\mathbb{C}}$. Then λ is called real-valued if $\lambda(\mathfrak{h}) \subseteq \mathbb{R}$, it is called purely imaginary-valued if $\lambda(\mathfrak{h}) \subseteq i\mathbb{R}$.* $\square$

Thus the assertion of our Main Lemma 2.7 could be expressed also as: If $\mathfrak{h}$ meets the interior of W then all roots with respect to $\mathfrak{h}_{\mathbb{C}}$ are either real-valued or purely imaginary-valued.

2.9. NOTATION. (i) We shall write $\mathfrak{g}^\lambda := \mathfrak{g}_\mathbb{C}^\lambda \cap \mathfrak{g}$ whenever λ is a real-valued root with respect to $\mathfrak{h}_\mathbb{C}$. Since no ambiguities are likely to occur we also address the vector space $\mathfrak{g}^\lambda$ as the *root space of the real-valued root* λ. The set of all real-valued roots will be denoted with Λ_r, and we write

$$\mathfrak{g}_{\Lambda_r} = \bigoplus_{\lambda \in \Lambda_r} \mathfrak{g}^\lambda.$$

(ii) If λ is a purely imaginary-valued root with respect to $\mathfrak{h}_\mathbb{C}$ then we write $\omega = -i\lambda$ for the *imaginary part* of λ and put $\mathfrak{g}^\omega = (\mathfrak{g}_\mathbb{C}^\lambda + \mathfrak{g}_\mathbb{C}^{-\lambda}) \cap \mathfrak{g}$. For a fixed strictly regular element $h \in \mathfrak{h}$ we define

$$\Omega_+ := \{\omega \in \mathfrak{h}^* \mid i\omega \text{ is a root with respect to } \mathfrak{h}_\mathbb{C} \text{ and } \omega(h) > 0\},$$
$$\mathfrak{g}^{\Omega_+} := \bigoplus_{\omega \in \Omega_+} \mathfrak{g}^\omega.$$

□

2.10. PROPOSITION. *Suppose that $\mathfrak{h}$ is a Cartan subalgebra of a Lie algebra $\mathfrak{g}$, such that all roots with respect to $\mathfrak{h}_\mathbb{C}$ are either real-valued or purely imaginary-valued. Then the following conclusions hold:*
  (i) $\mathfrak{g} = \mathfrak{h} \oplus \mathfrak{g}_{\Lambda_r} \oplus \mathfrak{g}^{\Omega_+}$.
  (ii) $[\mathfrak{g}_{\Lambda_r}, \mathfrak{g}^{\Omega_+}] = \{0\}$.
  (iii) $[\langle \mathfrak{g}_{\Lambda_r} \rangle, \langle \mathfrak{g}^{\Omega_+} \rangle] = \{0\}$.
  (iv) *The subalgebras $\langle \mathfrak{g}_{\Lambda_r} \rangle$ and $\langle \mathfrak{g}^{\Omega_+} \rangle$ generated by the vector subspaces $\mathfrak{g}_{\Lambda_r}$ and $\mathfrak{g}^{\Omega_+}$ are ideals in $\mathfrak{g}$.*
  (v) $\mathfrak{h} \cap \langle \mathfrak{g}_{\Lambda_r} \rangle \subseteq \bigcap_{\omega \in \Omega_+} \ker \omega$, *and* $\mathfrak{h} \cap \langle \mathfrak{g}^{\Omega_+} \rangle \subseteq \bigcap_{\lambda \in \Lambda_r} \ker \lambda$.

Proof. Assertion (i) follows directly from our hypotheses.

(ii) For any $\lambda \in \Lambda_r$ and $\omega \in \Omega_+$. Then $\lambda+i\omega$, $\lambda-i\omega \in \mathfrak{h}_\mathbb{C}^*$ fail to be roots by our hypothesis. But we have $[\mathfrak{g}^\lambda, \mathfrak{g}^\omega] = [\mathfrak{g}^\lambda, \mathfrak{g} \cap (\mathfrak{g}_\mathbb{C}^{i\omega} + \mathfrak{g}_\mathbb{C}^{-i\omega}+)] \subseteq \mathfrak{g}_\mathbb{C}^{\lambda+i\omega} + \mathfrak{g}_\mathbb{C}^{\lambda-i\omega} = \{0\}$.

Assertion (iii) follows from (ii) with the aid of the Jacobi identity; (iv) follows from (i) and (iii).

(v) Let $h \in \mathfrak{h} \cap \langle \mathfrak{g}_{\Lambda_r} \rangle$ and $\omega \in \Omega_+$. Let $x, y \in \mathfrak{g}^\omega$ be nonzero elements such that $[h, x] = \omega(h) \cdot y$ and $[h, y] = -\omega(h) \cdot x$. By (ii), $[h, x] \in [\mathfrak{g}_{\Lambda_r}, \mathfrak{g}^{\Omega_+}] = \{0\}$. Since $y \neq 0$ we have $\omega(h) = 0$. This proves the first assertion and the second is proved analogously. □

In the next section we shall further analyze the real-valued roots λ.

3. Root Spaces with Respect to Real-Valued Roots

In this section we transfer some handy results of EGGERT on Lie semialgebras to Lie wedges of reduced weakly exponential subsemigroups, these results are summarized in the following theorem. (Recall that a Lie subsemialgebra W of $\mathfrak{g}$ is said to be reduced in $\mathfrak{g}$ if its edge does not contain any non-zero ideals of $\mathfrak{g}$.)

3.1. THEOREM. (EGGERT) *Let W be a reduced semialgebra in a Lie algebra $\mathfrak{g}$, and suppose that $\mathfrak{h}$ is a Cartan subalgebra of $\mathfrak{g}$ which meets the interior of W. Let λ be a real-valued root. Then the following assertions hold:*
  (i) *the root space $\mathfrak{g}^\lambda$ is actually an eigenspace, that is, $[h, x] = \lambda(h)\cdot x$ for all $x \in \mathfrak{g}^\lambda$ and $h \in \mathfrak{h}$;*
  (ii) *Suppose that μ is a real-valued root with respect to $\mathfrak{h}$ such that $[\mathfrak{g}^\lambda, \mathfrak{g}^\mu] \neq \{0\}$. Then*
    (ii.a) $\mu = -\lambda$,
    (ii.b) *the vector subspaces $\mathfrak{g}^\lambda$, $\mathfrak{g}^{-\lambda}$ and $[\mathfrak{g}^\lambda, \mathfrak{g}^{-\lambda}]$ are one-dimensional, and*
    (ii.c) *their sum is a Lie subalgebra of $\mathfrak{g}$ which is isomorphic with $\mathfrak{sl}(2, \mathbb{R})$,*
    $$\mathfrak{g}^\lambda \oplus \mathfrak{g}^{-\lambda} \oplus [\mathfrak{g}^\lambda, \mathfrak{g}^{-\lambda}] \cong \mathfrak{sl}(2, \mathbb{R}).$$

Proof. This theorem is a combination of EGGERT's results Lemma 3.14 (assertion (i)), Lemma 3.10, Lemma 3.15, Theorem 3.19 and Theorem 3.20 (assertion (ii)) in [18] (p.49f). □

The material used here covers six pages in EGGERT's treatise; it is directly accessible to readers knowing the fundamental facts on Lie semialgebras discussed in [27].

Our principal goal is the following scholium parallel to 3.1, which we shall establish in several steps through the next four lemmas. In this proof we shall actually utilize the results in Theorem 3.1 above.

3.2. SCHOLIUM. *Consider the following two conditions:*
  (a) W *is a reduced subsemialgebra of $\mathfrak{g}$, and*
  (b) W *is the Lie wedge of a reduced weakly exponential subsemigroup of G.*
Then Theorem 3.1 remains true if we replace hypothesis (a) by hypothesis (b).
□

For technical convenience we first introduce a testing device designed especially for our present purposes.

3.3. LEMMA. *Let W be the tangent wedge of a weakly exponential subsemigroup S of a Lie group G and suppose that $\mathfrak{h}$ is a Cartan subalgebra of $\mathfrak{g} = L(G)$ which meets the interior of W. Furthermore, we let P be a set of real-valued roots with respect to $\mathfrak{h}$, write $\mathfrak{n}$ for the Lie subalgebra generated by the union of the subspaces $\mathfrak{g}^\rho$, $\rho \in P$, and define $\mathfrak{m} := \mathfrak{h} + \mathfrak{n}$. We suppose that $\mathfrak{n}$ is a nilpotent Lie algebra. Then the following assertions hold:*
  (i) $\mathfrak{m}$ *is a solvable maximal rank subalgebra of $\mathfrak{g}$ and $\mathrm{comp}(\mathfrak{m})$ is central in $\mathfrak{m}$. In particular, $M = \langle \exp \mathfrak{m} \rangle$ is a UMCS-group (IV.6.1).*

(ii) *The wedge* $\mathfrak{m} \cap W$ *is a semialgebra.*

Proof. (i) The root spaces $\mathfrak{g}^\rho$, $\rho \in P$, are invariant under the action of $\operatorname{ad}\mathfrak{h}$, hence the Lie algebra $\mathfrak{n} = \langle \mathfrak{g}^\rho \mid \rho \in P \rangle$ generated by them is an ideal of $\mathfrak{m} = \mathfrak{h} + \mathfrak{n}$. Since $\mathfrak{m}/\mathfrak{n} \cong \mathfrak{h}/(\mathfrak{h} \cap \mathfrak{n})$ is nilpotent it follows that $\mathfrak{m}$ is solvable.

If $x \in \operatorname{comp}(\mathfrak{m}) \cap \mathfrak{h}$ then the operator $\operatorname{ad}_{\mathfrak{m}} x$ is semisimple and $\operatorname{Spec}\operatorname{ad}_{\mathfrak{m}} x \subseteq \{\rho(x) \mid \rho \in P\}$, and has no non-zero eigenvalues. Hence it must be zero. Since every k with $\operatorname{ad} k$ semisimple lies in a Cartan subalgebra (cf., e.g., [5], chap VII, §2, n° 3, Proposition 10) and since all Cartan subalgebras of a solvable algebra are conjugate this means that $\operatorname{comp}(\mathfrak{m})$ is central in $\mathfrak{m}$.

The analytic subgroup $M = \langle \exp \mathfrak{m} \rangle$ of G is closed since $\mathfrak{m}$ is a maximal rank subalgebra. Also, M has a unique maximal compact subgroup which is central in M.

(ii) follows from Corollary IV.6.4 since $\mathfrak{m}$ meets the interior of W. □

3.4. NOTATION. Throughout the rest of this section we always suppose that W is the Lie wedge of a weakly exponential subsemigroup of G and that h is a strictly regular element of $\mathfrak{g} = \mathfrak{L}(G)$, contained in the interior of W. We consider real-valued roots with respect to the Cartan subalgebra $\mathfrak{h}$ generated by h. □

3.5. LEMMA. *Fix a strictly regular element* $h \in \mathfrak{h} \cap \operatorname{int}(W)$ *and a real-valued root* λ. *Then*
  (i) *the root space* $\mathfrak{g}^\lambda$ *is actually an eigenspace for* $\operatorname{ad} h$, *that is,* $[h, x] = \lambda(h) \cdot x$ *for all* $x \in \mathfrak{g}^\lambda$;
  (ii) $[\mathfrak{g}^\lambda, \mathfrak{g}^\mu] = \{0\}$ *whenever* μ *is a real-valued root such that* $\lambda(h)$ *and* $\mu(h)$ *have the same sign.*

Proof. Recall first that by Proposition 2.2(iii) the root spaces of $\mathfrak{h}$ are exactly the weight spaces of $\operatorname{ad} h$.

Let P be the set of all real-valued roots ρ such that $\lambda(h)$ and $\rho(h)$ have the same sign. Then the vector subspace $\mathfrak{n} := \bigoplus_{\rho \in P} \mathfrak{g}^\rho$ is a subalgebra of $\mathfrak{g}$. Moreover, $\mathfrak{n}$ is nilpotent (note that for $\rho, \rho' \in P$ we always have $[\mathfrak{g}^\rho, \mathfrak{g}^{\rho'}] \subseteq \mathfrak{g}^{\rho+\rho'}$ with $|(\rho + \rho')(h)| > \max(|\rho(h)|, |\rho'(h)|)$). By Lemma 3.3 the vector subspace $\mathfrak{m} := \mathfrak{h} \oplus \mathfrak{n}$ is a solvable maximal rank subalgebra of $\mathfrak{g}$ and $\mathfrak{m} \cap W$ is a semialgebra. Let $\mathfrak{j}$ be the maximal ideal of $\mathfrak{m}$ which is contained in $H(W)$.

(i) We have to show that the vector subspace $V = (\operatorname{ad} h - \lambda(h) \cdot \operatorname{id})\mathfrak{g}^\lambda$ of $\mathfrak{m}$ vanishes. But $W \cap \mathfrak{m}$ is a generating semialgebra in $\mathfrak{m}$, so Theorem 3.1(i) implies that V must lie in the kernel of the quotient map $\mathfrak{m} \to \mathfrak{m}/\mathfrak{j}$, that is, $V \subseteq \mathfrak{j}$, hence $V \subseteq H(W)$. Since this is true for any conjugate $\operatorname{Ad}(g)h$ of h with $\operatorname{Ad}(g)h \in \operatorname{int} W$, Occam's Razor 1.3 applies and we conclude that $V = \{0\}$.

(ii) By Theorem 3.1(ii) we know that $[\mathfrak{g}^\lambda, \mathfrak{g}^\mu] \subseteq \mathfrak{j} \subseteq H(W)$ for all $\lambda, \mu \in P$. Applying Occam's Razor 1.3, as in the proof of (i), we conclude that $[\mathfrak{g}^\lambda, \mathfrak{g}^\mu]$ must vanish. This establishes our claim. □

3.6. COROLLARY. *Let W be the tangent wedge of a reduced weakly exponential subsemigroup S in a Lie group G. Suppose that $\mathfrak{h}$ is a Cartan subalgebra of $\mathfrak{g}$ which meets the interior of W.*

If λ is a real-valued root on $\mathfrak{h}$ then $\mathfrak{g}^\lambda = \mathfrak{g}_\lambda$, i. e., the root space for λ is an eigenspace for all maps $\operatorname{ad} h'$ with $h' \in \mathfrak{h}$.

(This establishes assertion 3.1(i) for the weakly exponential case.)

Proof. By 3.5 the assertion holds for all strictly regular elements $h' \in \mathfrak{h} \cap \operatorname{int}(W)$. But since these elements span $\mathfrak{h}$ (they form a nonvoid open subset), the Corollary follows. □

3.7. LEMMA. *If λ, μ are real-valued roots with $\lambda \neq -\mu$ then $[\mathfrak{g}^\lambda, \mathfrak{g}^\mu] = \{0\}$.*

Proof. Let h be as in 3.5 and suppose that, to the contrary, there exists a pair of real-valued roots λ, μ with $\lambda \neq -\mu$ and $[\mathfrak{g}^\lambda, \mathfrak{g}^\mu] \neq \{0\}$. Then $\mu(h) \neq \lambda(h)$ since h is strictly regular. We know from Lemma 3.5 that $\mu(h)$ and $\lambda(h)$ have different signs, say, $\mu(h) < 0 < \lambda(h)$. We may assume that among such pairs ours renders the distance $\lambda(h) - \mu(h)$ minimal. Replacing W by $-W$ if $\lambda(h) < -\mu(h)$, we enforce that $\lambda(h) + \mu(h) > 0$, so that $\mu(h) < 0 < \mu(h) + \lambda(h) < \lambda(h)$.

We put $P = \{\lambda, \mu\}$ and claim that the associated $\mathfrak{n}$ is nilpotent, so that the Lemma 3.3 applies and yields that $W \cap \mathfrak{m}$ is a semialgebra.

By 3.5 we know already that the spaces $\mathfrak{g}^\lambda, \mathfrak{g}^\mu, [\mathfrak{g}^\lambda, \mathfrak{g}^\mu]$ are abelian subalgebras and that $[\mathfrak{g}^\lambda, [\mathfrak{g}^\lambda, \mathfrak{g}^\mu]] \subseteq [\mathfrak{g}^\lambda, \mathfrak{g}^{\lambda+\mu}] = \{0\}$, (note that $\lambda(h)$ and $(\lambda+\mu)(h)$ have the same sign).

Case (i): $\lambda + \mu \neq -\mu$. Then $[\mathfrak{g}^\mu, \mathfrak{g}^{\lambda+\mu}] \neq \{0\}$ would contradict the minimality of $\lambda(h) + \mu(h)$, so we must have $[\mathfrak{g}^\mu, [\mathfrak{g}^\lambda, \mathfrak{g}^\mu]] \subseteq [\mathfrak{g}^\mu, \mathfrak{g}^{\lambda+\mu}] = \{0\}$ and hence $\mathfrak{n} = \mathfrak{g}^\lambda + \mathfrak{g}^\mu + [\mathfrak{g}^\lambda, \mathfrak{g}^\mu]$. It follows that $[\mathfrak{g}^\lambda, \mathfrak{g}^\mu]$ is central in $\mathfrak{n}$ and hence $\mathfrak{n}$ is nilpotent.

Case (ii): $\lambda + \mu = -\mu$, so that $\lambda = -2\mu$. Then $\mathfrak{n}_1 := \mathfrak{g}^{-2\mu} + \mathfrak{g}^{-\mu} + \mathfrak{g}^\mu + [\mathfrak{g}^{-\mu}, \mathfrak{g}^\mu]$ is a subalgebra containing $\mathfrak{n}$; note that $\mathfrak{n}_1$ (and hence $\mathfrak{n}$) is nilpotent if $[\mathfrak{g}^{-\mu}, \mathfrak{g}^\mu]$ is central in $\mathfrak{n}_1$ (the subspaces $\mathfrak{g}^{-2\mu}$ and $\mathfrak{g}^{-\mu}$ commute). Since by 3.5 all root spaces to real-valued roots are eigenspaces, it suffices to show that $\mu([\mathfrak{g}^{-\mu}, \mathfrak{g}^\mu]) = \{0\}$. To see this, we observe that

$$[\mathfrak{g}^{-2\mu}, [\mathfrak{g}^{-\mu}, \mathfrak{g}^\mu]] = [[\mathfrak{g}^{-2\mu}, \mathfrak{g}^{-\mu}], \mathfrak{g}^\mu] + [\mathfrak{g}^{-\mu}, [\mathfrak{g}^{-2\mu}, \mathfrak{g}^\mu]]$$
$$= 0 + [\mathfrak{g}^{-\mu}, [\mathfrak{g}^{-2\mu}, \mathfrak{g}^\mu]] \subseteq [\mathfrak{g}^{-\mu}, \mathfrak{g}^{-\mu}]$$
$$= \{0\}.$$

Let $x \in [\mathfrak{g}^{-\mu}, \mathfrak{g}^\mu]$ and choose a non-zero eigenvector y of $\operatorname{ad} x$ in $\mathfrak{g}^{-2\mu}$. Then $0 = [x, y] = -2\mu(x) \cdot y$ and hence $\mu(x) = 0$. Thus $\mathfrak{n}$ is nilpotent.

Applying Lemma 3.3 we now see that $W \cap \mathfrak{m}$ is a semialgebra. As in the proof of (i) we conclude from 3.1(ii) that $H(W)$ must contain $[\mathfrak{g}^\lambda, \mathfrak{g}^\mu]$. Thus by Occam's Razor 1.3 $[\mathfrak{g}^\lambda, \mathfrak{g}^\mu] = \{0\}$. □

The following Lemma will finish the proof of the Scholium (see 3.2).

3.8. LEMMA. *Let λ and $-\lambda$ be real-valued roots such that $[\mathfrak{g}^\lambda, \mathfrak{g}^{-\lambda}] \neq \{0\}$. Then the vector subspaces $\mathfrak{g}^\lambda$, $\mathfrak{g}^{-\lambda}$ and $[\mathfrak{g}^\lambda, \mathfrak{g}^{-\lambda}]$ are one-dimensional, and $\mathfrak{g}^\lambda \oplus \mathfrak{g}^{-\lambda} \oplus [\mathfrak{g}^\lambda, \mathfrak{g}^{-\lambda}] \cong \mathfrak{sl}(2, \mathbb{R})$.*

Proof. (Cf. [18], p. 53). Suppose first that $\lambda([\mathfrak{g}^\lambda, \mathfrak{g}^{-\lambda}]) = \{0\}$. Then we define $P = \{\lambda, -\lambda\}$ and note that $\mathfrak{n} = \mathfrak{g}^\lambda + \mathfrak{g}^{-\lambda} + [\mathfrak{g}^\lambda, \mathfrak{g}^{-\lambda}]$ is nilpotent, since $[\mathfrak{g}^\lambda, \mathfrak{g}^{-\lambda}]$ is central in $\mathfrak{n}$ (because all real-valued root spaces are eigenspaces). Thus we conclude from Lemma 3.3 that $W \cap \mathfrak{m}$ is a semialgebra. Let $\mathfrak{j}$ be the largest ideal of $\mathfrak{m}$ which is contained in $H(W)$. Then we apply Theorem 3.1(ii) to the reduced semialgebra $(W \cap \mathfrak{m})/\mathfrak{j}$ in the quotient algebra $\mathfrak{m}/\mathfrak{j}$ and find that the vector space $[\mathfrak{g}^\lambda, \mathfrak{g}^{-\lambda}]$ must lie in the kernel of the quotient morphism $\mathfrak{m} \to \mathfrak{m}/\mathfrak{j}$, hence is contained in $\mathfrak{j}$ and therefore in $H(W)$. Thus, invoking Occam's Razor 1.3, we deduce $[\mathfrak{g}^\lambda, \mathfrak{g}^{-\lambda}] = \{0\}$, a contradiction.

It follows that there are points $x \in \mathfrak{g}^\lambda$ and $y \in \mathfrak{g}^{-\lambda}$ such that $\lambda([x, y]) \neq 0$. Then for any $x_1 \in \mathfrak{g}^\lambda$ we have $[x, x_1] = 0$ (by 3.5), hence

$$0 = [y, [x, x_1]] = [[y, x], x_1] + [x, [y, x_1]] = -\lambda([x, y]) \cdot x_1 - \lambda([x_1, y]) \cdot x,$$

and therefore x_1 is a real multiple of x. Thus $\mathfrak{g}^\lambda$ is one-dimensional. In the same way we prove that $\mathfrak{g}^{-\lambda}$ has dimension one; this also implies that $[\mathfrak{g}^\lambda, \mathfrak{g}^{-\lambda}] = \mathbb{R} \cdot [x, y]$ is one-dimensional.

By the classification of three-dimensional real Lie algebras this means that $\mathfrak{g}^\lambda + \mathfrak{g}^{-\lambda} + [\mathfrak{g}^\lambda, \mathfrak{g}^{-\lambda}]$ is isomorphic with $\mathfrak{sl}(2, \mathbb{R})$. □

4. Weakly Exponential Subsemigroups and Eggert Algebras

We have seen in the last sections that a Lie algebra containing the Lie wedge of a reduced weakly exponential subsemigroup satisfies a set of very stringent conditions: not only are the Cartan subalgebras abelian (by 1.4), but the assertions listed in 2.7 and 3.1 must be also valid. These properties are shared by Lie algebras containing reduced semialgebras, and their general significance were first recognized by ANSELM EGGERT in his work on the classification of semialgebras (cf. [18]). We therefore propose to call such Lie algebras *Eggert algebras*, in recognition of his contribution to the final classification of semialgebras. The present section is devoted to purely Lie algebra theoretic information on Eggert algebras.

4.1. DEFINITION. (I) Let $\mathfrak{h}$ be a Cartan subalgebra of a finite dimensional real Lie algebra $\mathfrak{g}$. Then the pair $(\mathfrak{g}, \mathfrak{h})$ is then called an *Eggert pair* if $\mathfrak{h}$ is abelian and $(\mathfrak{g}, \mathfrak{h})$ satisfies the following conditions:
  (i) Every root λ with respect to $\mathfrak{h}_\mathbb{C}$ is either real-valued, $\lambda(\mathfrak{h}) \subseteq \mathbb{R}$, or purely imaginary-valued, $\lambda(\mathfrak{h}) \subseteq i\mathbb{R}$.
  (ii) The root spaces with respect to real-valued roots λ are eigenspaces, $\mathfrak{g}^\lambda = \mathfrak{g}_\lambda$.
  (iii) If λ, μ are real-valued roots with $[\mathfrak{g}^\lambda, \mathfrak{g}^\mu] \neq \{0\}$ then
    (iii.a) $\mu = -\lambda$,

(iii.b) the vector subspaces $\mathfrak{g}^\lambda$, $\mathfrak{g}^{-\lambda}$ and $[\mathfrak{g}^\lambda, \mathfrak{g}^{-\lambda}]$ are one-dimensional, and
(iii.c) their sum is a Lie subalgebra of $\mathfrak{g}$ which is isomorphic with $\mathfrak{sl}(2, \mathbb{R})$,
$$\mathfrak{g}^\lambda \oplus \mathfrak{g}^{-\lambda} \oplus [\mathfrak{g}^\lambda, \mathfrak{g}^{-\lambda}] \cong \mathfrak{sl}(2, \mathbb{R}).$$
(II) A Lie algebra $\mathfrak{g}$ is called an *Eggert algebra* if it contains an abelian Cartan subalgebra $\mathfrak{h}$ such that $(\mathfrak{g}, \mathfrak{h})$ is an Eggert pair. □

4.2. THEOREM. *Let W be a Lie wedge in a Lie algebra $\mathfrak{g}$, and suppose that $\mathfrak{h}$ is a Cartan subalgebra of $\mathfrak{g}$ which meets the interior of W. Then $(\mathfrak{g}, \mathfrak{h})$ is an Eggert pair in each of the following cases:*
  (a) *W is a reduced subsemialgebra in $\mathfrak{g}$;*
  (b) *W is the Lie wedge of a reduced weakly exponential subsemigroup S in a Lie group G with Lie algebra $\mathfrak{g}$.*

Proof. In case (a) the assertion is well established through EGGERT's work; it is the combination of Theorem II.5.28 in [27] with EGGERT's results cited in 3.1. In case (b) it is a consequence of 2.7 and the Scholium 3.2. □

We decompose the set of all real-valued roots into those giving rise to a copy of $\mathfrak{sl}(2, \mathbb{R})$ and those which do not.

4.3. DEFINITION. *If λ is real-valued and $[\mathfrak{g}^\lambda, \mathfrak{g}^{-\lambda}] + \mathfrak{g}^\lambda + \mathfrak{g}^{-\lambda}$ is isomorphic to $\mathfrak{sl}(2, \mathbb{R})$ then λ is called a simple real root, otherwise a nonsimple real root.* □

Let us deal with the simple roots first. The following Lemma is a variant of EGGERT's Lemma 3.27.

4.4. LEMMA. *Let $(\mathfrak{g}, \mathfrak{h})$ be an Eggert pair and let λ be a simple real-valued root. Then*
  (i) *the vector subspace*
$$\mathfrak{s}_\lambda = \mathfrak{g}_\lambda \oplus \mathfrak{g}_{-\lambda} \oplus [\mathfrak{g}_\lambda, \mathfrak{g}_{-\lambda}]$$
  is an ideal of $\mathfrak{g}$ and isomorphic with $\mathfrak{sl}(2, \mathbb{R})$;
  (ii) *$\mathfrak{s}_\lambda$ commutes with $\mathfrak{g}^{\Omega+}$ and with all root spaces $\mathfrak{g}_\mu$ where μ is a real-valued root different from both λ and $-\lambda$;*
  (iii) *$\mathfrak{h}$ is the direct sum $\mathfrak{h} = [\mathfrak{g}_\lambda, \mathfrak{g}_{-\lambda}] \oplus \ker \lambda$.*
  (iv) *$\mathfrak{g}$ decomposes as a direct sum of ideals $\mathfrak{g} = \mathfrak{s}_\lambda \oplus \mathfrak{g}_1$, where*
$$\mathfrak{g}_1 = (\ker \lambda) \oplus \bigoplus_{\pm\lambda \neq \lambda' \in \Lambda_r} \mathfrak{g}_\lambda \oplus \mathfrak{g}^{\Omega+}$$
  is an Eggert algebra and $(\mathfrak{g}_1, \ker \lambda)$ is an Eggert pair.

Proof. Assertion (ii) follows 4.1(i), 2.10(ii), and property 4.1(iii) of Eggert pairs; assertion (i) is a consequence of (ii) and 4.1(iii). Assertion (iii) follows from the fact that $\lambda([\mathfrak{g}_\lambda, \mathfrak{g}_{-\lambda}]) \neq \{0\}$ and dimensional reasons. The last assertion follows from the decomposition $\mathfrak{g} = \mathfrak{h} \oplus \mathfrak{g}_{\Lambda_r} \oplus \mathfrak{g}^{\Omega+}$ and (i)–(iii). □

4.5. NOTATION. We write Λ^* for the set of all non-simple real roots and

$$\mathfrak{g}_{\Lambda^*} = \bigoplus_{\lambda \in \Lambda^*} \mathfrak{g}_\lambda,$$

where $\mathfrak{g}_\lambda = \{x \in \mathfrak{g} \mid (\forall h \in \mathfrak{h})\,[h,x] = \lambda(h)\cdot x\}$. □

Note that in the case of Eggert pairs we have $\mathfrak{g}^\lambda = \mathfrak{g}_\lambda$ for every real-valued root λ.

4.6. REMARKS. (i) Recall that a real Lie algebra $\mathfrak{m}$ is said to be *metabelian* if its commutator algebra $[\mathfrak{m}, \mathfrak{m}]$ is abelian. It is called *diagonally metabelian* (= specially metabelian in the terminology of [27]) if each of the operators $\operatorname{ad} m|[\mathfrak{m},\mathfrak{m}]$, $m \in \mathfrak{m}$, is diagonalizable (over the reals). A Lie algebra $\mathfrak{m}$ is diagonally metabelian if and only if it contains an abelian Cartan subalgebra $\mathfrak{h}$ such that the roots associated with $\mathfrak{h}$ are real-valued and satisfy $\mathfrak{g}^\lambda = \mathfrak{g}_\lambda$ as well as $[\mathfrak{g}^\lambda, \mathfrak{g}^\mu] = \{0\}$ for any pair of roots λ, μ (cf. [18]).

Thus if $(\mathfrak{g}, \mathfrak{h})$ is an Eggert pair then $\mathfrak{h} \oplus \mathfrak{g}_{\Lambda^*}$ is diagonally metabelian.

(ii) For any Lie algebra $\mathfrak{g}$ the Δ-*radical* of $\mathfrak{g}$ is defined as the intersection Δ of all hyperplane subalgebras. The Δ-radical is a characteristic ideal, and the structure of $\mathfrak{g}/\Delta$ is explicitly known by [30]. □

4.7. THEOREM. *Let $(\mathfrak{g}, \mathfrak{h})$ be an Eggert pair. Then $\mathfrak{g}$ is the ideal direct sum $\mathfrak{g} = \mathfrak{s} \oplus \mathfrak{d} \oplus \mathfrak{k}$, where*

(i) *$\mathfrak{s} = \mathfrak{s}_1 \oplus \cdots \oplus \mathfrak{s}_k$ is a direct sum of ideals isomorphic with $\mathfrak{sl}(2, \mathbb{R})$;*

(ii) *$\mathfrak{d} = \mathfrak{h}_1 \oplus \mathfrak{g}_{\Lambda^*}$ is a centerfree diagonally metabelian Lie algebra, where $\mathfrak{h}_1$ is a Cartan subalgebra of $\mathfrak{d}$;*

(iii) *$\mathfrak{k} = \mathfrak{h}_2 \oplus \mathfrak{g}^{\Omega^+}$ is a Lie algebra and $\mathfrak{h}_2$ is a Cartan subalgebra of $\mathfrak{k}$ which contains the center of $\mathfrak{g}$ and has only purely imaginary-valued roots on $\mathfrak{k}$; and*

(iv) *$\mathfrak{h} = (\mathfrak{h} \cap \mathfrak{s}) \oplus \mathfrak{h}_1 \oplus \mathfrak{h}_2$.*

Proof. By applying induction on Lemma 4.4 we find ideals $\mathfrak{s}_1, \mathfrak{s}_2, \ldots, \mathfrak{s}_k$, all isomorphic with $\mathfrak{sl}(2, \mathbb{R})$, such that $\mathfrak{g}$ is the ideal direct sum

$$\mathfrak{g} = \mathfrak{s}_1 \oplus \mathfrak{s}_2 \oplus \ldots \oplus \mathfrak{s}_k \oplus \mathfrak{g}_s, \text{ with}$$
$$\mathfrak{g}_s = \mathfrak{h}_s \oplus \mathfrak{g}_{\Lambda^*} \oplus \mathfrak{g}^{\Omega^+},$$
$$\mathfrak{h}_s = \{x \in \mathfrak{h} \mid \lambda(x) = 0 \text{ for all simple real-valued roots } \lambda\}.$$

Furthermore, $(\mathfrak{g}_s, \mathfrak{h}_s)$ is an Eggert pair without simple real-valued roots. To finish the proof it suffices to show the assertion under the assumption that there are no simple real-valued roots. Thus we suppose $(\mathfrak{g}, \mathfrak{h}) = (\mathfrak{g}_s, \mathfrak{h}_s)$.

We define:

$$\mathfrak{h}_0 = \bigcap_{\omega \in \Omega^+} \ker \omega, \qquad \mathfrak{h}_2 = \bigcap_{\lambda \in \Lambda^*} \ker \lambda.$$

Note that the sum of $\mathfrak{h}_0$ and $\mathfrak{h}_2$ is the annihilator of $\{0\} = \mathrm{span}(\Lambda^*) \cap \mathrm{span}(i\Omega^+)$ in $\mathfrak{h}$ and therefore equals $\mathfrak{h}$. Furthermore, $[\mathfrak{h}_2, \mathfrak{g}_{\Lambda^*}] = \{0\}$ and $\mathfrak{h}_0 \cap \mathfrak{h}_2$ is the center $\mathfrak{z}$ of $\mathfrak{g}$.

Now we pick an arbitrary vector space complement $\mathfrak{h}_1$ of $\mathfrak{h}_0 \cap \mathfrak{h}_2$ in $\mathfrak{h}_0$. Then we define

$$\mathfrak{d} := \mathfrak{h}_1 \oplus \mathfrak{g}_{\Lambda^*} \quad \text{and} \quad \mathfrak{k} := \mathfrak{h}_2 \oplus \mathfrak{g}^{\Omega^+}.$$

Since $\mathfrak{h}_1 \cap \mathfrak{h}_2 = \{0\}$ we see that $\mathfrak{d}$ is a centerfree diagonally metabelian Lie algebra.

From 4.1(iii), 2.10(iii), and the Jacobi identity it follows that $[\mathfrak{g}_{\Lambda^*}, [\mathfrak{g}^{\Omega^+}, \mathfrak{g}^{\Omega^+}]] = \{0\}$, so $[\mathfrak{g}^{\Omega^+}, \mathfrak{g}^{\Omega^+}] \cap \mathfrak{h} \subseteq \mathfrak{h}_2$. Thus $\mathfrak{k}$ is a Lie algebra and $\mathfrak{h}_2$ is a Cartan subalgebra of $\mathfrak{k}$. Moreover, $[\mathfrak{d}, \mathfrak{k}] = \{0\}$. This proves the asserted ideal direct sum decomposition of $\mathfrak{g}$.

The center $\mathfrak{z}$ of $\mathfrak{g}$ decomposes into a direct sum $\mathfrak{z} = (\mathfrak{s} \cap \mathfrak{z}) \oplus (\mathfrak{d} \cap \mathfrak{z}) \oplus (\mathfrak{k} \cap \mathfrak{z})$. Since $\mathfrak{s}$ and $\mathfrak{d}$ are centerfree this implies that $\mathfrak{z} = \mathfrak{k} \cap \mathfrak{z}$, that is, $\mathfrak{z} \subseteq \mathfrak{k}$. Since the center of a Lie algebra is contained in every Cartan subalgebra this implies $\mathfrak{z} \subseteq \mathfrak{h}_2$. $\square$

We recall from [27], p. 136ff. that the *base ideal* of a Lie algebra $\mathfrak{g}$ is the sum of all one dimensional ideals. A *base root* φ is a linear map $\mathfrak{g} \to \mathbb{R}$ such that the set $\mathfrak{g}_\varphi = \{x \in \mathfrak{g} : \forall g \in \mathfrak{g}\, [g, x] = \varphi(g) \cdot x\}$ is not singleton. The space $\mathfrak{g}_\varphi$ is called the *base root space for φ*.

4.8. THEOREM. *Let $\mathfrak{g} = \mathfrak{s} \oplus \mathfrak{d} \oplus \mathfrak{k}$ be the decomposition of Theorem 4.7. Let $\mathfrak{z}$ be the center of $\mathfrak{g}$. Then the following conditions are satisfied:*

(i) *The sum $\mathfrak{m} := \mathfrak{z} \oplus \mathfrak{g}_{\Lambda^*}$ is the base ideal.*

(ii) *The root spaces $\mathfrak{g}^\lambda = \mathfrak{g}_\lambda$ for $\lambda \in \Lambda^*$ are ideals in $\mathfrak{g}$; they may be identified with the base root spaces of $\mathfrak{m}$ for non-zero base roots.*

(iii) *The commutator algebra of $\mathfrak{g}$ is*

$$\mathfrak{g}' = \mathfrak{s} \oplus \mathfrak{g}_{\Lambda^*} \oplus \langle \mathfrak{g}^{\Omega^+} \rangle$$

(iv) *The Lie algebra $\langle \mathfrak{g}^{\Omega^+} \rangle$ generated by $\mathfrak{g}^{\Omega^+}$ is the Δ-radical Δ of $\mathfrak{g}$, so the $\Delta \subseteq \mathfrak{k}$.*

(v) *The solvable radical $\mathfrak{r}$ of $\mathfrak{g}$ decomposes as $\mathfrak{r} = \mathfrak{d} \oplus (\mathfrak{r} \cap \mathfrak{k})$ and $\mathfrak{r} \cap \mathfrak{k} = (\mathfrak{r} \cap \mathfrak{h}_2) + (\mathfrak{r} \cap \langle \mathfrak{g}^{\Omega^+} \rangle)$.*

Proof. (i) By definition we have $\mathfrak{g}_{\Lambda^*} = \bigoplus_{\lambda \in \Lambda^*} \mathfrak{g}_\lambda$ with $[h, x] = \lambda(h) \cdot x$ for each $x \in \mathfrak{g}_\lambda$. Furthermore, $[\mathfrak{s} + \mathfrak{g}^{\Omega^+}, \mathfrak{g}_{\Lambda^*}] = \{0\}$ since the decomposition of 4.7 is an ideal direct sum. Thus $\mathfrak{m} = \mathfrak{z} \oplus \mathfrak{g}_{\Lambda^*}$ is an abelian ideal which is the sum of one-dimensional ideals. If, conversely, $\mathfrak{a}$ is a one dimensional ideal of $\mathfrak{g}$, then it is, in particular, spanned by a common eigenvector of $\mathfrak{h}$ and is therefore contained in $\mathfrak{g}_{\Lambda^*}$.

(ii) By the definition of Eggert algebras and of the vector space $\mathfrak{g}_{\Lambda^*}$ we have $\mathfrak{g}^\lambda = \mathfrak{g}_\lambda$ for $\lambda \in \Lambda^*$. Every $\lambda : \mathfrak{h} \to \mathbb{R}$, $\lambda \in \Lambda^*$ extends to a base root φ which vanishes on $\mathfrak{s} + \mathfrak{g}^{\Omega^+}$. Accordingly, the function $\varphi \mapsto \varphi|\mathfrak{h}$ is a bijection from the set of base roots to Λ^*, and the base root space $\mathfrak{g}_\varphi$ is precisely the root space $\mathfrak{g}_{\varphi|\mathfrak{h}}$.

Assertion (iii) follows by straightforward calculation.

(iv) By the results of [30], the hyperplane subalgebras of an algebra of the form $\mathfrak{s} \oplus \mathfrak{g}_{\Lambda^\bullet}$ intersect in $\{0\}$, and this assertion remains correct for any homomorphic image. In particular, it applies to $\mathfrak{g}/\langle \mathfrak{g}^{\Omega+}\rangle$. Thus the intersection of all hyperplane subalgebras of $\mathfrak{g}$ is contained in $\langle \mathfrak{g}^{\Omega+}\rangle$.

Conversely, if $\mathfrak{e}$ is a hyperplane subalgebra of $\mathfrak{g}$, then $\mathfrak{e} \cap (\mathfrak{h} \oplus \mathfrak{g}^{\Omega^+})$ either equals $\mathfrak{h} \oplus \mathfrak{g}^{\Omega^+}$ or is a hyperplane subalgebra therein. But any such contains $\mathfrak{g}^{\Omega+}$ since the algebra $\mathfrak{h} \oplus \mathfrak{g}^{\Omega+}$ has no real roots. Therefore, $\mathfrak{g}^{\Omega+}$ is contained in Δ. Since Δ is a subalgebra, $\langle \mathfrak{g}^{\Omega+}\rangle \subseteq \Delta$ follows.

(v) We first recall that $\mathfrak{g}$ is the ideal direct sum $\mathfrak{g} = \mathfrak{s} \oplus \mathfrak{d} \oplus \mathfrak{k}$, where $\mathfrak{s}$ is semisimple and $\mathfrak{d}$ is solvable. Thus $\mathfrak{r} \cap \mathfrak{s} = \{0\}$ and $\mathfrak{d} \subseteq \mathfrak{r}$, so $\mathfrak{r}$ decomposes as $\mathfrak{r} = \mathfrak{d} \oplus (\mathfrak{r} \cap \mathfrak{k})$. We write $\mathfrak{r}_*$ for the solvable radical in $\langle \mathfrak{g}^{\Omega+}\rangle$. (Recall that $\langle \mathfrak{g}^{\Omega+}\rangle$ is an ideal of $\mathfrak{g}$, by 2.10.) Since $\mathfrak{r}_*$ is a characteristic ideal we have $\mathfrak{r}_* = \mathfrak{r} \cap \langle \mathfrak{g}^{\Omega+}\rangle$. Since $\mathfrak{h}$ is abelian we know that $\mathfrak{k}/\langle \mathfrak{g}^{\Omega+}\rangle$ is abelian, so $\mathfrak{k}/\mathfrak{r}_*$ is reductive and the assertion follows. $\square$

We conclude this section with an application of these results to semigroups in Lie groups.

4.9. MAIN LEMMA. *Let G be a Lie group containing a reduced weakly exponential subsemigroup S with Lie wedge W. Suppose that $\mathfrak{h}$ is a Cartan algebra meeting the interior of W and that $\mathfrak{g} = \mathfrak{s} \oplus \mathfrak{d} \oplus \mathfrak{k}$ is a decomposition according to Theorem 4.7. Assume the following hypothesis:*

(comp) *$\mathfrak{k}$ has a unique largest compactly embedded subalgebra.*

Then the following conclusions hold:

(i) *W is a semialgebra.*

(ii) *W decomposes in the form*

$$W = (W \cap \mathfrak{s}) \oplus W_0, \quad W_0 = (W \cap (\mathfrak{d} \oplus \mathfrak{k})).$$
$$W \cap \mathfrak{s} = (W \cap \mathfrak{s}_1) \oplus \cdots \oplus (W \cap \mathfrak{s}_n)$$

where the $\mathfrak{s}_j \cong \mathfrak{sl}(2, \mathbb{R})$ are the simple components of $\mathfrak{s}$.

(iii) *Set $\mathfrak{d}' = [\mathfrak{d}, \mathfrak{d}] = \mathfrak{g}_{\Lambda^\bullet}$,*
$W_{inv} = \overline{W_0 + \mathfrak{d}'}$, *and*
$W_{sec} = \bigcap \{L_p(W_0) \mid p \in C^1(W_0) \text{ and } \mathfrak{d}' \not\subseteq T_p(W_0)\}$.
Then $\mathfrak{d}'$ is an ideal of $\mathfrak{g}$ such that $\mathfrak{d}' \oplus \mathfrak{z}$ with the center $\mathfrak{z}$ of $\mathfrak{g}$ is the base ideal of $\mathfrak{g}$. The wedge W_{inv} is the smallest invariant wedge containing W_0, the wedge W_{sec} is an intersection of half-space semialgebras containing $\Delta = \mathfrak{k}'$, and $W_0 = W_{inv} \cap W_{sec}$.

Proof. (i) We let $\mathfrak{c}$ denote the largest compactly embedded subalgebra of $\mathfrak{k}$. Now $[\mathfrak{s} + \mathfrak{d}, \mathfrak{k}] = \{0\}$ and $\text{comp}\,\mathfrak{d} = \{0\}$. Then $\text{comp}\,\mathfrak{g} = (\text{comp}\,\mathfrak{s}) \oplus \mathfrak{c}$. But $\text{comp}_G(\mathfrak{g}) \subseteq \text{comp}\,\mathfrak{g}$. Then the Main Lemma IV.7.5 applies and shows that W is a semialgebra.

(ii) In view of (i) this is now a consequence of EGGERT's Classification Theorem, Part I, [18], p. 76, 4.20.

(iii) By (i) again this follows from EGGERT's Small Intersection Theorem [18], p. 79, 4.29. □

We note that in the proof of the Main Lemma the principal results of EGGERT's dissertation [18] enter into the stream of our argument.

CHAPTER 8

ROOTS AND ROOT SPACES

1. Roots and Root Spaces

In the following $\mathfrak{g}$ is a real Lie algebra and $\mathfrak{g}_{\mathbb{C}}$ its complexification. Let $\mathfrak{h}$ be a Cartan subalgebra in $\mathfrak{g}$ and $\mathfrak{h}_{\mathbb{C}}$ its complexification. For any root λ we let $\mathfrak{g}_{\mathbb{C}}^\lambda$ denote the root space of λ, i.e, the nilspace of $\operatorname{ad} X - \lambda(X)\cdot\mathbf{1}$ for all $X \in \mathfrak{h}_{\mathbb{C}}$.

1.1. HYPOTHESIS. We suppose that $\mathfrak{h}$ is abelian and that there exists an imaginary-valued root (see VII.2.8) $\lambda\colon \mathfrak{h}_{\mathbb{C}} \to \mathbb{C}$. $\square$

Let $\kappa : \mathfrak{g}_{\mathbb{C}} \to \mathfrak{g}_{\mathbb{C}}$ denote complex conjugation $\kappa(X + i\cdot Y) = X - i\cdot Y$ for $X, Y \in \mathfrak{g}$. Then κ is a real Lie algebra involution satisfying $\kappa(c\cdot X) = \overline{c}\cdot\kappa(X)$.

1.2. DEFINITION. (i) We define $\overline{\lambda}\colon \mathfrak{h}_{\mathbb{C}} \to \mathfrak{h}_{\mathbb{C}}$ by $\overline{\lambda}(X) = \overline{\lambda(\kappa(X))}$. Then $\overline{\lambda}$ is a root which, according to our hypothesis $\lambda(\mathfrak{h}) \subseteq i\cdot\mathbb{R}$, satisfies $\overline{\lambda}(X) = -\lambda(X)$ for all $X \in \mathfrak{h}$. Consequently, $\lambda + \overline{\lambda}$ vanishes on $\mathfrak{h}$ and hence on $\mathfrak{h}_{\mathbb{C}}$, i.e., $\lambda + \overline{\lambda} = 0$. Thus

(1) $$[\mathfrak{g}_{\mathbb{C}}^\lambda, \mathfrak{g}_{\mathbb{C}}^{\overline{\lambda}}] \subseteq \mathfrak{g}_{\mathbb{C}}^0 = \mathfrak{h}_{\mathbb{C}}.$$

(ii) We write $\omega = -i\cdot\lambda$ and $\mathfrak{g}^\omega = \mathfrak{g} \cap (\mathfrak{g}_{\mathbb{C}}^\lambda \oplus \mathfrak{g}_{\mathbb{C}}^{\overline{\lambda}})$. We say that $\mathfrak{g}^\omega$ is the *real root space attached to* λ. $\square$

1.3. LEMMA. (i) $\mathfrak{g}_{\mathbb{C}}^\lambda$ *contains an eigenvector* $v \ne 0$.
(ii) *The elements* $X := \frac{1}{2}\cdot(v + \kappa(v))$ *and* $IX := \frac{1}{2i}\cdot(v - \kappa(v))$ *are contained in* $\mathfrak{g}^\omega$, *and* $Q(X) := [IX, X] \in \mathfrak{h}$.
(iii) $\mathfrak{n} := \mathbb{R}\cdot X + \mathbb{R}\cdot IX + \mathbb{R}\cdot Q(X)$ *is a subalgebra with* $[\mathfrak{h}, \mathfrak{n}] \subseteq \mathfrak{n}$.
(iv) $\mathfrak{m}(\lambda, v) := \mathfrak{h} + \mathfrak{n}$ *is a maximal rank subalgebra.*
(v) *We have the following cases:*

(S_-) $\omega(Q(X)) < 0$, *then* $\mathfrak{n} \cong \mathfrak{so}(3)$,
(S_+) $\omega(Q(X)) > 0$, *then* $\mathfrak{n} \cong \mathfrak{sl}(2, \mathbb{R})$
(N) $\omega(Q(X)) = 0$, *and* $Q(X) \ne 0$, *then* $\mathfrak{n}$ *is the Heisenberg algebra,*
(A) $Q(X) = 0$, *then* $\mathfrak{n} \cong \mathbb{R}^2$.

In cases (N) and (A) we select a vector $Y \in \mathfrak{h}$ with $\omega(Y) \neq 0$, we set $Y = 0$ in cases (S_+) and (S_-). Let $\mathfrak{b} := \mathbb{R} \cdot Y + \mathfrak{n}$. Then we have

$$\mathfrak{b} \cong \begin{cases} \mathfrak{so}(3), & \text{in case } (S_-), \\ \mathfrak{sl}(2, \mathbb{R}), & \text{in case } (S_+), \\ \mathfrak{osc}, & \text{in case } (N), \\ \mathfrak{mot}, & \text{in case } (A). \end{cases}$$

(vi) $\mathfrak{m}(\lambda, v) = \mathfrak{a} \oplus \mathfrak{b}$ with an abelian ideal $\mathfrak{a}$ of $\mathfrak{m}(\lambda, v)$.

Proof. (i) This familiar fact is seen easily as follows: There is a smallest natural number n such that $(\operatorname{ad} Z - \lambda(Z) \cdot 1)^n |_{\mathfrak{g}_\mathbb{C}^\lambda} = 0$ for all $Z \in \mathfrak{h}_\mathbb{C}$. Thus there is an $X_0 \in \mathfrak{g}_\mathbb{C}^\lambda$ and a $Z \in \mathfrak{h}_\mathbb{C}$ such that $v := (\operatorname{ad} Z - \lambda(Z) \cdot 1)^{n-1} X_0 \neq 0$. This is the required eigenvector.

(ii) The first assertion follows from the definitions and the fact that X and IX are clearly κ-invariant. Next we compute

$$Q(X) = \frac{1}{2i}[v, \kappa(v)] \in [\mathfrak{g}_\mathbb{C}^\lambda, \mathfrak{g}_\mathbb{C}^{\overline{\lambda}}] \cap \mathfrak{g} \subseteq \mathfrak{h}_\mathbb{C} \cap \mathfrak{g} = \mathfrak{h}.$$

(iii) We note $[Z, X] = \omega(Z) \cdot IX$ and $[Z, IX] = -\omega(Z) \cdot X$ for $Z \in \mathfrak{h}$. This holds, in particular, for $Z = Q(X)$.

(iv) This is immediate from (iii).

(v) and (vi) See [27], p. 230ff. and Definition VI.2.1. □

1.4. DEFINITION. *We say that the algebra* $\mathfrak{m}(\lambda, v) = \mathfrak{a} \oplus \mathfrak{b}$ *constructed in* 1.3 *is associated with* λ *and* v. □

1.5. LEMMA. *Suppose that* $\mathfrak{g} = \mathfrak{a} \oplus \mathfrak{b}$ *with an abelian ideal* $\mathfrak{a}$ *and an ideal* $\mathfrak{b} \cong \mathfrak{sl}(2, \mathbb{R})$, *and that* W *is an invariant wedge in* $\mathfrak{g}$. *Then*
  (i) $W \cap \mathfrak{b}$ *is the projection of* W *into* $\mathfrak{b}$ *along* $\mathfrak{a}$,
  (ii) *if* $W \not\subseteq \mathfrak{a}$, *then* W *contains a nonzero compact element of* $\mathfrak{b}$.

Proof. (i): See [27], p. 122, Lemma II.3.23.

(ii) follows from (i) and the fact that any nonzero invariant wedge in $\mathfrak{sl}(2, \mathbb{R})$ contains nonzero compact elements. ([27], p. 109, Theorem II.3.7.) □

1.6. LEMMA. *Let G denote a connected Lie group and S a reduced weakly exponential subsemigroup. Suppose that $\mathfrak{h}$ is a Cartan subalgebra meeting the interior of $W = \mathfrak{L}(S)$, that λ is an imaginary-valued root, and that $v \neq 0$ is an eigenvector in $\mathfrak{g}_\mathbb{C}^\lambda$. We write $\mathfrak{m} = \mathfrak{m}(\lambda, v)$.*
Then we have the following conclusions:
  (i) $\mathfrak{h} = \mathfrak{a} \oplus (\mathfrak{h} \cap \mathfrak{b})$.
  (ii) *Either* $\mathfrak{b} \cong \mathfrak{so}(3)$, *or else* $\mathfrak{b} \cong \mathfrak{sl}(2, \mathbb{R})$.
In the following conclusions we suppose that $\mathfrak{b} \cong \mathfrak{sl}(2, \mathbb{R})$.

(iii) $\mathrm{int}_{\mathfrak{b}}(\mathfrak{b} \cap W)$ *contains no compact element of* $\mathfrak{b}$.
(iv) $W \cap \mathfrak{m}$ *is not invariant in* $\mathfrak{m}$.
(v) $W \cap \mathfrak{m} = (W \cap \mathfrak{a}) \oplus (W \oplus \mathfrak{b})$.

Proof. First we note by our Main Lemma VII.1.4 that all Cartan subalgebras of $\mathfrak{g}$ are abelian, since S is reduced.

(i) Since $\mathfrak{h}$ is its own centralizer and $\mathfrak{a}$ is central in $\mathfrak{m}$ we have $\mathfrak{a} \subseteq \mathfrak{h}$. Then the assertion follows.

(ii) and (iii) The analytic subgroup M generated by $\mathfrak{m}$ is closed since $\mathfrak{m}$ has maximal rank. The Maximal Rank Theorem II.4.3 shows $S \cap M = \overline{\exp(W \cap \mathfrak{m})}$. By way of contradiction we assume now that either $\mathfrak{b}$ is not simple (i.e., that we have one of the cases (A) and (N)), or that $\mathfrak{b} \cong \mathfrak{sl}(2, \mathbb{R})$ and $\mathrm{int}_{\mathfrak{b}}(\mathfrak{b} \cap W)$ contains a compact element of $\mathfrak{b}$.

Then the Testing Theorem VI.2.5 applies and shows that $\mathfrak{m}' \subseteq H(W)$. Since W is assumed to be reduced, we apply Occam's Razor VI.1.3 as follows: Pick any regular element h of $\mathfrak{g}$ in $\mathfrak{h} \cap \mathrm{int}\, W$ and set $V = \mathfrak{m}'$. Then condition (Occ) of VI.1.3 is satisfied, and Theorem VI.1.3 shows that $\mathfrak{b}' = \mathfrak{m}' = \{0\}$. This is impossible for $\mathfrak{b} = \mathfrak{mot}, \mathfrak{osc}$, or $\mathfrak{sl}(2, \mathbb{R})$.

Now we assume $\mathfrak{b} \cong \mathfrak{sl}(2, \mathbb{R})$. The Lie wedge $W \cap \mathfrak{m}$ is a semialgebra and $\mathfrak{m} \cap \mathrm{int}\, W \neq \emptyset$.

(iv) Assume that $\mathfrak{m} \cap W$ is invariant in $\mathfrak{m}$. Since $\mathfrak{h} \cap \mathrm{int}\, W \neq \emptyset$ we have $\mathfrak{m} \cap W \not\subseteq \mathfrak{a}$. Then by Lemma 1.5 there is a compact element of $\mathfrak{b}$ in W, which is impossible by (iii).

(v) By (iv) $\mathfrak{m} \cap W$ is not invariant in $\mathfrak{m}$. Then a result of EGGERT's ([18], Theorem 4.15, p. 73) proves the assertion. $\square$

1.7. LEMMA. *Let G denote a connected Lie group and S a reduced weakly exponential subsemigroup with Lie wedge W. Suppose that $\mathfrak{h}$ is a Cartan subalgebra and λ is an imaginary-valued root w.r.t. $\mathfrak{h}$, and that $P \neq 0$ is an eigenvector in $\mathfrak{g}_{\mathbb{C}}^{\lambda}$. We write $\mathfrak{m} = \mathfrak{m}(\lambda, P) = \mathfrak{a} \oplus \mathfrak{b}$ as in 1.3.*

Suppose further that the following hypotheses are satisfied:
(a) $\mathfrak{h} \cap \mathrm{int}\, W \neq \emptyset$.
(b) $\mathfrak{h}$ *is compactly embedded in* $\mathfrak{m}$.
Then $\mathfrak{b} \cong \mathfrak{so}(3)$.

Proof. By 1.6(ii) $\mathfrak{b} \cong \mathfrak{sl}(2, \mathbb{R})$ or $\mathfrak{b} \cong \mathfrak{so}(3)$. We assume the former and derive a contradiction.

By hypothesis (a) there is an $x \in \mathfrak{h} \cap \mathrm{int}\, W$. We write $x = x_{\mathfrak{a}} + x_{\mathfrak{b}}$ with $x_{\mathfrak{a}} \in \mathfrak{a}$ and $x_{\mathfrak{b}} \in \mathfrak{b}$. By 1.6(i), $\mathfrak{h} = \mathfrak{a} \oplus (\mathfrak{h} \cap \mathfrak{b})$, and thus $x_{\mathfrak{b}} \in \mathfrak{h}$. Also, by 1.6(v), $\mathfrak{m} \cap W = (\mathfrak{a} \cap W) \oplus (\mathfrak{b} \cap W)$. Since $x \in \mathrm{int}\, W$, we have $x_{\mathfrak{b}} \in \mathrm{int}_{\mathfrak{b}}(\mathfrak{b} \cap W)$. Thus $x_{\mathfrak{b}} \in \mathrm{int}_{\mathfrak{b}}(\mathfrak{b} \cap \mathfrak{h} \cap W)$.

By (b) $x_{\mathfrak{b}} \in \mathfrak{h}$ is a compact element in $\mathfrak{m}$ hence in $\mathfrak{b}$. This contradicts 1.6(iii). $\square$

Note that $\mathfrak{h}$ is compactly embedded in $\mathfrak{m}$ if it is compactly embedded in $\mathfrak{g}$.

1.8. DEFINITION. Suppose that $\mathfrak{h}$ is a compactly embedded Cartan subalgebra of a Lie algebra $\mathfrak{g}$. Then $\mathfrak{k}(\mathfrak{h})$ denotes the unique maximal compactly embedded subalgebra containing $\mathfrak{h}$. (See [27], III.5.6, p. 217.) We say that a root λ with respect to $\mathfrak{h}_{\mathbb{C}}$ is *noncompact* iff the space $\mathfrak{g}^{\omega} := \mathfrak{g} \cap (\mathfrak{g}_{\mathbb{C}}^{\lambda} + \mathfrak{g}_{\mathbb{C}}^{-\lambda})$, $\omega = -i\lambda|\mathfrak{h}$ (see [27], p. 226) is **not** contained in $\mathfrak{k}(\mathfrak{h})$. □

1.9. LEMMA. *Let G denote a connected Lie group and S a reduced weakly exponential subsemigroup. Suppose that the following hypotheses are satisfied:*
  (a) *$\mathfrak{h}$ is a Cartan subalgebra meeting the interior of $W = \mathfrak{L}(S)$.*
  (b) *No root λ with respect to $\mathfrak{h}_{\mathbb{C}}$ satisfies $\lambda(\mathfrak{h}) \subseteq \mathbb{R}$.*
Then we have the following conclusions:
  (i) *$\mathfrak{g}$ is reductive.*
  (ii) *$\mathfrak{h}$ is compactly embedded.*

Proof. Recall that the Cartan subalgebras of $\mathfrak{g}$ are abelian since S is reduced (VII.1.4).

(i) By our Main Lemma VII.2.7 all roots are either real- or imaginary-valued. By hypothesis (a) $\lambda(\mathfrak{h}) \subseteq i\mathbb{R}$. If the radical $\mathfrak{r}$ of $\mathfrak{g}$ were not central, then we would find a root λ and and eigenvector $v \in \mathfrak{g}_{\mathbb{C}}^{\lambda} \cap \mathfrak{r}_{\mathbb{C}}$. We set $X = \frac{1}{2} \cdot (v + \kappa(v))$ and $IX = \frac{1}{2} \cdot (v - \kappa(v))$ as in Lemma 1.3. Then X and IX are both in $\mathfrak{r}$ and thus $\mathfrak{m}(\lambda, v)' = \mathfrak{n} = \mathbb{R} \cdot X + \mathbb{R} \cdot IX + \mathbb{R} \cdot [IX, X] \subseteq \mathfrak{r}$. But this contradicts Lemma 1.6(ii). Hence $\mathfrak{r} = \mathfrak{z}(\mathfrak{g})$ and $\mathfrak{g}$ is reductive.

(ii) Since in a reductive Lie algebra all $\operatorname{ad} X$ with $X \in \mathfrak{h}$ are semisimple we conclude from $\lambda(\mathfrak{h}) \in i\mathbb{R}$ that $\mathfrak{h} \subseteq \operatorname{comp}(\mathfrak{g})$. Hence $\mathfrak{h}$ is compactly embedded. □

1.10. MAIN LEMMA. *Let G denote a connected Lie group and S a reduced weakly exponential subsemigroup. Suppose that the following hypotheses are satisfied:*
  (a) *$\mathfrak{h}$ is a Cartan subalgebra meeting the interior of $W = \mathfrak{L}(S)$.*
  (b) *No root λ with respect to $\mathfrak{h}_{\mathbb{C}}$ satisfies $\lambda(\mathfrak{h}) \subseteq \mathbb{R}$.*
Then $\mathfrak{g}$ is compact.

Proof. By Lemma 1.9(ii), $\mathfrak{h}$ is compactly embedded. Let $\mathfrak{k}(\mathfrak{h})$ be the largest compactly embedded subalgebra containing $\mathfrak{h}$ (see 1.8). We must show $\mathfrak{g} = \mathfrak{k}(\mathfrak{h})$. For this purpose we show that there are no noncompact roots (see 1.8). Let us assume, on the contrary, that λ is a noncompact root w.r.t. $\mathfrak{h}_{\mathbb{C}}$. Pick the eigenvector v as in 1.3 and construct $\mathfrak{m} = \mathfrak{m}(\lambda, v) = \mathfrak{a} \oplus \mathfrak{b}$. By [27], p. 232, Corollary III.6.16, $\mathfrak{m}$ is not a compact Lie algebra. From 1.6(ii) we know $\mathfrak{b} = \mathfrak{sl}(2, \mathbb{R})$ since λ is a noncompact root. Since $\mathfrak{h}$ is compactly embedded the hypotheses of 1.7 are satisfied and therefore $\mathfrak{b} \cong \mathfrak{so}(3)$. This is a contradiction. □

1.11. COROLLARY. *Let G denote a connected Lie group and S a weakly exponential subsemigroup. Let $\mathfrak{j}$ be the largest ideal of $\mathfrak{g}$ contained in $\mathfrak{L}(S)$ Suppose that the following hypotheses are satisfied:*
  (a) *$\mathfrak{h}$ is a Cartan subalgebra meeting the interior of $W = \mathfrak{L}(S)$.*
  (b) *No root λ with respect to $\mathfrak{h}_\mathbb{C}$ satisfies $\lambda(\mathfrak{h}) \subseteq \mathbb{R}$.*
Then $\mathfrak{g}/\mathfrak{j}$ is compact.

Proof. Define $\mathfrak{g}^\# := \mathfrak{g}/\mathfrak{j}$, $G^\# := G/J$, and $\mathfrak{h}^\# = \frac{\mathfrak{h}+\mathfrak{j}}{\mathfrak{j}}$. Further $W^\# = W/\mathfrak{j}$ and $S^\# = S/J$. Then by IV.4.2(iii) the semigroup $S^\#$ is weakly exponential and
  ($\mathrm{a}^\#$) $\mathfrak{h}^\#$ is a Cartan subalgebra of $\mathfrak{g}^\#$ meeting the interior of $W^\# = \mathfrak{L}(S^\#)$.
  ($\mathrm{b}^\#$) No root λ with respect to $\mathfrak{h}_\mathbb{C}^\#$ satisfies $\lambda(\mathfrak{h}^\#) \subseteq \mathbb{R}$.
  ($\mathrm{c}^\#$) $S^\#$ is reduced.
Now our Main Lemma 1.10 applies to $G^\#$ and shows that $\mathfrak{g}^\#$ is compact. This is the assertion. □

2. The Answer

2.1. THEOREM. *Let G be a connected Lie group with a reduced weakly exponential subsemigroup S. Then the following assertions hold:*
(i) *$\mathfrak{g}$ is an ideal direct sum*
$$\mathfrak{g} = \mathfrak{s}_1 \oplus \cdots \oplus \mathfrak{s}_k \oplus \mathfrak{d} \oplus \mathfrak{k}$$
such that $\mathfrak{s}_j \cong \mathfrak{sl}(2,\mathbb{R})$, $j = 1,\ldots,k$, $\mathfrak{d}$ is diagonally metabelian without center, and $\mathfrak{k}$ is a compact ideal.
(ii) *$\mathfrak{k}$ contains the Δ-radical and the center of $\mathfrak{g}$.*
(iii) *The radical of $\mathfrak{g}$ is $\mathfrak{d} \oplus \mathfrak{z}(\mathfrak{k})$.*
(iv) *There is a unique Levi complement $\mathfrak{s}_1 \oplus \cdots \oplus \mathfrak{s}_k \oplus \mathfrak{k}'$.*
(v) *$W = \mathfrak{L}(S)$ is a Lie semialgebra.*

Proof. (i) From Theorem VII.4.2 we know that $\mathfrak{g}$ is an Eggert algebra and by Theorem VII.4.7 that $\mathfrak{g}$ is the ideal direct sum $\mathfrak{g} = \mathfrak{s} \oplus \mathfrak{d} \oplus \mathfrak{k}$, where
  (α) $\mathfrak{s} = \mathfrak{s}_1 \oplus \cdots \oplus \mathfrak{s}_k$ is a direct sum of ideals isomorphic with $\mathfrak{sl}(2,\mathbb{R})$;
  (β) $\mathfrak{d} = \mathfrak{h}_1 \oplus \mathfrak{g}_{\Lambda^\bullet}$ is a centerfree diagonally metabelian Lie algebra; and
  (γ) $\mathfrak{k} = \mathfrak{h}_2 \oplus \mathfrak{g}^{\Omega^+}$, where $\mathfrak{h}_2$ is a Cartan subalgebra of $\mathfrak{k}$ with imaginary-valued roots, containing the center of $\mathfrak{g}$.
Let $\mathfrak{h}_0$ denote a Cartan subalgebra of $\mathfrak{s}$ not containing any compact elements. The Eggert decomposition is constructed in such a way that $\mathfrak{h} := \mathfrak{h}_0 \oplus \mathfrak{h}_1 \oplus \mathfrak{h}_2$ is a Cartan algebra of $\mathfrak{g}$ meeting int W.

The subalgebra $\mathfrak{m} = \mathfrak{h} + \mathfrak{k} = \mathfrak{h}_0 \oplus \mathfrak{h}_1 \oplus \mathfrak{k}$ is a maximal rank subalgebra meeting int W. Let $M = \langle \exp \mathfrak{m} \rangle$. Then by the Maximal Rank Theorem II.4.3 $W \cap \mathfrak{m}$ is the Lie wedge of a weakly exponential semigroup $S \cap M$. Let $\mathfrak{j}$ denote the largest ideal of $\mathfrak{m}$ contained in $W \cap \mathfrak{m}$ and J the group generated by it. We apply Corollary 1.11 and find that $\mathfrak{m}/\mathfrak{j}$ is compact. We claim that $\mathfrak{j} \cap \mathfrak{k} = \{0\}$. When this claim is established we know that $\mathfrak{k} \cong \frac{\mathfrak{k}+\mathfrak{j}}{\mathfrak{j}}$ is compact, and this will finish the proof of (i). We formulate the proof of this claim in the following Lemma, for which we briefly interrupt the proof of Theorem 2.1.

2.2. LEMMA. *Let $\mathfrak{g}$ be an ideal direct sum $\mathfrak{a} \oplus \mathfrak{b}$ and $\mathfrak{e}$ a subalgebra of $\mathfrak{g}$ not containing any nonzero ideal of $\mathfrak{g}$. Suppose that $\mathfrak{g}_1 = \mathfrak{a} \oplus \mathfrak{b}_1$ with a subalgebra $\mathfrak{b}_1$ of $\mathfrak{b}$ and that $\mathfrak{j}$ is an ideal of $\mathfrak{g}_1$ contained in $\mathfrak{e}$. Then $\mathfrak{j} \cap \mathfrak{a} = \{0\}$*

Proof. The normalizer of $\mathfrak{j} \cap \mathfrak{a}$ contains $\mathfrak{a}$ because $\mathfrak{j}$ is an ideal in $\mathfrak{a} \oplus \mathfrak{b}_1$, and it contains $\mathfrak{b}$ since $\mathfrak{b}$ centralizes $\mathfrak{a}$. Hence $\mathfrak{j} \cap \mathfrak{a}$ is an ideal in $\mathfrak{g}$. But $\mathfrak{a} \cap \mathfrak{j} \subseteq \mathfrak{j} \subseteq \mathfrak{e}$. Hence $\mathfrak{j} \cap \mathfrak{a} = \{0\}$. □

We now continue the proof of Theorem 2.1:
(ii) By Theorem VII.4.8(iv) the ideal $\mathfrak{k}$ contains the Δ-radical $\langle \mathfrak{g}^{\Omega^+} \rangle$.

(iii) The radical $\mathfrak{r}$ adapts to the direct ideal decomposition and therefore is $\mathfrak{d} \oplus (\mathfrak{r} \cap \mathfrak{k})$. Since $\mathfrak{k}$ is a compact Lie algebra by (i), its radical is its center. Thus $\mathfrak{r} = \mathfrak{d} \oplus \mathfrak{z}(\mathfrak{k})$.

(iv) Again since $\mathfrak{g} = \mathfrak{s} \oplus \mathfrak{d} \oplus \mathfrak{k}$ is an ideal direct sum, Levi complements for $\mathfrak{r} = \mathfrak{d} \oplus \mathfrak{z}(\mathfrak{k})$ adapt to the decomposition. The Levi complement of $\mathfrak{z}(\mathfrak{k})$ in $\mathfrak{k}$ is $\mathfrak{k}'$. Thus there is a unique Levi complement $\mathfrak{s} \oplus \mathfrak{k}'$.

(v) This follows at once from (i) and our Main Lemma VII.4.9(i).
The proof is now complete. □

2.3. THEOREM. *Let G be a Lie group and S a weakly exponential subsemigroup with Lie wedge W. Then W is a Lie semialgebra.*

Proof. The theorem follows from Theorem 2.1 and Proposition IV.4.8. □

We shall use the following lemma below.

2.4. LEMMA. *Let $\mathfrak{g} = \mathfrak{s} \oplus \mathfrak{d} \oplus \mathfrak{k}$ be an Eggert algebra decomposed as in* Theorem VII.4.7. *Suppose that $\mathfrak{e}$ is a subalgebra satisfying*
  (a) *$\mathfrak{e}$ is ideal free.*
  (b) *$\mathfrak{e} = (\mathfrak{e} \cap \mathfrak{s}_1) \oplus \cdots \oplus (\mathfrak{e} \cap \mathfrak{s}_k) \oplus (\mathfrak{e} \oplus (\mathfrak{d} \oplus \mathfrak{k}))$.*
  (c) *$\mathfrak{e} \cap \mathfrak{k} = \{0\}$.*
  (d) *$\mathfrak{k}$ is a compact algebra.*
Then $\mathfrak{e}$ is diagonally metabelian (cf. Remark V.4.6.).

Suppose, in addition, that the following condition is also satisfied:
  (e) *For all $j = 1, \ldots, k$ we have $\dim(\mathfrak{s}_j \cap \mathfrak{e}) \leq 1$.*
Then $\mathfrak{e} \cap \mathfrak{s}$ is central in $\mathfrak{e}$.

Finally, suppose that we have the following condition:
  (f) *$\mathfrak{e} \cap (\mathfrak{d} \oplus \mathfrak{k}) = I \cap J$ where I is an ideal and J is an intersection of hyperplane subalgebras.*
Then $\mathfrak{e} \cap (\mathfrak{d} \oplus \mathfrak{k}) \subseteq \mathfrak{d} \oplus \mathfrak{z}(\mathfrak{k})$.

Proof. By (a) we have $\mathfrak{e} \cap \mathfrak{s}_j \neq \mathfrak{s}_j$, $j = 1, \ldots, k$. Thus $\mathfrak{e} \cap \mathfrak{s}_j$ is at most two-dimensional and thus is diagonally metabelian. By (b), therefore, $\mathfrak{e} \cap \mathfrak{s}$ is diagonally metabelian. By (b) again, the subalgebra $\mathfrak{e}$ is now diagonally metabelian iff $\mathfrak{e} \cap (\mathfrak{d} \oplus \mathfrak{k})$ is diagonally metabelian. Thus we may assume that $\mathfrak{g} = \mathfrak{d} \oplus \mathfrak{k}$. Since $\mathfrak{e} \cap \mathfrak{k} = \{0\}$ by (c), then $\mathfrak{e}$ is the graph of a morphism $\varphi \colon D \to \mathfrak{k}$, $D \subseteq \mathfrak{d}$, namely, $\mathfrak{e} = \{d \oplus \varphi(d) \mid d \in D\}$. Any subalgebra of $\mathfrak{d}$ is diagonally metabelian. But $\mathfrak{e} \cong D$. Hence $\mathfrak{e}$ is diagonally metabelian.

If condition (e) is also satisfied, then $\mathfrak{e} \cap \mathfrak{s}$ is abelian and hence central in $\mathfrak{e}$ by (b).

Finally, we assume (f). Now $\mathfrak{k}' \subseteq J$ because $\mathfrak{k}'$, being a compact semisimple Lie algebra, does not contain any hyperplane subalgebra. Thus $J = \mathfrak{v} \oplus \mathfrak{k}'$ where $\mathfrak{v}$ is an intersection of hyperplane subalgebras. Furthermore, I is an ideal of the ideal direct sum $\mathfrak{d} \oplus \mathfrak{z} \oplus \mathfrak{k}'$, where $\mathfrak{d} \oplus \mathfrak{z}$ is the radical and $\mathfrak{k}'$ the unique Levi complement of $\mathfrak{d} \oplus \mathfrak{z} \oplus \mathfrak{k}'$. Hence $I = \mathfrak{a} \oplus \mathfrak{j}$ where $\mathfrak{a}$ is an ideal in $\mathfrak{d} \oplus \mathfrak{z}$ and $\mathfrak{j}$ an ideal in $\mathfrak{k}'$. Now

$$\mathfrak{e} \cap (\mathfrak{d} \oplus \mathfrak{k}) = I \cap J = (\mathfrak{a} \oplus \mathfrak{j}) \cap (\mathfrak{v} \oplus \mathfrak{k}') = (\mathfrak{a} \cap \mathfrak{v}) \oplus (\mathfrak{j} \cap \mathfrak{k}'),$$

where $\mathfrak{j} = \mathfrak{j} \cap \mathfrak{k}'$ is an ideal of $\mathfrak{d} \oplus \mathfrak{z} \oplus \mathfrak{k}'$, and thus of $\mathfrak{g}$. By (a) it follows that $\mathfrak{j} = \{0\}$. Thus $\mathfrak{e} \cap (\mathfrak{d} \oplus \mathfrak{k}) = \mathfrak{a} \cap \mathfrak{v} \subseteq \mathfrak{d} \oplus \mathfrak{z}$ as asserted. Hence (ii) is proved. □

For a better understanding of the following theorem we note that the semialgebras in $\mathfrak{sl}(2, \mathbb{R})$ are classified in [27], pp.109–110 as follows:

Let $\mathfrak{g} = \mathfrak{sl}(2, \mathbb{R})$. Then every plane $[X, \mathfrak{g}]$ for $k(X) = 0$ is a tangent plane to the double cone $\mathcal{D}^0$ (see IV.1.1) and bounds two half space semialgebras in $\mathfrak{g}$. Every semialgebra in $\mathfrak{g}$ is the intersection of such half-space semialgebras. Recall here that the intersection of any family of semialgebras is a semialgebra. Also recall V.1.7 which states that the semialgebras in $\mathfrak{g}$ which are Lie wedges of weakly exponential semigroups are contained in $\overline{\mathcal{D}^+}$.

We retain the notation of Theorem 2.1, we recall $\mathfrak{d}' = [\mathfrak{d}, \mathfrak{d}] = \mathfrak{g}_{\Lambda^*}$, and for a wedge W in $\mathfrak{g}$ we set, firstly, $W_0 = W \cap (\mathfrak{d} \oplus \mathfrak{k})$ and, secondly,

$$W_{\mathrm{inv}} = \overline{W_0 + \mathfrak{d}'}, \quad \text{and} \quad W_{\mathrm{sec}} = \bigcap \{L_p(W_0) \mid p \in C^1(W_0) \text{ and } \mathfrak{d}' \not\subseteq T_p(W_0)\}.$$

2.5. THEOREM. *Let G be a connected Lie group with a reduced weakly exponential subsemigroup S with Lie wedge W. Decompose $\mathfrak{g}$ as in Theorem 2.1(i). Then the following conclusions hold:*

(i) *There is a canonical decomposition*

$$W = (\mathfrak{s}_1 \cap W) \oplus \cdots \oplus (\mathfrak{s}_k \cap W) \oplus (W \cap (\mathfrak{d} \oplus \mathfrak{k}))$$

for which the following information is available:

   (α) $W_0 = W_{\mathrm{inv}} \cap W_{\mathrm{sec}}$, *further W_{inv} is the smallest invariant wedge containing W_0, and W_{sec} is an intersection of half-space semialgebras containing $\Delta = \mathfrak{k}'$.*

   (β) *Under the isomorphism $\mathfrak{s}_j \cong \mathfrak{sl}(2, \mathbb{R})$, the semialgebras $\mathfrak{s}_j \cap W$ correspond to semialgebras contained in $\overline{\mathcal{D}^+} \subseteq \mathfrak{sl}(2, \mathbb{R})$.*

   (γ) $\dim H(W \cap \mathfrak{s}_j) \leq 1$.

(ii) $H(W)$ *is diagonally metabelian, $H(W) \cap \mathfrak{k} = \{0\}$, and we have an ideal direct sum*

$$H(W) = (H(W) \cap \mathfrak{s}) \oplus H(W_0)$$

such that $H(W) \cap \mathfrak{s}$ is abelian and $H(W_0) \subseteq \mathfrak{d} \oplus \mathfrak{z}$ with $\mathfrak{z} = \mathfrak{z}(\mathfrak{k})$ being the center of $\mathfrak{g}$.

(iii) $\mathrm{comp}(\mathfrak{g}) \cap H(W) = \{0\}$.

Proof. (i) The decomposition of W and Conclusion (α) follows from Theorem 2.1 and Proposition IV.4.8. Now we prove (β): We consider a Cartan subalgebra $\mathfrak{h}$ of $\mathfrak{g}$ meeting int W. The projection of $\mathfrak{h}$ into $\mathfrak{s}_j$ is $\mathfrak{h} \cap \mathfrak{s}_j$ and meets $\mathrm{int}_{\mathfrak{s}_j}(\mathfrak{s}_j \cap W)$. Now $\mathfrak{m} := \mathfrak{s}_j + \mathfrak{h}$ is a test algebra (VI.2.1) containing a Cartan subalgebra. Let M be the Lie subgroup with Lie algebra $\mathfrak{m}$. Then the semigroup $M \cap S$ is weakly exponential by Theorem II.4.2. It is reduced by construction. Now M is a test group (VI.2.1) with Lie algebra $\mathfrak{m} = \mathfrak{a} \oplus \mathfrak{b}$ with $\mathfrak{b} = \mathfrak{s}_j \cong \mathfrak{sl}(2,\mathbb{R})$. Since W is reduced, $\mathfrak{b} = \mathfrak{s}_j \not\subseteq W$ and therefore $W \cap \mathfrak{m}$ does not contain $\mathfrak{m}' = \mathfrak{b}$. Thus by Definition VI.2.4 and the Testing Theorem VI.2.5, the wedge $W \cap \mathfrak{b}$ does not contain any nonzerio compact element of $\mathfrak{b} = \mathfrak{s}_j \cong \mathfrak{sl}(2,\mathbb{R})$. This completes the proof of (β). Finally (γ) follows from the observation that for any $\mathfrak{w}$ is a semialgebra of $\mathfrak{sl}(2,\mathbb{R})$ contained in $\overline{\mathcal{D}^+}$ we have $\dim H(\mathfrak{w}) \leq 1$. This finishes the proof of (i).

(ii) We shall apply Lemma 2.4 with $\mathfrak{e} = H(W)$. For this we must verify the hypotheses (a)–(f) of 2.4.

—(a) follows from the fact that S is reduced.

—(b) follows from Part (i) above.

—(c): By Corollary IV.6.7, $\mathfrak{m} \cap W$ is invariant in $\mathfrak{m}$. Now $\mathfrak{k}$ is a compact ideal and $W \cap \mathfrak{k}$ is invariant in $\mathfrak{k}$. Since $[\mathfrak{s} \oplus \mathfrak{d}, \mathfrak{k}] = \{0\}$ it follows that $W \cap \mathfrak{k}$ is invariant in $\mathfrak{g}$. Thus $\mathfrak{e} \cap \mathfrak{k}$ is an ideal. Now (c) follows from (a).

—(e) follows from (iγ) above.

—(f) is a consequence of $W_0 = W_{\mathrm{inv}} \cap W_{\mathrm{sec}}$ in (i).

Proof of (iii). We know that $W = (W \cap \mathfrak{s}) \oplus W_0$. Thus $H(W) = (H(W) \cap \mathfrak{s}) \oplus (H(W) \cap (\mathfrak{d} \oplus \mathfrak{k}))$. Also $\mathrm{comp}(\mathfrak{g}) = \mathrm{comp}(\mathfrak{s}) \oplus \mathfrak{k}$. Therefore

$$H(W) \cap \mathrm{comp}(\mathfrak{g}) =$$
$$\{H(W) \cap \mathfrak{s} \cap \mathrm{comp}(\mathfrak{g})\} \oplus \{(H(W) \cap (\mathfrak{d} \oplus \mathfrak{k})) \cap \mathfrak{k}\}$$
$$\{H(W) \cap \mathrm{comp}(\mathfrak{s})\} \oplus \{H(W) \cap \mathfrak{k}\}.$$

Now $H(W) \cap \mathrm{comp}(\mathfrak{s}) = \bigoplus_{j=1}^{k}(H(W) \cap \mathrm{comp}(\mathfrak{s}_j)) = \{0\}$ by (i). Also, $H(W) \cap \mathfrak{k} = \{0\}$ by (ii). Thus (iii) is proved. $\square$

2.6. COROLLARY. *Let G be a connected Lie group containing a weakly exponential closed subsemigroup S, with Lie wedge W. We suppose that G is the smallest closed subgroup containing S. Then $\mathrm{comp}(\mathfrak{g}) \cap H(W)$ is invariant under all inner automorphisms of $\mathfrak{g}$.*

Proof. We denote by A the analytic subgroup generated by S and by N the largest normal subgroup of G contained in S. Note that any closed normal subgroup of A_{Lie} is normal in G, since by hypothesis A is dense in G. By Proposition IV.4.2 the semigroup S/N is weakly exponential, so by Theorem

II.5.8 and the preceding remark it is reduced in A_{Lie}/N. We apply Theorem 2.5 to A_{Lie}/N and S/N, and conclude that $\text{comp}(\mathfrak{a}/\mathfrak{n}) \cap H(W/\mathfrak{n}) = \{0\}$. Now

$$\frac{\text{comp}(\mathfrak{g}) + \mathfrak{n}}{\mathfrak{n}} \subseteq \frac{\text{comp}(\mathfrak{a}) + \mathfrak{n}}{\mathfrak{n}} \subseteq \text{comp}(\frac{\mathfrak{a}}{\mathfrak{n}}),$$

hence $\text{comp}(\mathfrak{g}) \cap H(W) \subseteq \mathfrak{n}$. Since $\mathfrak{n} \subseteq H(W)$ we have $\text{comp}(\mathfrak{g}) \cap H(W) = \text{comp}(\mathfrak{g}) \cap \mathfrak{n}$. Thus $\text{comp}(\mathfrak{g}) \cap H(W)$ is invariant under all inner automorphisms of $\mathfrak{g}$, as the intersection of the invariant subsets $\mathfrak{n}$ and $\text{comp}(\mathfrak{g})$. □

2.7. COROLLARY. *Let G be a connected Lie group containing a weakly exponential closed subsemigroup S, with Lie wedge W. Let A be the subgroup generated by S. Then G, A and A_{Lie} induce the same topology on S.*

Proof. Since A and $\overline{A}$ indice the same topology on S we may assume that $G = \overline{A}$. By Corolloray 2.6 the set $\text{comp}(G) \cap H(W)$ is invariant under all inner automorphisms of $\mathfrak{g}$. Thus Proposition II.5.10 applies and establishes the assertion. □

We shall say that a semialgebra which is an intersection of half-space semialgebras in a Lie algebra is an *intersection algebra*. Similarly we say that a subsemigroup of a Lie group which is an intersection of half-space semigroups is an *intersection semigroup*.

2.8. COROLLARY. *Let G be a connected Lie group with a reduced weakly exponential subsemigroup S with Lie wedge W and suppose that G does not contain any compact normal simple subgroup. Then S is an intersection semigroup.*

Proof. Under the present hypotheses, W_{inv} is a trivial wedge in $\mathfrak{d} \oplus \mathfrak{z}$ and therefore an intersection algebra. Thus $W_0 = W_{\text{inv}} \cap W_{\text{sec}}$ is an intersection algebra. Also, since all algebras $W \cap \mathfrak{s}_j$ are intersection algebras by Proposition V.1.6 and V.1.7 also $W_Q = (W \cap \mathfrak{s}_1) \oplus \cdots \oplus (W \cap \mathfrak{s}_j)$ is an intersection algebra. □

2.9. COROLLARY. *Let G be a connected Lie group with a reduced weakly exponential subsemigroup S with Lie wedge W. If $\mathfrak{g}$ is solvable, then $\mathfrak{g}$ is metabelian.* □

2.10. COROLLARY. *Let G be a connected Lie group with a reduced weakly exponential subsemigroup S with Lie wedge W. If $\mathfrak{g}$ is reductive, then there are simple Lie algebras $\mathfrak{s}_1, \ldots, \mathfrak{s}_k$, all isomorphic to $\mathfrak{sl}(2, \mathbb{R})$, and compact simple Lie algebras $\mathfrak{k}_j$, $j = 1, \ldots, p$ such that $\mathfrak{g}$ is an ideal direct sum*

$$\mathfrak{g} = \mathfrak{s}_1 \oplus \cdots \oplus \mathfrak{s}_k \oplus \mathfrak{k}_1 \oplus \cdots \oplus \mathfrak{k}_p \oplus \mathbb{R}^q.$$

If $\mathfrak{g}$ is semisimple then $\mathfrak{g} = \mathfrak{s}_1 \oplus \cdots \oplus \mathfrak{s}_k$. □

2.11. PROPOSITION. *Under the hypotheses and with the notation of* Theorem 2.5 *the following sets are closed exponential subsemigroups:*
  (i) $\exp W_{\text{inv}}$,
  (ii) $\exp W_{\text{sec}}$,
  (iii) $\exp(W_{\text{inv}} \cap W_{\text{sec}})$, *and this set equals* $\exp W_{\text{inv}} \cap \exp W_{\text{sec}}$.

Proof. According to Theorem 2.1 we write $\mathfrak{g} = \mathfrak{s} \oplus \mathfrak{d} \oplus \mathfrak{z} \oplus \mathfrak{k}'$ where $\mathfrak{z}$ is the center of $\mathfrak{g}$ and $\mathfrak{k}' = [\mathfrak{k}, \mathfrak{k}]$ is a compact semisimple ideal. Note $\mathfrak{k} = \mathfrak{z} \oplus \mathfrak{k}'$. By Theorem 2.5(i) we know that $W = (W \cap \mathfrak{s}) \oplus W_0$, $W_0 = W_{\text{inv}} \cap W_{\text{sec}} \subseteq \mathfrak{d} \oplus \mathfrak{z} \oplus \mathfrak{k}'$, where W_{inv} is invariant and contains $\mathfrak{d}'$, and where W_{sec} contains $\mathfrak{k}' = \Delta$ and is an intersection of half-space semialgebras. Also, $W \cap \mathfrak{s} = (W \cap \mathfrak{s}_1) \oplus \cdots \oplus (W \cap \mathfrak{s}_k)$ with $W \cap \mathfrak{s}_j \subseteq \overline{\mathcal{D}^+}$.

(i) Now we set $S_{\text{inv}} = \overline{\langle \exp_G W_{\text{inv}} \rangle}$. Let $D = \exp_G \mathfrak{d}$. Since $\mathfrak{d}$ is centerfree diagonally metabelian, the analytic normal subgroup D is closed. Then $D' = [D, D] \subseteq S_{\text{inv}}$. (Note that the commutator group of a diagonally metabelian connected Lie group is also closed.) Then $\mathcal{L}(S_{\text{inv}}/D') = W_{\text{inv}}/\mathfrak{d}'$ and $DZK'/D' \cong Z \times K'$ has a compact Lie algebra. Since $S_{\text{inv}} = \overline{\langle \exp_G W_{\text{inv}} \rangle}$ we have $S_{\text{inv}}/D' = \overline{\langle \exp_G(W_{\text{inv}}/\mathfrak{d}') \rangle}$. By the Theorem IV.6.8 of MITTENHUBER and NEEB we have $S_{\text{inv}}/D' = \exp_G(W_{\text{inv}}/\mathfrak{d}')$. Let $\mathfrak{h}_\mathfrak{d}$ denote a Cartan subalgebra of $\mathfrak{d}$. Since D' is exponential, $[\mathfrak{d}, \mathfrak{z} + \mathfrak{k}'] = \{0\}$, and $\mathfrak{d}$ is diagonally metabelian, the hypotheses of Lemma IV.4.4 are satisfied with DZK' in place of G, with D' in place of N, and with $\mathfrak{h}_\mathfrak{d} \oplus \mathfrak{z} \oplus \mathfrak{k}'$ in place of $\mathfrak{c}$. Thus Lemma IV.4.4 shows that $S_{\text{inv}} = \exp_G W_{\text{inv}}$.

(ii) Next we set $S_{\text{sec}} = \overline{\langle \exp_G W_{\text{sec}} \rangle}$. Let $K' = \langle \exp_G \mathfrak{k}' \rangle$. Since $\mathfrak{k}'$ is compact semisimple, K' is a compact group. We know $K' \subseteq S_{\text{sec}}$. Then as in the preceding paragraph we see that $S_{\text{sec}}/K' = \overline{\langle \exp_G(W_{\text{sec}}/\mathfrak{k}') \rangle}$ and note that $DZK'/K' \cong D \times Z$ is a metabelian exponential group which is diffeomorphic with $\mathfrak{d} \oplus \mathfrak{z}$ under $\exp_G$. By [27], p. 101, Corollary II.2.42 every Lie semialgebra in the Lie algebra of such a group is mapped diffeomorphically under its exponential function onto a closed subsemigroup. Hence $S_{\text{sec}}/K' = \exp_G(W_{\text{sec}}/\mathfrak{k}')$. Since K' is exponential and centralizes DZ we can again apply IV.4.4 with $G = DZK'$ in place of G, with K' in place of N, and with $\mathfrak{d} \oplus \mathfrak{z}$ in place of $\mathfrak{c}$. We conclude that $S_{\text{sec}} = \exp_G W_{\text{sec}}$.

(iii) We claim that $S_{\text{inv}} \cap S_{\text{sec}} = \exp_G(W_{\text{inv}} \cap W_{\text{sec}})$, which will prove the assertion. The right side is clearly contained in the left, and in order to prove the reverse containment we pick an element $s \in S_{\text{inv}} \cap S_{\text{sec}}$. Since $s \in S_{\text{inv}}$ we find a $w_0 \in W_{\text{inv}}$ such that $s = \exp_G w_0$. Likewise we find a $w_1 \in W_{\text{sec}}$ with $s = \exp_G w_1$. Since W_{inv} and W_{sec} are contained in $\mathfrak{d} \oplus \mathfrak{z} \oplus \mathfrak{k}'$ we can write uniquely $w_0 = d_0 + z_0 + k_0$ and $w_1 = d_1 + z_1 + k_1$ with $d_j \in \mathfrak{d}$, $z_j \in \mathfrak{z}$, and $k_j \in \mathfrak{k}'$. From $\exp_G w_0 = s = \exp_G w_1$ we conclude $\exp_G d_0 = \exp_G d_1$ and $\exp_G z_0 = \exp_G z_1$. These equations yield $d_0 = d_1$, $z_0 = z_1$. Since $\mathfrak{k}' \subseteq W_{\text{sec}}$ we thus have $w_0 = z_0 + d_0 + k_0 = z_1 + d_1 + k_1 - (k_1 - k_0) = w_1 - (k_1 - k_0) \in W_{\text{sec}}$. Since also $w_0 \in W_{\text{inv}}$ we have $w_0 \in W_{\text{inv}} \cap W_{\text{sec}}$, and this proves the claim.

We have shown that the closed subsemigroup $S_0 := S_{\text{inv}} \cap S_{\text{sec}}$ is exponential, agreeing, as it were, with $\exp_G W_0$. □

2.12. Theorem. *Let G be a connected Lie group with a reduced weakly exponential subsemigroup S with Lie wedge W.*

(i) *The simply connected covering group $\widetilde{G}$ of G contains a closed exponential subsemigroup $\widetilde{S}$ such that $\widetilde{S} = \overline{\exp_{\widetilde{G}} W}$.*

(ii) *If $p \colon \widetilde{G} \to G$ is the universal covering homomorphism then $S = p(\widetilde{S})$.*

(iii) *S is exponential.*

Proof. (i) From Theorem 2.1, $\widetilde{G}$ may be identified with the (internal) direct product $\widetilde{Q}\widetilde{D}\widetilde{Z}\widetilde{K'} = \widetilde{Q} \times \widetilde{D} \times \widetilde{Z} \times \widetilde{K'}$ where $\widetilde{Q} \cong \widetilde{\mathrm{Sl}}(2,\mathbb{R})^k$, $\widetilde{D} \cong D$ is an exponential centerfree metabelian Lie group (diffeomorphic to $\mathfrak{d}$ under $\exp_{\widetilde{G}}$), $\widetilde{Z} \cong \mathbb{R}^n$, and $\widetilde{K'}$ is a simply connected semisimple compact group. We set $\widetilde{S}_Q = \overline{\langle \exp_{\widetilde{G}}(W \cap \mathfrak{s}) \rangle} = \overline{\langle \exp_{\widetilde{G}}(W \cap \mathfrak{s}_1) \rangle} \times \cdots \times \overline{\langle \exp_{\widetilde{G}}(W \cap \mathfrak{s}_k) \rangle}$. By Proposition V.1.7, since $W \cap \mathfrak{s}_j \subseteq \mathcal{D}^+$, we have $\overline{\langle \exp_{\widetilde{G}}(W \cap \mathfrak{s}_j) \rangle} = \exp_{\widetilde{G}}(W \cap \mathfrak{s}_j)$. Thus $\widetilde{S}_Q$ is an exponential subsemigroup of $\widetilde{Q}$.

We set $\widetilde{S}_{\mathrm{inv}} = \exp_{\widetilde{G}} W_{\mathrm{inv}}$, $\widetilde{S}_{\mathrm{sec}} = \exp_{\widetilde{G}} W_{\mathrm{sec}}$, and $\widetilde{S}_0 = \widetilde{S}_{\mathrm{inv}} \cap \widetilde{S}_{\mathrm{sec}}$. We know from Proposition 2.11 that $\widetilde{S}_0$ is an exponential semigroup satisfying $\widetilde{S}_0 = \exp_{\widetilde{G}} W_0$. Now we set $\widetilde{S} = \widetilde{S}_Q \times \widetilde{S}_0$ in $\widetilde{G} = \widetilde{Q} \times \widetilde{D}\widetilde{Z}\widetilde{K'}$. Then $\widetilde{S}$ is a closed subsemigroup of $\widetilde{G}$ with $\mathcal{L}(\widetilde{S}) = (W \cap \mathfrak{s}) \oplus W_0 = W$ and $\exp_{\widetilde{G}} W = \exp_{\widetilde{G}}(W \cap \mathfrak{s}) \times \exp_{\widetilde{G}}(W_{\mathrm{inv}} \cap W_{\mathrm{sec}}) = \widetilde{S}_Q \times (\widetilde{S}_{\mathrm{inv}} \cap \widetilde{S}_{\mathrm{sec}}) = \widetilde{S}$. Thus $\widetilde{S} = \overline{\exp_{\widetilde{G}} W} = \overline{\langle \exp_{\widetilde{G}} W \rangle}$ and we conclude $p(\widetilde{S}_0) = p(\exp_{\widetilde{G}} W_0) = \exp_G W_0 = S_0$. In particular,

(b*) $p(\widetilde{S}_0)$ is closed.

(ii) We want to apply Proposition V.2.5 with $G_1 = Q$, $G_2 = DZK'$, $T_1 = S_Q$, $T_2 = S_0$. Hypothesis (a) is satisfied. As S is reduced, $\mathrm{int}\, W \neq \emptyset$ (see IV.4.7) and thus hypothesis (c) of V.2.5 is satisfied, too. From (b*) we know that $p(\widetilde{S}_0)$ is closed in G. This verifies hypothesis (b) of V.2.5. Therefore, V.2.5 applies and shows that $S = p(\widetilde{S}_Q)p(\widetilde{S}_0) = p(\widetilde{S}_Q)S_0 = p(\widetilde{S})$ is closed.

(iii) From (ii) we get $p(\widetilde{S}) = p(\exp_{\widetilde{G}} W) = \exp_G W$. □

2.13. Proposition. *For a connected Lie group G the following statements are equivalent:*

(i) *G supports a reduced weakly exponential subsemigroup which is invariant.*

(ii) *G supports a reduced exponential subsemigroup which is invariant.*

(iii) *G is a noncompact group with compact Lie algebra $\mathfrak{g}$.*

Proof. The implications (ii)$\Rightarrow$(i) is trivial; (iii)$\Rightarrow$(ii) follows from Proposition V.5.1. (i)$\Rightarrow$(ii) is known from Theorem 2.12 above.

(i)$\Rightarrow$(iii) Suppose that S is an invariant reduced subsemigroup of G. Then W is an invariant wedge in $\mathfrak{g}$ and therefore the ideal $\mathfrak{s}$ in the decomposition $\mathfrak{g} = \mathfrak{s} \oplus \mathfrak{d} \oplus \mathfrak{k}$ of VII.4.7 must vanish, by 2.5. Thus, in the notation of Theorem 2.5, $W = W_0 = W_{\mathrm{inv}}$. Since by 2.5(ii) W_{inv} contains $\mathfrak{d}' = [\mathfrak{d},\mathfrak{d}]$, and since S is reduced, $\mathfrak{d}'$ must vanish. But by its very construction, $\mathfrak{d}$ is center free, so $\mathfrak{d} = \{0\}$ and we conclude $\mathfrak{g} = \mathfrak{k}$, that is, $\mathfrak{g}$ is compact. □

2.14. THEOREM. *Let G be a Lie group and S a weakly exponential subsemigroup which does not contain any nondegenerate subgroups which are invariant under inner automorphisms implemented by elements of S. Then S is exponential.*

Proof. Let A be the analytic subgroup $\langle S \cup S^{-1} \rangle$. Then by the Intrinsic Embedding Theorem II.5.8 the semigroup S is weakly exponential in the topology of A_{Lie} and has inner points in A_{Lie}. By hypothesis it is in fact reduced since it does not contain any nontrivial subgroups normal in A. Then it is exponential by Theorem 2.12. □

2.15. COROLLARY. *Any weakly exponential subsemigroup without nontrivial units in a Lie group is exponential.* □

CHAPTER 9

APPENDIX: THE HYPERSPACE OF A LOCALLY COMPACT SPACE

Let Γ_{mn} denote the set of all m-dimensional vector subspaces of $\mathbb{R}^n$. The orthogonal group $O(n)$ operates transitively on Γ_{mn}, and if we identify $\mathbb{R}^m$ with an element of Γ_{mn}, then the isotropy group at this element is $O(m) \times O(n-m)$ considered as a subgroup of $O(n)$. The natural bijection $O(n)/(O(m) \times O(n-m)) \to \Gamma_{mn}$ allows us to consider Γ_{mn} as a compact manifold. Manifolds of this kind are called GRASSMANN manifolds. They are well-known and much studied objects.

BOURBAKI used Haar measures to equip the set of all closed subgroups of a locally compact group G with a compact Hausdorff topology which he then also described in alternative ways. ([2], p. 174ff., p. 206ff.)

F. HAUSDORFF defined a metric on the set of closed subsets of a compact metric space and noted that his metric defines a compact Hausdroff topology on this set. Later VIETORIS noted that in fact the set of all closed subsets of an arbitrary compact Hausdorff space X could be given a compact Hausdorff topology in a more or less natural fashion. The resulting space is frequently called the *hyperspace of* X and it plays a role in the topology of continua. A further step was taken by J. M. G. FELL in [20]. He constructed a compact Hausdorff topology on the space of closed subsets of a not necessarily Hausdorff locally quasicompact space. (His result is recovered in 1.8 below.)

In recent years, it turned out to be surprisingly useful to investigate the space of closed subgroups of a Lie group and to develop this tool in a manner parallel to the space of all subalgebras of the corresponding Lie algebra [32, 70]. We have seen this in Chapter I, II.3.7.

The field which provides the exact tools needed to associate with a locally compact topological space X a compact Hausdorff topology τ on the set $\Gamma(X)$ of its closed subspaces is the area of *continuous lattices*. Furthermore, the theory of continuous lattices makes us understand that this works only for locally compact spaces. Indeed $\Gamma(X)$ is a distributive lattice, whose opposite lattice is the lattice $\mathcal{O}(X)$ of open sets of X. These lattices are complete lattices (i.e., permit the formation of arbitrary infs and sups); but $\mathcal{O}(X)$ is, for a Hausdorff space X at least, a continuous lattice if and only if X is locally compact. Therefore, the lattice theoretical counterpiece of locally compact spaces is the theory of distributive continuous lattices. There is a considerable body of literature on them, but it may not be easy to penetrate to the relevant portions.

1. Continuous Lattices, the Lawson Topology, and Hyperspaces

A subset U of a partially ordered set L is called an *upper* set if

$$(\forall u \in U, x \in L) \quad (u \leq x) \Rightarrow (x \in U).$$

For any subset $S \subseteq L$ the upper set $\uparrow S := \{y \in L : (\exists s \in S) \ s \leq y\}$ is called the *upper set generated by* S. We write $\uparrow x$ instead of $\uparrow \{x\}$. *Lower* sets and the set $\downarrow S$ are defined dually.

A subset D of L is *directed* if every nonempty finite subset has an upper bound in D.

We begin with a complete lattice L and we define two topologies σ and ω. The open sets of σ are all upper sets, those of ω are lower sets, but the two topologies are not defined dually to each other.

σ is the set of all upper sets U such that for every directed subset $D \subseteq L$ the relation $\sup D \in U$ implies $D \cap U \neq \emptyset$. Clearly, σ is closed under arbitrary unions and finite intersections, and $L \setminus \downarrow x \in \sigma$ for all x; hence σ is a topology called the *Scott-topology* on L.

Because of $\uparrow S \cap \uparrow T = \uparrow\{s \vee t : s \in S, t \in T\}$ and $\uparrow S \cup \uparrow T = \uparrow(S \cup T)$, the set of all subsets $\uparrow F$ generated by finite subsets F of L is a sublattice of the powerset of X. In particular, it is a basis for the collection of closed sets of a topology, namely, ω. This topology is called the *weak topology*.

1.1. DEFINITION. The topology $\lambda = \sigma \vee \omega$ is called the LAWSON-*topology* of L. □

We say a space is *quasicompact* if it has the Heine-Borel property, and *compact* if it is, in addition, Hausdorff. A topological space X is *locally quasicompact* if for every $x \in X$ and every open neighborhood V of x there is an open neighborhood U of x and a quasicompact subspace Q such that $U \subseteq Q \subseteq V$.

1.2. PROPOSITION. *On every complete lattice L, the Lawson topology is a quasicompact topology in which all singleton sets are closed.*

Proof. See [21], p. 146. □

In particular, if X is an arbitrary topological space, then the topology $L = \mathcal{O}(X)$ carries a quasicompact topology, as does, as a consequence, the lattice $\Gamma(X) \cong L^{op}$ of closed subsets of X. The point is, however, that for a particular class of complete lattices the Lawson topology is Hausdorff.

In order to describe this class we consider an auxiliary transitive relation on a complete lattice L: We write $x \ll y$ if for all directed subsets $D \subseteq L$ the relation $y \leq \sup D$ implies the existence of some $d \in D$ with $x \leq d$. We also write $\Uparrow x = \{y \in L : x \ll y\}$ and $\Downarrow y = \{x \in L : x \ll y\}$.

1.3. DEFINITION. A lattice L is said to be a *continuous lattice* if for each element x we have $x = \sup {\downarrow} x$. □

There is a simple but very effective tool for continuous lattices, called the *Interpolation Lemma*:

1.4. LEMMA. *If $a \ll b$ in a continuous lattice then there is an element x with $a \ll x \ll b$.*

Proof. [21], p. 47. □

As a consequence, in every continuous lattice L the sets ${\Uparrow} x$ are members of σ, and it is not difficult to verify directly from the definitions that $\{{\Uparrow} x : x \in L\}$ is a basis for the topology σ. [Indeed if $x \in U \in \sigma$ then $x = \sup {\downarrow} x$ and ${\downarrow} x$ is directed. Hence there is a $u \in {\downarrow} x \cap U$, and then $x \in {\Uparrow} u \subseteq U$.] Therefore, the Lawson topology of a continuous lattice has a basis of sets of the form ${\Uparrow} x \cap (L \backslash {\uparrow} F)$ where $x \in L$ and $F \subseteq L$ is finite. We may reformulate this

1.5. PROPOSITION. *In a continuous lattice L, the Lawson topology has a basis consisting of sets of the form*
$$V(x_0, \ldots, x_n) := {\Uparrow} x_0 \setminus ({\uparrow} x_1 \cup \cdots \cup {\uparrow} x_N), \quad x_0, \ldots x_N \in L. \ \square$$

If $y_0 \leq x_0$ and $x_j \leq y_j$ for $j = 1, \ldots, n$ then
$$V(x_0, \ldots x_n) \leq V(y_0, \ldots, y_n).$$

We illustrate these concepts by looking at their impact on topology.

1.6. PROPOSITION. *Suppose that X is a topological space and $L = \mathcal{O}(X)$. Then*
  (i) *For $U \ll V$ in L to hold it is sufficient that there is a quasicompact subset Q of X such that $U \subseteq Q \subseteq V$. If X is locally quasicompact, this condition is also necessary.*
  (ii) *If X is locally quasicompact then L is a continuous lattice*
  (iii) *If X is Hausdorff regular and L is continuous, then X is locally compact.*

Proof. [21], p. 40 for (i). Condition (ii) is immediate from (i). For (iii) see [21], p. 42. □

Condition (iii) is in fact a weak form of the much stronger condition
  (iii′) *If X is a sober space, i.e. a T_0-space in which every irreducible closed set is a singleton closure, and if L is a continuous lattice, then X is locally quasicompact.*

For the concept of sobriety and this condition we refer to [21], p. 79, p. 251, p. 259. There are topological T_0-spaces X for which $\mathcal{O}(X)$ is a continuous lattice and which fail to be locally quasicompact so badly that every quasicompact subset has empty interior. (See [21], p. 265, 5.22.)

Now we make our definitions formal:

1.7. DEFINITION. Let X be a locally quasicompact topological space. Then $\Gamma(X)$ is the lattice of all closed subsets equipped with the compact Hausdorff topology τ for which the function $A \mapsto X \setminus A \colon (\Gamma(X), \tau) \to (\mathcal{O}(X), \lambda)$ is a homeomorphism. The topological space $(\Gamma(X), \tau)$ is called the hyperspace of X. □

Now we wish to describe a basis for the topology τ on the hyperspace. Let $A_0, \ldots A_n$ be arbitrary closed subsets of X. By 1.5 the sets

$$W(A_0, \ldots, A_n) = \{A \in \Gamma(X) : X \setminus A \in \Uparrow(X \setminus A_0) \setminus (\uparrow(X \setminus A_1) \cup \cdots \cup \uparrow(X \setminus A_n))\}$$

form a basis of τ. This needs some conversion. We have $X \setminus A \in \Uparrow(X \setminus A_0)$ iff there is a quasicompact subset Q of X such that $X \setminus A_0 \subseteq Q \subseteq X \setminus A$. This is tantamount to $X \setminus A_0 \subseteq Q$ and $Q \cap A = \emptyset$ Further, $X \setminus A \notin \uparrow(X \setminus A_j)$ iff $X \setminus A_j \not\subseteq X \setminus A$ iff $A \not\subseteq A_j$ iff $A \cap (X \setminus A_j) \neq \emptyset$. We rewrite this in terms of the open sets $U_j = X \setminus A_j$. Thus:

1.8. PROPOSITION. (i) *The topology of the hyperspace $\Gamma(X)$ of a locally quasicompact space has a basis consiting of all sets of the form $W(U_0, \ldots, U_n)$, where the U_j are open sets in X and where*

$$W(U_0, \ldots, U_n) = \\ \{A \in \Gamma(X) : (\exists Q \text{ quasicompact in } X) \quad (U_0 \subseteq Q) \text{ and } (A \cap Q = \emptyset) \\ \text{ and } (A \cap U_j \neq \emptyset, \quad j = 1, \ldots, n)\}.$$

(ii) *The relations $V_0 \subseteq U_0$ and $U_j \subseteq V_j$, $j = 1, \ldots, n$ imply $W(U_0, \ldots, U_n) \subseteq W(V_0, \ldots, V_n)$.* □

This description of $\Gamma(X)$ was first given by FELL [20].

1.9. PROPOSITION. *Let X be a locally quasicompact space whose topology has a countable basis. Then the topology of $\Gamma(X)$ has a countable basis and is therefore a compact metrizable space.*

Proof. We use the result and the notation of 1.8. If $\mathcal{B}$ is a countable basis of $\mathcal{O}(X)$ then

$$\{W(U_0, \ldots, U_n) : \quad n = 1, 2, \ldots, \quad U_j \in \mathcal{B}\}$$

is a countable set $\mathcal{C}$. Suppose that $A \in W(V_0, \ldots, V_n)$ for $V_j \in \mathcal{O}(X)$, $j = 0, \ldots, n$. Then there is a quasicompact set Q such that $V_0 \subseteq Q$ and $Q \cap A = \emptyset$, and there are elements $a_j \in A \cap V_j$ for $j = 1, \ldots n$. Since A is closed, every point $q \in Q$ has an open neighborhood $W(q)$ not meeting A. Then by the local quasicompactness of X, every point q has a basic open neigborhood $U_0(q) \in \mathcal{B}$ such that there is a quasicompact set $Q(q)$ such that $U_0(q) \subseteq Q(q) \subseteq W(q)$. By the quasicompactness of Q, there is a finite sequence $q_1, \ldots, q_k$ such that $Q \subseteq U_0(q_1) \cup \cdots \cup U_0(q_k)$. Thus if we set $U_0^j := U_0(q_j)$ and $Q_j := Q(q_j)$ we have a finite sequence of sets $U_0^1, \ldots U_0^k \in \mathcal{B}$ and a sequence of quasicompact subsets

$Q_1, \ldots Q_k$ such that $U_0^j \subseteq Q_j$ and $Q_j \cap A \subseteq W(q_j) \cap A = \emptyset$ for $j = 1, \ldots, k$ and that $V_0 \subseteq Q \subseteq U_0^1 \cup \cdots \cup U_0^k$. Further, there are basic open neighborhoods $U_j \in \mathcal{B}$ of a_j, such that $a_j \in U_j \subseteq V_j$ for $j = 1, \ldots, n$. Now

$$A \in W(U_0^1, U_1, \ldots U_n) \cap \cdots \cap W(U_0^k, U_1, \ldots U_n) \subseteq W(V_0, \ldots V_n)$$

in view of 1.8(ii), and thus $\mathcal{C}$ is a basis of the topology of $\Gamma(X)$ in view of 1.8(i). □

If X is a Hausdorff space then Q is closed and thus $C := \overline{U_0}$ is compact and one may take $Q = C$ in the definition of $W(U_0, \ldots U_j)$. This set really depends on C. For an $(n+1)$-tuple $(C; U_1, \ldots, U_n)$ consisting of a compact set C and open sets U_j let us write $W(C; U_1, \ldots, U_n) =$

$$\{A \in \gamma(X) : (A \cap C = \emptyset) \text{ und } (A \cap U_j \neq \emptyset, \quad j = 1, \ldots, n)\}.$$

There are more sets $W(C; U_1, \ldots, U_n)$ than there are sets $W(U_0, \ldots, U_n)$ with $\overline{U_0}$ compact. However, if $A \in W(C; U_1, \ldots, U_n)$ then in view of the local compactness of X and the compactness of C we find an open neighborhood U_0 of C such that $Q := \overline{U_0}$ is compact and satisfies $Q \cap A = \emptyset$. Thus $A \subseteq W(U_0, \ldots, U_n) \subseteq W(C; U_1, \ldots, U_n)$. This shows that we have the following specialisation of 1.8 to the class of Hausdorff spaces:

1.10. COROLLARY. *If X is a Hausdorff space then the compact Hausdorff topology of the hyperspace $\Gamma(X)$ has a basis of sets*
$W(C; U_1, \ldots, U_n)$, C *compact in X,* $U_1, \ldots, U_n$ *open in X.* □

We note that

$$((D \subseteq C) \text{ and } (U_j \subseteq V_j, j = 1, \ldots, n)) \Rightarrow (W(C; U_1, \ldots, U_n) \subseteq W(D; V_1, \ldots, V_n)),$$

that
$$W(C; U_1, \ldots, U_n) = W(C; U_1) \cap \cdots \cap W(C; U_n),$$

and that, consequently, the family of sets $W(C; U)$ forms a subbasis of τ as C ranges through the compact and U through the open sets of X.

If X is compact Hausdorff, then Corollary 1.10 shows that we have recovered the classical Vietoris topology which people have used since the twenties. In this special case, however, experience shows that much can be said for viewing the Vietoris topology as being the topology associated with a uniform structure (which is unique since τ is a compact Hausdorff topology): Indeed since X is compact Hausdorff its topology derives from a unique uniform structure. If $\mathcal{U} \subseteq X \times X$ is an entourage of this uniform structure on X then we set

$$\mathcal{U}^* = \{(A, B) \subseteq \Gamma(X) \times \Gamma(Y) : A \subseteq \mathcal{U}(B) \text{ and } B \subseteq \mathcal{U}(A)\}.$$

Then the filter basis of all $\mathcal{U}^*$ as $\mathcal{U}$ ranges through the entourages of the uniform structure of X generates a Hausdorff uniform structure on $\Gamma(X)$ whose associated topology is τ. [Indeed if $A \in W(C;U)$ then $A \cap C = \emptyset$ and $A \cap U \neq \emptyset$ and we claim that we find an entourage $\mathcal{U} \subseteq X \times X$ such that $\mathcal{U}(A) \subseteq W(C;U)$; indeed let $u \in A \cap U$ and select $\mathcal{U}$ so that $\mathcal{U}(A) \cap C = \emptyset$ and $\mathcal{U}(u) \subseteq U$; this entourage will satisfy the requirements. Thus $(\Gamma(X), \text{uniform topology}) \to (\Gamma(X), \tau)$ is continuous and then the compactness of the uniform topology proves the equality of the uniform topology and τ.]

If, moreover, (X, d) is a compact metric space and A and B are closed subsets of X we set

$$D(A, B) = \max\{\sup\nolimits_{a \in A} \inf\nolimits_{b \in B} d(a,b), \sup\nolimits_{b \in B} \inf\nolimits_{a \in A} d(a,b)\}.$$

This is the Hausdorff metric, and we recognize rather directly that the uniform structure derived from this metric is the uniform structure which gave us the Vietoris topology. Thus the topology τ indeed generalizes the classical hyperspace topologies. The classical definitions have no chance to work in the absence of Hausdorff separation. Thus, no doubt, the definition via the Lawson topology of continuous lattices is conceptually the most appropriate definition. Condition (iii) of Proposition 1.6 explains why—at least inside the class of Hausdorff topological spaces—we have no chance to find reasonable compact Hausdorff hyperspace topologies outside the subclass of locally compact spaces. *The class of locally (quasi)compact topological spaces is the correct domain for a theory of Hausdorff compact hyperspaces.*

2. Continuous Functions

We have to know how continuous functions $f: X \to Y$ affect hyperpsaces and continuous functions between them. To put it more formally: we have to understand in which way $X \mapsto \Gamma(X)$ is a functor. For instance, we certainly want to know for a locally compact Hausdorff space X how the hyperspace (X, τ) is related to the Vietoris hyperspace of the one-point-compactification $X \cup \{\infty\}$ which is, so to speak, a classical object. More generally: For a closed, or an open subspace, or, more aptly, any locally compact subspace Y of a locally compact Hausdorff space X, we certainly want to know how the hyperspace $\Gamma(Y)$ is related to $\Gamma(X)$. In all of these cases we are dealing with inclusion maps $f: Y \to X$.

While the definition of the hyperspace $\Gamma(X)$ in Section 1 was a relatively straightforward matter, the question of the functioriality is, perhaps somewhat unexpectly, more delicate. A closer inspection of the lattice theoretical background, however, again will completely clarify what is going on; it just needs a little patience.

Let us return for the moment to the class of continuous lattices which we have already recognized as being the platform on which we deal with locally compact spaces in a lattice theoretical fashion. To each continuous lattice L we have associated a compact Hausdorff space $\Lambda(L) = (L, \lambda)$, namely, the set of elements with the Lawson topology. Since $\lambda = \sigma \vee \omega$, a function $g: L_1 \to L_2$ between continuous lattices will certainly induce a λ-continuous function if it induces an ω-continuous function and a σ-continuous function.

First, ω: If g preserves arbitrary infs, then for $y \in L_2$ we set $x = \inf g^{-1}(\uparrow y)$. Then $g(x) = \inf g(g^{-1}(\uparrow y))$. Since $g(g^{-1}(\uparrow y)) \subseteq \uparrow y$ we have $y \leq g(x)$, i.e., $g(x) \in \uparrow y$. Thus $x = \min g^{-1}(\uparrow y)$ and if $x' \in g^{-1}(\uparrow y)$, then $x \leq x'$. If g preserves arbitrary infs it preserves infs of two comparable elements and is, therefore monotone. If $x \leq x'$, then $y \leq g(x) \leq g(x')$ whence $x' \in g^{-1}(\uparrow y)$. Hence

$$g^{-1}(\uparrow y) = \uparrow x.$$

Since the sets $\uparrow y$, respectively, $\uparrow x$ are subbasic sets of the weak topologies of L_2, respectively, L_1, the ω-continuity of g follows.

Secondly, σ: If g preserves sups of directed sets (or, as one says briefly, directed sups), then a quick verification shows that for V Scott-open in L_2 the set $g^{-1}(V)$ is Scott-open in L_1.

Conclusion: *If g preserves arbitrary infs and directed sups, then it is Lawson continuous.*

Such maps are sometimes called $\bigwedge$-$\bigvee$-maps. The unit interval $[0,1]$ is a continuous lattice and the natural topology is the Lawson topology. There are numerous continuous self-maps which neither preserve infs nor sups. It is, therefore, not feasible to inspect the converse implication; the Lawson topologies are too fine. A function between complete lattices is Scott continuous iff if preserves directed sups. In [21], p. 143 it is explained what one can expect as consequences of ω-continuity. There are many very good reasons why the $\bigwedge$-$\bigvee$-maps play a central role in continuous lattice theory: Continuous lattices may be singled out from the class of complete lattices as those which satisfy infinitary equations involving arbitrary infs and directed sups. A subset of a large cube $[0,1]^J$ is a continuous lattice if and only if it is closed under pointwise infs and pointwise directed sups; indeed *every* continuous lattice occurs in this fashion. These aspects are discussed in detail in [21].

The class of continuous lattices and the $\bigwedge$-$\bigvee$-maps form a category **CL** and the class of compact Hausdorff spaces and continuous functions between them forms a category **comp**. A functor $F: \mathbf{A} \to \mathbf{B}$ is *faithful* if $F(f) = F(g)$ implies $f = g$. Our observations congel to the following summary:

2.1. PROPOSITION. *The assignment which associates with any continuous lattice L the compact Hausdorff space $(\Lambda(L), \lambda)$ and with any $\bigwedge$-$\bigvee$-map between continuous lattices the same function as λ-continuous function is a faithful functor $\Lambda: \mathbf{CL} \to \mathbf{comp}$.* $\square$

If L denotes a continuous lattice and $x_0 \in L$, then $\uparrow x_0$ is a continuous lattice and the function $x \mapsto x \wedge x_0 : L \to L$ is a $\bigwedge$-$\bigvee$-retraction of L with image $\uparrow x_0$. As an immediate consequence we note

2.2. PROPOSITION. *Suppose that X is a locally quasicompact space and X_0 a closed subspace. Then the subspace $\Gamma(X;X_0)$ of all closed subsets of the hyperspace $\Gamma(X)$ containing X_0 is a closed hence compact subset, and the function*
$$A \mapsto A \cup X_0 : \Gamma(X) \to \Gamma(X;X_0)$$
is a continuous retraction. □

Now we consider the category of topological spaces and continuous maps. We associate with each topological space X the complete lattice $\mathcal{O}(X)$. (Recall that if X is locally quasicompact, then the lattice $\Gamma(X)$ of closed subsets is none other than the opposite lattice $\mathcal{O}(X)^{op} \cong \Gamma(X)$. Then $\mathcal{O}(X)$ is a continuous lattice). With a continuous function $f: X_1 \to X_2$ we associate the function $\mathcal{O}(f): \mathcal{O}(X_2) \to \mathcal{O}(X_1)$ given by $\mathcal{O}(f)(V) = f^{-1}(V)$. The function $\mathcal{O}(f)$ preserves arbitrary unions and intersections. As far as open sets are concerned, this means that it preserves arbitrary sups and finite infs. We say that it is a $\wedge$-$\vee$-map. All of this applies, in particular, to the full subcategory **Locp** of locally quasicompact topological spaces and continuous maps.

Let us denote by **cframe** the category of continuous distributive lattices and $\wedge$-$\vee$-maps. Then the assignment $\mathcal{O}$ is a contravariant functor from the category **Locp** to the category **cframe**. If we denote with **cframe**^{op} the opposite category (sometimes called the category of *locally compact locales*), then we have a functor $\mathcal{O}: \textbf{Locp} \to \textbf{cframe}^{op}$.

Now the problem with our candidate Γ for a contravariant functor **Locp** $\to$ **comp** becomes apparent. Replacing temporarily Γ by $\mathcal{O}$ (which is really no restriction of generality, as we know), we clearly want to compose $\mathcal{O}$ with Λ. But we can't because **cframe** is not a subcategory of **CL** even though the class of objects of theses categories agree. We therefore must determine that class of continuous functions $f: X_1 \to X_2$ between topological spaces for which $\mathcal{O}(f): \mathcal{O}(X_2) \to \mathcal{O}(X_3)$ preserves arbitrary, not only finite infs.

This is the case if and only if for each family $\{V_j : j \in J\}$ of open subsets of X_2 we have
$$\text{int} \bigcap_{j \in J} f^{-1}(U_j) = f^{-1}\left(\bigcap_{j \in J} U_j\right).$$

We note that $\text{int} \bigcap_{j \in J} f^{-1}(U_j) = \text{int } f^{-1}\left(\bigcap_{j \in J} U_j\right)$. On each topological space we have a transitive and reflexive relation $\preceq$ given by $x \preceq y$ iff $\overline{\{x\}} \subseteq \overline{\{y\}}$ iff $x \in \overline{\{y\}}$. Then $\preceq$ is equality if and only if all singletons are closed; a fortiori this is true if X is Hausdorff. We say that a set is *saturated* if it is an upper set w.r.t. the relation $\preceq$. Then *a subset of a topological space is saturated if and only if it is the intersection of a collection of open sets*. In a Hausdorff space every subset is saturated. If T is an arbitrary subset, then the intersection of the family $\mathcal{V}$ of all open subsets V of X containing T is the smallest saturated subset containing T, the saturation T^*.

Notice that on account of the continuity of f, for any subset S of X_2, we have $\text{int } f^{-1}(S) \supseteq f^{-1}(\text{int } S)$. As a consequence of this and the preceding remarks we record:

2.3. LEMMA. *For a continuous function $f: X_1 \to X_2$ between topological spaces the function $\mathcal{O}(f)$ preserves arbitrary sups if and only if for each saturated subset S of X_2 we have*
(∗) $$\operatorname{int} f^{-1}(S) \subseteq f^{-1}(\operatorname{int} S). \quad \square$$

2.4. DEFINITION. A function $f: X_1 \to X_2$ between topological spaces is called *open* if $f(U)^* \in \mathcal{O}(X_2)$ for all $U \in \mathcal{O}(X_1)$. $\square$

A sufficient condition for the openness of f is that open sets have open images, and if X_2 is Hausdorff this is also necessary.

2.5. PROPOSITION. *For a continuous function $f: X_1 \to X_2$ between topological spaces the function $\mathcal{O}(f)$ preserves arbitrary infs if and only if it is open.*

Proof. We shall invoke Lemma 2.3 and show that (∗) holds iff f is open.

First suppose that f is open and let S be any saturated subset of X_2. Set $U = \operatorname{int} f^{-1}(S)$. Then U is open. Now $f(U) \subseteq ff^{-1}(S) \subseteq S$, whence $f(U)^* \subseteq S$ since S is saturated. As f is open, $f(U)^*$ is open and thus $f(U) \subseteq f(U)^* \subseteq \operatorname{int} S$. Thus $U \subseteq f^{-1}(\operatorname{int} S)$ and thus (∗) holds.

Conversely, assume (∗) and let U be an open subset of X_1. Define $S = f(U)^*$. Then $U \subseteq \operatorname{int} f^{-1}(S) \subseteq f^{-1}(\operatorname{int} S)$ (by (∗)), and this implies $f(U) \subseteq \operatorname{int} S$. Then $\operatorname{int} S \in \mathcal{V}$, and thus $S = \bigcap_{f(U) \subseteq V \in \mathcal{O}(X_2)} V \subseteq \operatorname{int} S \subseteq S$. Thus $S = f(U)^*$ is open. $\square$

Now suppose that $f: X_1 \to X_2$ is continuous and open. Then
$$\mathcal{O}(f): \mathcal{O}(X_2) \to \mathcal{O}(X_1)$$
is a $\bigwedge$-$\bigvee$-map. Hence $\mathcal{O}(f)$ is a continuous map
$$\Lambda(\mathcal{O}(X_2)) \to \Lambda(\mathcal{O}(X_1)).$$
If we set $\Gamma(f)(A) = f^{-1}(A)$, then $\Gamma(f): \Gamma(X_2) \to \Gamma(X_1)$ is continuous. Thus we have seen:

2.6. THEOREM. *Every continuous and open function $f: X_1 \to X_2$ between topological spaces yields a continuous function $\Gamma(f): \Gamma(X_2) \to \Gamma(X_1)$ between compact hyperspaces.* $\square$

In other words, the hyperspace assignment Γ is a contravariant functor from the full category of locally quasicompact topological spaces and continuous and open maps into the category of compact Hausdorff spaces.

2.7. COROLLARY. *Suppose that U is an open subset of a locally quasicompact topological space X. Then*
  (i) *$A \mapsto A \cap U: \Gamma(X) \to \Gamma(U)$ is a surjective continuous function beetween compact spaces, and*
  (ii) *$B \mapsto B \cup (X \setminus U): \Gamma(U) \to \Gamma(X; X \setminus U)$ is a homeomorphism.*

Proof. Conclusion (i) is a consequence of Theorem 2.6 which we apply to the continuous open inclusion map $f: U \to X$. In order to prove (ii) we first recall from 2.2 that $A \mapsto A \cup (X \setminus U): \Gamma(X) \to \Gamma(X; (X \setminus U))$ is a continuous retraction. Since U is open in X, a subset B of U is closed in U iff $B \cup (X \setminus U)$ is closed in X. Thus the function $A \mapsto A \cap U: \Gamma(X; X \setminus U) \to \Gamma(U)$ is bijective; it is continuous as a restriction of the continuous function in (i). Because of the compactness of $\Gamma(X, X \setminus U)$ it is a homeomorphism and the function in (ii) is its inverse. $\square$

An interesting special case is the issue of the one point compactification of a locally compact Hausdorff space:

2.8. COROLLARY. *Let X be a locally compact Hausdorff space and $X \cup \{\infty\}$ its one-point-compactification. Then the function*
$$A \mapsto A \cap X: \Gamma(X \cup \{\infty\}) \to \Gamma(X)$$
is a surjective continuous function, and
$$A \mapsto A \cup \{\infty\}: \Gamma(X) \to \Gamma(X \cup \{\infty\}; \{\infty\})$$
is a homeomorphism. $\square$

This corollary permits us to link the hyperspace of a locally compact Hausdorff space to the classical Vietoris topology on the hyperspace of a compact space. In fact, it is possible to define the hyperspace topology in this fashion. (See [2], [70].) The question lingers whether this is an adequate approach; we recall that the one-point-compactification very quickly looses its power in the absence of Hausdorff separation.

The following is a useful consequence of 2.7(i):

2.9. COROLLARY. *Suppose that $\mathcal{U}$ is an open cover of X. Then the function*
$$A \mapsto (U \cap A)_{U \in \mathcal{U}}: \Gamma(X) \to \prod_{U \in \mathcal{U}} \Gamma(U)$$
is a homeomorphism onto the image.

Proof. By Corollary 2.7(i) and the definition of the product, the function is continuous. Since $\mathcal{U}$ is a cover, it is injective. Since Γ is compact, it is a homeomorphism onto its image. $\square$

We notice that by 2.7(i), the projection of the image onto any factor $\Gamma(U)$ is surjective. In universal algebra one says that such a morphism $\Gamma(X) \to \prod_{U \in \mathcal{U}} \Gamma(U)$ is a subdirect product representation.

We have seen that open functions play an important role in making Γ a contravariant functor. Now we use closed maps in order to produce a covariant functor.

2.10. DEFINITION. A function $f: X_1 \to X_2$ between topological spaces is called *closed* if $f(A)$ is closed for each closed subset $A \subseteq X_1$. It is called *proper* if it is continuous and closed, and if it has the additional property that $f^{-1}(Q)$ is quasicompact for each quasicompact saturated subset $Q \subseteq X_2$. □

Recall that in any Hausdorff space, every subset is saturated. The definition therefore reduces to the usual one in this case.

2.11. THEOREM. ('The Proper Mapping Theorem') *Let $f: X_1 \to X_2$ denote a proper map between topological spaces. Then the function*

$$\gamma: \Gamma(X_1) \to \Gamma(X_2), \quad \gamma(A) = f(A),$$

between the corresponding hyperspaces is continuous.

Proof. Suppose that $U_0, \ldots U_n$ are open in X_2. Note that $A \in \Gamma(X_1)$ is contained in $\gamma^{-1}(W(U_0, \ldots, U_n))$ iff $f(A) \in W(U_0, \ldots, U_n)$ iff there is a quasicompact $Q \subseteq X_2$ such that $U_0 \subseteq Q$, $f(A) \cap Q = \emptyset$, and $f(A) \cap U_j \neq \emptyset$ for $j = 1, \ldots, n$. In view of the fact that $f(A)$ is closed and the saturation of a set is the intersection of all of its open neighborhoods, these last $n + 2$ conditions are equivalent to $A \cap f^{-1}(Q^*) = \emptyset$ and $A \cap f^{-1}(U_j) \neq \emptyset$ for $j = 1, \ldots, n$. Now $f^{-1}(Q^*)$ is quasicompact since the saturation of a quasicompact set is compact and since f is proper, and $f^{-1}(U_j)$ is open since f is continuous. Thus, since $U_0 \subseteq Q$ implies $f^{-1}(U_0) \subseteq f^{-1}(Q)$, we see that $f(A) \in W(U_0, \ldots, U_n)$ is equivalent to $A \in W(f^{-1}(U_0), \ldots, f^{-1}(U_n))$. Hence

$$\gamma^{-1}(W(U_0, \ldots, U_n)) = W(f^{-1}(U_0), \ldots, f^{-1}(U_n)).$$

In view of 1.8 this shows the continuity of γ. □

2.12. COROLLARY. *Let Y be a closed subspace of a locally quasicompact space X. Then the inclusion function $\Gamma(Y) \to \Gamma(X)$ is continuous. I.e., we may consider $\Gamma(Y)$ as a closed subspace of $\Gamma(X)$.*

Proof. The inclusion function $Y \to X$ is proper. Hence the inclusion function $\Gamma(Y) \to \Gamma(X)$ is continuous by 2.11. Since $\Gamma(Y)$ is compact and $\Gamma(X)$ is Hausdorff, we may consider $\Gamma(Y)$ as a compact (hence closed) subspace of $\Gamma(X)$. □

If we let **open**, respectively, **prop** denote the category of locally quasicompact topological spaces and continuous open, respectively, proper maps and define a function $\Gamma^{\#}: \textbf{prop} \to \textbf{comp}$ by $\Gamma^{\#}(X) = \Gamma(X)$ on objects and $\Gamma^{\#}(f)(A) = f(A)$ on morphisms, then we obtain at once the following reformulation and complementation of 2.11:

2.13. COROLLARY. *There are well-defined functors*
Γ: **open** $\to$ **comp**^{op} *and* $\Gamma^\#$: **prop** $\to$ **comp**. $\square$

If G is a locally compact topological group and H a compact subgroup, then the quotient map $G \to G/H$ is a morphism in **open** $\cap$ **prop**.

Let $f: X_1 \to X_2$ be a proper map between locally quasicompact spaces. We note that for a closed subset A of X_1 and a closed subset B of X_2 we have $f(A) \subseteq B$ iff $A \subseteq f^{-1}(B)$, i.e., $\Gamma^\#(f)(A) \subseteq B$ iff $A \subseteq \Gamma(f)(B)$. This observation places our formalism into the general frame of Galois connections. Indeed, if L_1 and L_2 are two partially ordered sets, then a pair (g, d) of functions $d: L_1 \to L_2$ and $g: L_2 \to L_1$ is called a *Galois connection* if for all $x \in L_1$ and $y \in L_2$ the relations $x \leq g(y)$ and $d(x) \leq y$ are equivalent. We call g the *upper adjoint* and d the *lower adjoint*. For the general background on Galois connections see e.g. [21], p. 18ff. We have identified $\Gamma^\#(f)$ as the lower adjoint of $\Gamma(f)$. There is a lattice theoretical background behind all of this which, on the level of the lattices $\mathcal{O}(X)$ rather than the opposite lattices $\Gamma(X)$, is discussed in [21], p. 262 ff. For further information see [48], p.257.

There is a function $s: X \to \Gamma(X)$ given by $s(x) = \overline{\{x\}}$. In order to clarify its continuity we note first that for any open subset U of X we have $s(x) \cap U \neq \emptyset$ iff $x \in U$, and for a quasicompact set Q we have $s(x) \cap Q = \emptyset$ iff $Q \subseteq X \setminus \overline{\{x\}}$ iff $x \notin Q^*$. If $S \subseteq X$ we let S^q denote the intersection of all quasicompact subsets of X containing S. If X is Hausdorff, then X^q is closed and is $\overline{S}$ if S is relatively compact and is X otherwise. Then for any collection $U_0, \ldots, U_n$ in $\mathcal{O}(X)$, we have
$$s^{-1}(W(U_0, \ldots, U_n)) = (X \setminus (U_0)^q) \cap U_1 \cap \cdots \cap U_n,$$
$$s((X \setminus (U_0)^q) \cap U_1 \cap \cdots U_n) = (\text{im } s) \cap W(U_0, \ldots, U_n).$$

The interior of this set is $(X \setminus \overline{(U_0)^q}) \cap U_1 \cap \cdots \cap U_n$. The unit interval with its Scott topology is a quasicompact and locally quasicompact T_0-space X in which $\overline{(U_0)^q} = X$ for any nonempty open set U_0. For such spaces $s^{-1}(W(U_0, \ldots, U_n)) = \emptyset$ whenever $U_0 \neq \emptyset$. In such spaces s has little chance of being continuous. However, if X is a locally compact Hausdorff space, then
$$s^{-1}(W(C; U_1, \ldots, U_n)) = (X \setminus C) \cap U_1 \cap \cdots \cap U_n,$$
$$s((X \setminus C) \cap U_1 \cap \cdots \cap U_n) = (\text{im } s) \cap W(C; U_1, \ldots, U_n).$$

Therefore:

2.14. PROPOSITION. *If X is a locally compact Hausdorff space, then*
$$x \mapsto \overline{\{x\}} : X \to \Gamma(X)$$
is a continuous embedding of X into its hyperspace. $\square$

3. The Use of Nets

In many applications one likes to work with convergence rather than open or closed sets. This means handling nets. One may restrict one's attention to sequences if one has the second axiom of countability available. By 1.9 this is the case if $\mathcal{O}(X)$ has a countable basis.

Generally we may utilize the fact that the topology τ on $\Gamma(X)$ is the image topology of the Lawson topology λ on $\mathcal{O}(X)$ under the map $U \mapsto X \setminus U$. On a complete lattice we always have the so-called lim-inf convergence.

3.1. DEFINITION. Let $(x_j)_{j \in J}$ denote a net in a complete lattice L. We write $x = \underline{\lim}_{j \in J} x_j$ if and only if $x = \bigvee_{k \in J} \bigwedge \{x_j : k \leq j\}$. □

We recall that a function $\alpha: K \to J$ between directed sets defines a subnet $(y_k)_{k \in K}$, $y_k := x_{\alpha(j)}$ of $(x_j)_{j \in J}$ if α is cofinal, i.e.,

$$(\forall j \in J)(\exists k \in K)(\forall m \in K)\,(k \leq m) \Rightarrow (j \leq \alpha(m)).$$

3.2. LEMMA. *In any complete lattice L, we consider the statements*
(1) $x = \lambda\text{-}\lim_{j \in J} x_j$,
(2) $x = \underline{\lim}_{k \in K} y_k$ *for all subnets* $(y_k)_{k \in K}$ *of* $(x_j)_{j \in J}$,
(3) $x = \underline{\lim}_{j \in J} y_k$.
Then (1) $\Rightarrow$ (2) $\Rightarrow$ (3). *If L is continuous, then all three statements are equivalent.*

Proof. (1)$\Rightarrow$(2): First assume that $x = \lambda\text{-}\lim_{j \in J} x_j$. Then certainly $x = \lambda\text{-}\lim_{k \in K} y_k$ for each subnet of $(y_k)_{k \in K}$ of $(x_j)_{j \in J}$. Set $y_k^* = \bigwedge \{y_m : k \leq m\}$. Then $x = \underline{\lim}_{k \in K} y_k$ means $x = \bigvee_{k \in K} y_k^*$. We claim that $y_k^* \leq x$ for all $k \in K$. Indeed if not, then $y_k^* \notin \downarrow x$ for some k, and then $k \leq m$ implies $y_m \in \uparrow y_k^*$ which is incompatible with λ-convergence of y_m to x. Thus $(y_k^*)k \in K$ is a nondecreasing net in $\downarrow x$. If $a \notin \downarrow x$ is arbitrary, then $L \setminus \downarrow a$ is a σ-open neighborhood of x and thus a λ-open neighborhood of x. Thus eventually, $y_k \notin \downarrow a$. Thus $x' := \bigvee_{k \in K} y_k^* \leq x$ is not contained in $\downarrow a$. Since a was arbitrary in $L \setminus \uparrow x$, we have $x \leq x'$. Hence $x = x'$. Thus $x = \underline{\lim}_{k \in K} y_k$.

(2)$\Rightarrow$(3): Trivial.

(3)$\Rightarrow$(1): Next assume that L is a continuous lattice and that $x = \underline{\lim}_{j \in J} x_j$. This first implies $y_k \leq x$ for all k and thus that for every $u \in L$ with $u \notin \downarrow x$ we eventually have $y_k \notin \uparrow u$, and thus eventually $x_j \notin \uparrow u$. Hence given $u_1, \ldots, u_n \notin \downarrow x$, eventually $y_k \in L \setminus (\uparrow u_1 \cup \cdots \cup \uparrow u_n)$. If $u_0 \ll x$ then by the Interpolation Lemma for Continuous Lattices (1.4) $\uparrow u_0$ is Scott open, and thus $x = \bigvee_{k \in J} y_k$ implies that there is a k with $y_k \in \uparrow u_0$, and thus $y_m \in \uparrow u_0$ for all $k \leq m$. Thus given $u_0 \ll x$ and $u_1, \ldots, u_n \notin \downarrow x$ the relation $x = \underline{\lim}_{j \in J} x_j$ implies that eventually $x_j \in V(u_0, \ldots, u_n)$. Hence $x = \lambda\text{-}\lim_{j \in J} x_j$. □

One can use lim-inf-convergence in the form of Condition (2) in 3.2 to create a topology on L relative to which lim-inf-convergence is topological convergence. This line of thought is discussed in [21], p.158ff.

Now we consider a locally quasicompact space X and apply Lemma 3.2 to the continuous lattice $L = \mathcal{O}(X)$.

3.3. PROPOSITION. *Let X be a locally quasicompact space, A a closed subset and $(A_j)_{j \in J}$ a net of closed subsets of X. Then the following statements are equivalent:*
   (1) $A = \lim_{j \in J} A_j$ *in the compact hyperspace* $\Gamma(X)$.
   (2) $A = \bigcap_{k \in K} \overline{\bigcup_{k \leq m} B_j}$ *for every subnet* $(B_k)_{k \in K}$ *of* $(A_j)_{j \in J}$.
   (3) $A = \bigcap_{k \in J} \overline{\bigcup_{k \leq j} A_j}$. □

3.4. COROLLARY. *Suppose that $A = \lim_{j \in J} A_j$ and $B = \lim_{j \in J} B_j$ in the hyperspace of a locally quasicompact space X. If $A_j \subseteq B_j$ eventually, then $A \subseteq B$.*

Proof. We represent A and B as in 3.3(3) and derive the claim at once. □

3.5. COROLLARY. *Suppose that $A = \lim_{j \in J} A_j$ in the hyperspace of a locally compact Hausdorff space X and and $a = \lim_{j \in J} a_j$ in X. If $a_j \in A_j$ eventually, then $a \in A$.*

Proof. By 2.14, the function $x \mapsto \{x\}: X \to \Gamma(X)$ is a topological embedding. Then the assertion follows from 3.4. □

3.6. COROLLARY. *Suppose that $A = \lim_{j \in J} A_j$ in the hyperspace of a locally quasicompact space X and suppose that all A_j are eventually contained in some quasicompact closed subspace Y of X. Then A is quasicompact and for any open neighborhood U of A there is a j_0 such that $j_0 \leq j$ implies $A_j \subseteq U$.*

Proof. By 2.11 we may assume that $Y = X$ is quasicompact. Using the local quasicompactness of X we find an open set V and a quasicompact set Q such that $A \subseteq V \subseteq Q \subseteq U$. We set $B_k = \overline{\bigcup_{k \leq j} A_j}$. Since X is quasicompact and closed subsets of quasicompact sets are quasicompact, the collection $\{B_k : k \in J\}$ is a filterbasis of quasicompact closed sets. Its intersection then is closed and quasicompact. Furthermore, $\{B_k \setminus U : k \in J\}$ is a downwards directed family of closed quasicompact sets with empty intersection. It therefore must contain an empty member B_{j_0}. Then $j_0 \leq j$ implies $A_j \subseteq B_{j_0} \subseteq U$. □

3.7. PROPOSITION. *Suppose that $A = \lim_{j \in J} A_j$ in the hyperspace $\Gamma(X)$ of a locally quasicompact space X and that $(a_1, \ldots, a_p) \in A^p$ for some natural number p. Then there is a subnet $(A_{j(\alpha)})_{\alpha \in \Omega}$ and elements $(a_{1\alpha}, \ldots, a_{p\alpha}) \in A_{j(\alpha)}^p$ such that $j \leq \alpha(j)$ and $a_k = \lim a_{k\alpha}$ for $k = 1, \ldots, p$.*

Proof. We let $\Omega = J \times \mathcal{U}(a_1) \times \cdots \times \mathcal{U}(a_p)$ where $\mathcal{U}(a)$ denote the filter basis of open neighborhoods of $a \in X$. We consider the partial order $(i, V_1, \ldots, V_p) \leq (j, U_1, \ldots, U_p)$ iff $i \leq j$ and $U_k \subseteq V_k$ for $k = 1, \ldots, p$. For each $\alpha = (j, U_1, \ldots, U_p)$ we find a $j(\alpha) \geq j$ such that $A_{j(\alpha)} \in W(\emptyset, U_1, \ldots, U_p)$; in particular, this implies the existence of some $a_{k\alpha} \in A_{j(\alpha)} \cap U_k$ for $k = 1, \ldots, p$. If U_k^0 is an arbitrary open neighborhood of a_k, then we pick any $j_0 \in J$ and set $\alpha_0 = (j_0, U_1^0, \ldots, U_p^0)$.

Then $\alpha_0 \leq \alpha = (j, U_1, \ldots, U_p)$ implies that $a_{j(\alpha)} \in A_{j(\alpha)} \subseteq W(\emptyset, U_1, \ldots, U_p) \subseteq W(\emptyset, U_1^0, \ldots, U_p^0)$, i.e, $a_{kj(\alpha)} \in U_k^0$. Hence $a_k = \lim_{\alpha \in \Omega} a_{kj(\alpha)}$ for $k = 1, \ldots, p$. □

This proposition remains relevant for $p = 1$. Corollary 3.5 and Proposition 3.7 characterize the points of $A = \lim A_j$ in a locally compact Hausdorff space as those which are of the form $a = \lim b_k$ with $b_k \in B_k$ for a subnet $(B_k)_{k \in K}$ of $(A_j)_{j \in J}$.

3.8. COROLLARY. *Suppose that $A = \lim_{j \in J} A_j$ in the hyperspace of a locally compact Hausdorf space X and suppose that all A_j are connected and eventually contained in some compact subspace Y of X. Then A is connected.*

Proof. Suppose not. Then the compact space A (see 3.6) decomposes into the disjoint union of two nonempty closed compact subspaces A_1 and A_2. In the locally compact space X every compact space has a basis of compact neighborhoods. If C_j ranges through the compact neighborhoods of A_k, $k = 1, 2$, then $C_1 \cap C_2$ is a down directed collection of compact sets with empty intersection. Hence for some C_1 and C_2 we have $C_1 \cap C_2 = \emptyset$. By 3.6 there is a j_0 such that $j_0 \leq j$ implies $A_j \subseteq C_1 \cup C_2$. If $a_k \in A_k$, $k = 1, 2$, then by 3.7 there exists some $j \geq j_0$ and elements $a'_k \in A_j \cap C_k$ for $k = 1, 2$. This contradicts the connectivity of A_j and proves the claim. □

This is a well known result concerning the hyperspaces of compact spaces proved here in a fashion adapted to our approach. If X is the planar set $\bigl([0, \infty[\times \{0, 1\}\bigr) \cup \bigl(\{1, 2, 3, \ldots\} \times [0, 1]\bigr)$, then the sequence $A_n = \bigl([0, n] \times \{0, 1\}\bigr) \cup \bigl(\{n\} \times [0, 1]\bigr)$ is a sequence of compact connected sets converging to the disconnected set $[0, \infty[\times \{0, 1\}$.

Corollary 2.9 allows us to formulate characterisation theorem for the convergence of nets in hyperspaces:

3.9. PROPOSITION. *Let $\mathcal{U}$ be an open cover of a locally quasicompact space X and let $(A_j)_{j \in J}$ be a net of closed subsets of X. Then for any closed subset A of X the following statements are equivalent:*
(1) $A = \lim A_j$ in $\Gamma(X)$.
(2) $A \cap U = \lim(A_j \cap U)$ in $\Gamma(U)$ for all $U \in \mathcal{U}$.

Proof. This is immediate from 2.9. □

4. Applications to Topological Algebras

4.1. PROPOSITION. *Let X be a locally compact space and $n \in \mathbb{N}$. Suppose that $f: X^n \to X$ is a continuous map. Then the set*

$$\Gamma(X; f) := \{Y \in \Gamma(X) \mid f(Y^n) \subseteq Y\}$$

is a closed subspace of $\Gamma(X)$. In particular, if X is a locally compact universal algebra then the set of all closed subalgebras is a closed subspace of $\Gamma(X)$.

Proof. Let $(Y_j)_{j \in J}$ be a net in $\Gamma(X; f)$ which converges to some closed subset $Y \in \Gamma(X)$ and pick $p \in Y^n$. We have to show that $f(p) \in Y$. Invoking Corollary 3.8 we find points $p_j \in Y_j^n$ with $p = \lim_j p_j$, and since f is continuous this implies that $f(p) = \lim_j f(p_j)$. By the definition of $\Gamma(X; f)$ we have $f(p_j) \in Y_j$, for every $j \in J$. Thus Corollary 3.5 applies and we see that $f(p) = \lim_j f(p_j) \in \lim_j Y_j = Y$. □

4.2. REMARK. The case of unary operations ($n = 0$) is covered by Proposition 2.2. □

4.3. COROLLARY. *If G is a locally compact group, then the set of closed subgroups, respectively, subsemigroups is closed in $\Gamma(G)$ and therefore is a compact Hausdorff space.* □

4.4. PROPOSITION. *Let G be a locally compact topological group, H a closed subgroup, and $\kappa: G \to G/H$ the induced quotient map. Then the following assertions hold:*
  (i) *$\Gamma(\kappa)$ is a homeomorphic embedding of $\Gamma(G/H)$ into $\Gamma(G)$. In particular, the H-saturated closed subsets of G form a closed subspace of $\Gamma(G)$.*
  (ii) *If H is compact then the map*

$$\gamma(\kappa): \Gamma(G) \to \Gamma(G/H), \quad A \mapsto \kappa(A),$$

is continuous and $\Gamma(\kappa) \circ \gamma(\kappa)$ is a continuous retraction.

Proof. (i) follows from Theorem 2.6, (ii) from Theorem 2.11. □

REFERENCES

1. A. Borel, *Linear Algebraic Groups*, W. A. Benjamin, New York, N. Y., 1969.
2. N. Bourbaki, *Intégration, Chap. 8*, Hermann, Paris, 1961.
3. _____, *Groupes et algèbres de Lie, Chap. 1*, Hermann, Paris, 1960.
4. _____, *Groupes et algèbres de Lie, Chap. 2 et 3*, Hermann, Paris, 1972.
5. _____, *Groupes et algèbres de Lie, Chap. 7 et 8*, Hermann, Paris, 1975.
6. _____, *Variétés différentielles et analytiques*, Diffusion CCLS, Paris, 1982.
7. P. B. Chen and T. S. Wu, *On exponential groups*, J. Pure Appl. Algebra **93** (1994), 169–178.
8. C. Chevalley, *Théorie des groupes de Lie III*, Hermann, Paris, 1955.
9. L. Corwin and M. Moskowitz, *A note on the exponential map of a real or p-adic Lie group*, J. of Pure Appl. Alg. **96** (1994), 113–132.
10. J. Dixmier, *L'application exponentielle dans les groupes de Lie résolubles*, Bull. Soc. Math. France **85** (1957), 113–121.
11. _____, *Sur les espaces localement quasi-compacts*, Can. J. Math., **20** (1968), 1093–1100.
12. N. Dörr, *A note on the oscillator group* (1992), 31–38.
13. D. Ž. Djoković, *On the exponential map in classical Lie groups* (1980), 76–88.
14. D. Ž. Đoković, *The interior and the exterior of the image of the exponential map in classical Lie groups* (1988), 90–109.
15. _____, *The exponential image of simple complex Lie groups of exceptional type* (1988), 101–111.
16. D. Ž. Đoković and Nguyen Q. Thăńg, *On the exponential map of almost simple real algebraic groups*, J. of Lie Theory **5** (1995), 275–291.
17. _____, *Lie groups with dense exponential image*, Math. Z., to appear.
18. A. Eggert, *Zur Klassifikation von Semialgebren*, (Dissertation, TH Darmstadt 1991) Mitt. Math. Sem. Giessen, vol. 204, Universität Giessen, Giessen, 1991.
19. R. Engelking, *General Topology*, Heldermann Verlag, Berlin, 1989.
20. J. M. G. Fell, *A Hausdorff topology for the closed subsets of a locally compact non-Hausdorff space* (1962), 472–476.
21. G. Gierz, K. H. Hofmann, K. Keimel, J. D. Lawson, M. Mislove, and D. Scott, *A Compendium of Continuous Latices*, Springer, Berlin etc., 1980.
22. V. M. Gichev, *On the structure of Lie algebras admitting and invariant cone*, Semigroups in Algebra, Geometry, and Analysis (K. H. Hofmann, J. D. Lawson, and E. B. Vinberg, eds.), De Gruyter Expositions in Mathematics 20, De Gruyter, Berlin etc., 1995, pp. 107–120.

23. M. Goto, *Index of the exponential map of a semialgebraic group*, J. Math. Soc. Japan **29** (1977), 161–163.
24. S. Helgason, *Differential Geometry, Lie Groups, and Symmetric Spaces*, Academic Press, Boston etc, 1978.
25. H. Heyer, *Probability Measurs on Locally Compact Groups*, Springer, Berlin etc, 1977.
26. J. Hilgert and K. H. Hofmann,, *Semigroups in Lie groups, semialgebras in Lie algebras* (1985), 481–504.
27. J. Hilgert, K. H. Hofmann, and J. D. Lawson,, *Lie groups, convex cones and semigroups*, Oxford Science Publications, Clarendon Press, Oxford, 1989.
28. J. Hilgert, K.-H. Neeb, *Lie Semigroups and their Applications*, Lecture Notes in Mathematics, vol. 1552, Springer, Berlin etc., 1993.
29. G. Hochschild, *The Structure of Lie Groups*, Holden-Day, San Francisco etc, 1965.
30. K. H. Hofmann, *Hyperplane subalgebras of real Lie algebras* (1990), 207–224.
31. _____, *A memo on singularities*, Memorandum circulated at Technische Hochschule Darmstadt 1990.
32. _____, *A memo on the exponential function and regular points* (1992), 24–37.
33. _____, *Near Cartan algebras and groups* (1992), 135–151.
34. _____, *The hyperspace of a locally compact space*, Preprint THD, Darmstadt 1994.
35. K. H. Hofmann and J. D. Lawson,, *Foundations of Lie semigroups*, Springer-Verlag, Berlin etc, 1983, pp. 128–201.
36. _____, *Divisible subsemigroups of Lie groups* (1983), 427–437.
37. K. H. Hofmann, J. D. Lawson, and J. S. Pym, *The Analytical Theory of Semigroups: Trends and Developments*, De Gruyter Expositions in Mathematics 1, De Gruyter, Berlin etc., 1990.
38. K. H. Hofmann, J. D. Lawson, and W. A. F. Ruppert, *On finiteness theorems and procupine varieties in Lie algebras* (1993), 49–63.
39. _____, *Weyl groups are finite—and other finitness properties of Cartan subalgebras*, Math. Nachr. **179** (1996), to appear.
40. K. H. Hofmann, J. D. Lawson, and E. B. Vinberg, *Semigroups in Algebra, Geometry, and Analysis*, De Gruyter Expositions in Mathematics 20, De Gruyter, Berlin etc., 1995.
41. K. H. Hofmann and P. S. Mostert, *Elements of Compact Semigroups*, Charles E. Merrill, Columbus, Ohio, 1966.
42. K. H. Hofmann and A. Mukherjea,, *On the density of the image of the exponential function*, Math. Ann. **234** (1978), 263–273.
43. K. H. Hofmann and W. A. F. Ruppert, *On the interior of subsemigroups of Lie groups*, Trans. Amer. Math. Soc. **324** (1991), 169–179.
44. _____, *The divisibility problem for subsemigroups of Lie groups*, Seminar Sophus Lie **1** (1991), 205–213.
45. _____, *The structure of Lie groups which support closed divisible subsemigroups*, Semigroups with Applications (J. M. Howie, W. D. Munn, and H.-J. Weinert, eds.), World Scientific, Singapore, 1992, pp. 11–30.

46. _____, *On Porcupine Varieties in Lie Algebras*, Math. Annalen, **298** (1994), 403–425.
47. _____, *Lie groups and exponential Lie subsemigroups*, Semigroups in Algebra, Geometry, and Analysis (K. H. Hofmann, J. D. Lawson, and E. B. Vinberg, eds.), De Gruyter Expositions in Mathematics 20, De Gruyter, Berlin etc., 1995, pp. 159–198.
48. K. H. Hofmann and F. Watkins, *The spectrum as a functor*, Continuous Lattices, Proceedings Bremen 1979, Lecture Notes in Math., (B. Banaschewski and R.-E. Hoffmann, eds.), vol. 871, Springer-Verl., Berlin etc., 1981, pp. 249–263.
49. W. Jaworski, *The density of the image of the exponential function and spacious locally compact groups*, J. of Lie Theory **5** (1995), 129–134.
50. V. Jurdevic and I. Kupka, *Control systems on semisimple Lie groups and their homogeneous spaces* (1981), 151–179.
51. H.-L. Lai, *Surjectivity of exponential map on semisimple Lie groups*, J. Math. Soc. **29** (1977), 303-325.
52. _____, *On the singularity of the exponential map on a Lie group*, Proc. Amer. Math. Soc. **62** (1977), 334–336.
53. _____, *Index of the exponential map of a center-free complex simple Lie group*, Osaka J. Math. **15** (1978), 553–560.
54. _____, *Index of the exponential map on a complex simple Lie group*, Osaka J. Math. **15** (1978), 561–567.
55. _____, *Corrections and supplements to "Index of the exponential map on a complex simple Lie group"*, Osaka J. Math. **17** (1980), 525–530.
56. _____, *Index of a simple adjoint group*, Bull. Inst. Math. Acad. Sinica **8** (1980), 603–608.
57. _____, *Exponential map on a simple group of classical type*, Bull. Inst. Math. Acad. Sinica **10** (1982), 417–430.
58. L. Markus, *Exponentials in algebraic matrix groups*, Adv. Math. **11** (1973), 351–367.
59. M. McCrudden, *On n-th roots and infinitely divisible elements in a connected Lie group* (1981), 293–299.
60. D. Mittenhuber, *Spacious Lie groups*, J. of Lie Theory **5** (1995), 135–146.
61. D. Mittenhuber and K.-H. Neeb, *On the exponential function of an invariant Lie semigroup* (1992), 21–30.
62. _____, *Remarks on our paper "On the exponential function of an invariant Lie semigroup"* (1993), 119–120.
63. Z. I. Moskalenko, *Exponential groups and ML-groups*, Ukrain. Mat. Zhurnal **28** (1976), 501–510.
64. M. Moskowitz, *The surjectivity of the exponential map for certain Lie groups*, Ann. Mat. Pura Appl., Ser. 4 **166** (1994), 129–143.
65. _____, *Correction to "The surjectivity of the exponential map for certain Lie groups"*, Ann. Mat. Pura. Appl., to appear.
66. _____, *Exponentiality of algebraic groups*, submitted.
67. M. Moskowitz and M. Wüstner, *Exponentiality of certain real solvable Lie groups*, submitted.
68. K.-H. Neeb, *Semigroups in the Universal Covering of* Sl(2) (1991), 33-43.

69. _____, *On the foundations of Lie semigroups* (1992), 165–189.
70. _____, *A topology on the space of closed subsets*, Interdepartmental Memo 1992, 16pp.
71. _____, *Invariant Subsemigroups of Lie Groups*, Memoirs pf the Amer. Math. Soc., vol. 499, American Mathematical Scoiety, Providence R. I., 1993.
72. _____, *Weakly exponential Lie groups*, J. of Algebra **179** (1996), 331–361.
73. M. Nishikawa, *Exponential image in the real general linear group*, Bull. Fukuoka Univ. Ed. III **26** (1976), 35–44.
74. _____, *Exponential image in the real special linear group*, Bull. Fukuoka Univ. Ed. III **28** (1978), 1–6.
75. _____, *On the exponential map of the group $O(p,q)_0$*, Mem. Fac. Sci. Kyushu Univ. Ser. A **37** (1983), 63–69.
76. _____, *Exponential image and conjugacy classes in the group $O(3,2)$*, Hiroshima Math. J. **14** (1984), 311–332.
77. M. Saito, *Sur certains groupes de Lie résolubles* (1957), 1–11, and 157–168.
78. L. San Martin, *Nonreversibility of subsemigroups of semisimple Lie groups* (1992), 376–387.
79. J. Tits, *Tabellen zu den einfachen Liegruppen und ihren Darstellungen,*, Lecture Notes in Math., 40, Springer, Berlin etc, 1967.
80. M. Wüstner, *Beiträge zur Struktur Theorie auflösbarer Lie Gruppen*, (Dissertation, TH Darmstadt 1995), TH Darmstadt, 1995.
81. _____, *A connected complex simple centerfree Lie group whose exponential function is not surjective*, J. of Lie Theory **5** (1995), 203–205.
82. _____, *On the surjectivity of the exponential function of complex algebraic, complex semisimple, and complex splittable Lie groups*, J. of Alg., to appear.
83. _____, *On the surjectivity of the exponential function of solvable Lie groups*, Math. Nachrichten, to appear.
84. _____, *On the exponential function of real splittable and real semisimple Lie groups*, in preparation.

FACHBEREICH MATHEMATIK, TECHNISCHE HOCHSCHULE DARMSTADT, SCHLOSSGARTENSTRASSE 7, D-64289 DARMSTADT, GERMANY
E-mail address: hofmann@mathematik.th-darmstadt.de

INSTITUT FÜR MATHEMATIK UND ANGEWANDTE STATISTIK, UNIVERSITÄT FÜR BODENKULTUR, GREGOR MENDELSTRASSE 33, A-1180 WIEN, ÖSTERREICH
E-mail address: ruppert@edv1.boku.ac.at

INDEX

Absolutely closed 14f, 98
algebraic hull 41d-44
ascending central series 50
'$ax + b$' group 14f

Base ideal 138d
base root 138d
Bernoulli numbers 76

Campbell-Hausdorff 2, 31, 76–83, 98
Cartan
— adapted 52d
— compact 73d
— dense 38d, 39d, 41–43
— finite 48, 52d, 55
— reductive 56d, 60, 62, 64, 66–68
— subalgebras, intersections of 52–54
Chronogeometry 1
Classsification Problem 4f, 6
codimension 65–74, 87f, 89, 92, 113
— of porcupine variety 67–74
compact
— element 104, 118, 142–143
— Lie algebra 107, 144–146, 149–151
compactly embedded subset 54d, 70f, 73f, 86f, 89–90, 143–144
confinement 76–78
Confluence Theorem 78
continuous lattice 147, 149d, 152
covering homomorphism 100f, 151

Delta radical 137, 145f
diagonalizable 8, 104
diagonally metabelian 8, 137f, 145–147, 150

directed set 148d
dispersion 79d–82
divisible 21d, 4, 20–24, 86

Edge of the wedge 12d
Eggert 3, 128
— algebra 135d, 146
— pair 135d, 137
— 's Classification Theorem 139
— 's Small Intersection Theorem 140
— 's Theorem 132d
exp-compact 25d
exp-regular 24d, 25
exponential 3, 13d, 84f, 92, 97, 108, 151f
extensions of mot and osc, 111d, 114, 124

Fitting
— decomposition 38, 41
— one-component 25, 38f, 42

Galois connection 164
Gamma-limit of one-parameter semigroups 28
Gamma-topology 25–29
— and nets 165
generalized Weyl groups 44d–47
Geometric
— Control Theory 1
— Semigroup Theory 1
Grassmann manifold 147
g-reductive 55d–60, 68
group of units 6, 8

Half-space
— in mot 14
— semialgebra 3, 75d, 84, 105–107, 147
— subsemigroup 103
Heisenberg algebra 39f, 103, 142

hyperplane subalgebra 3, 139, 146–147
hyperspace 7, 25, 147–162, 150d
— topology 25, 156

Imaginary valued root 130d–140, 143, 145
Interior Point Theorem 32d
interior points 31–37
intersection
— semialgebra 6, 147–149d
— semigroup 149d
Intrinsic Embedding Theorem 33, 34d, 124
intrinsic Lie group topology 31, 34, 85, 89
invariant Lie wedge 3, 6, 12d, 75, 91, 142, 147–152

Lawson topology 148f, 152f
lean subset 87d–89, 92–94, 113f
Lie semialgebra 2, 75d, 80, 86, 88, 91, 97f, 103, 105, 132–140, 143–152
Lie semigroup 1ff, 13d
Lie wedge 1ff, 12d, 12f, 30–37, 126–140, 142–152
lim-inf-convergence 165
local dimension 71d
local semigroup 75d
locally
— direct product 115–118
— divisible 79
— exponential 2, 6
lower set 65d

Master Examples 13d–17, 29, 95ff
maximal rank 29d, 30, 31
— subalgebra 54, 64, 69, 130, 132–133, 141, 145

173

— subgroup 7, 29d
Maximal Rank Theorem 30d, 90, 140, 145
maximally almost periodic 91d
metabelian 8, 14, 39, 137, 149
— algebra 39
motion
— algebra 40, 102–105, 107–108, 111, 116, 142–143
— group 24, 95ff, 102

Negative Examples 10, 17–20
net convergence 165
nilspace 25, 29, 34
noncompact root 144
nonsimple real root 136d–140

Occam's Razor 126f, 127d, 130, 133, 135, 143
one point compactification 158
one-parameter subsemigroups 2
oscillator
— algebra 18d, 103–105, 107–108, 111, 116–117, 142–143
— group 18d, 95f, 104–107

Porcupine
— set 61d
— variety 7, 61–74, 63d
proper continuous map 26, 163d
Proper Mapping Theorem 163d

Rank 29, 40, 65, 67
real valued root 130d–140
reduced

— Lie semialgebra 85d
— subsemigroup of a Lie group 5, 85d, 94f, 107, 114, 128–140
— weakly exponential 144–152
reductive Lie algebra 105, 144
reductively embedded subset 54d–57, 64, 67, 69
regular
— compact element 124
— element in a Lie algebra 24d–26, 69, 118, 143
— point in a Lie semigroup 28f
root 43, 48, 128–140, 141–152
noncompact —, 144
nonsimple —, 136d–140
real valued —, 130d–140
simple real valued —, 136d–140
imaginary valued —, 130d–140, 143, 145
— space 53

Saturated set 160d
Scholium 132d, 134
Scott topology 148d
semisimple element 62d
simple real root 136d–140
sl(2,R) 3, 8, 14f, 17f, 35, 92–94, 95–101, 110, 120–121, 132, 135–142, 145–149
slim center 116d-117
so(3) 15f, 141f
sober space 149d
spectrum of ad W 128–130
splittable hull 42d
standard double cone 95d
standard Lorentzian cone 17, 19, 104–105
standard pair 19d
strictly regular 128d–140

subtangent vector 12d
support 48–51

Tangent object 12d
test
— algebra 105–108
— group 105–108
— semigroups 112d, 111–114
— wedges 112d
Testing Theorem 112, 119, 143, 148
topological algebras 168

UMCS-groups 90d, 91, 132
unique maximal compact subgroup (UMCS) 90d
upper set 148d

Vietoris topology 157f, 162

Weak topology 148d
weakly exponential 4, 13d, 34, 84f, 87, 89, 91f, 97f, 103, 105, 112, 119f, 126–140
Weyl group 7, 10, 38–47, 58
generalized —, 44d–47

Zariski closure 61–63, 67–73, 87, 88, 113
Zariski-topology 38, 42, 61f, 66, 130

Editorial Information

To be published in the *Memoirs*, a paper must be correct, new, nontrivial, and significant. Further, it must be well written and of interest to a substantial number of mathematicians. Piecemeal results, such as an inconclusive step toward an unproved major theorem or a minor variation on a known result, are in general not acceptable for publication. *Transactions* Editors shall solicit and encourage publication of worthy papers. Papers appearing in *Memoirs* are generally longer than those appearing in *Transactions* with which it shares an editorial committee.

As of July 31, 1997, the backlog for this journal was approximately 8 volumes. This estimate is the result of dividing the number of manuscripts for this journal in the Providence office that have not yet gone to the printer on the above date by the average number of monographs per volume over the previous twelve months, reduced by the number of issues published in four months (the time necessary for preparing an issue for the printer). (There are 6 volumes per year, each containing at least 4 numbers.)

A Copyright Transfer Agreement is required before a paper will be published in this journal. By submitting a paper to this journal, authors certify that the manuscript has not been submitted to nor is it under consideration for publication by another journal, conference proceedings, or similar publication.

Information for Authors and Editors

Memoirs are printed by photo-offset from camera copy fully prepared by the author. This means that the finished book will look exactly like the copy submitted.

The paper must contain a *descriptive title* and an *abstract* that summarizes the article in language suitable for workers in the general field (algebra, analysis, etc.). The *descriptive title* should be short, but informative; useless or vague phrases such as "some remarks about" or "concerning" should be avoided. The *abstract* should be at least one complete sentence, and at most 300 words. Included with the footnotes to the paper, there should be the 1991 *Mathematics Subject Classification* representing the primary and secondary subjects of the article. This may be followed by a list of *key words and phrases* describing the subject matter of the article and taken from it. A list of the numbers may be found in the annual index of *Mathematical Reviews*, published with the December issue starting in 1990, as well as from the electronic service e-MATH [**telnet e-MATH.ams.org** (or **telnet 130.44.1.100**). Login and password are **e-math**]. For journal abbreviations used in bibliographies, see the list of serials in the latest *Mathematical Reviews* annual index. When the manuscript is submitted, authors should supply the editor with electronic addresses if available. These will be printed after the postal address at the end of each article.

Electronically prepared papers. The AMS encourages submission of electronically prepared papers in $\mathcal{A}_{\mathcal{M}}\mathcal{S}$-TeX or $\mathcal{A}_{\mathcal{M}}\mathcal{S}$-LaTeX. The Society has prepared author packages for each AMS publication. Author packages include instructions for preparing electronic papers, the *AMS Author Handbook*, samples, and a style file that generates the particular design specifications of that publication series for both $\mathcal{A}_{\mathcal{M}}\mathcal{S}$-TeX and $\mathcal{A}_{\mathcal{M}}\mathcal{S}$-LaTeX.

Authors with FTP access may retrieve an author package from the Society's Internet node **e-MATH.ams.org** (130.44.1.100). For those without FTP access, the author package can be obtained free of charge by sending e-mail to pub@ams.org (Internet) or from the Publication Division, American Mathematical Society, P.O. Box 6248, Providence, RI 02940-6248. When requesting an author package, please specify $\mathcal{AMS}$-TeX or $\mathcal{AMS}$-LaTeX, Macintosh or IBM (3.5) format, and the publication in which your paper will appear. Please be sure to include your complete mailing address.

Submission of electronic files. At the time of submission, the source file(s) should be sent to the Providence office (this includes any TeX source file, any graphics files, and the DVI or PostScript file).

Before sending the source file, be sure you have proofread your paper carefully. The files you send must be the EXACT files used to generate the proof copy that was accepted for publication. For all publications, authors are required to send a printed copy of their paper, which exactly matches the copy approved for publication, along with any graphics that will appear in the paper.

TeX files may be submitted by email, FTP, or on diskette. The DVI file(s) and PostScript files should be submitted only by FTP or on diskette unless they are encoded properly to submit through e-mail. (DVI files are binary and PostScript files tend to be very large.)

Files sent by electronic mail should be addressed to the Internet address pub-submit@ams.org. The subject line of the message should include the publication code to identify it as a Memoir. TeX source files, DVI files, and PostScript files can be transferred over the Internet by FTP to the Internet node e-math.ams.org (130.44.1.100).

Electronic graphics. Figures may be submitted to the AMS in an electronic format. The AMS recommends that graphics created electronically be saved in Encapsulated PostScript (EPS) format. This includes graphics originated via a graphics application as well as scanned photographs or other computer-generated images.

If the graphics package used does not support EPS output, the graphics file should be saved in one of the standard graphics formats—such as TIFF, PICT, GIF, etc.—rather than in an application-dependent format. Graphics files submitted in an application-dependent format are not likely to be used. No matter what method was used to produce the graphic, it is necessary to provide a paper copy to the AMS.

Authors using graphics packages for the creation of electronic art should also avoid the use of any lines thinner than 0.5 points in width. Many graphics packages allow the user to specify a "hairline" for a very thin line. Hairlines often look acceptable when proofed on a typical laser printer. However, when produced on a high-resolution laser imagesetter, hairlines become nearly invisible and will be lost entirely in the final printing process.

Screens should be set to values between 15% and 85%. Screens which fall outside of this range are too light or too dark to print correctly.

Any inquiries concerning a paper that has been accepted for publication should be sent directly to the Editorial Department, American Mathematical Society, P. O. Box 6248, Providence, RI 02940-6248.

Editors

This journal is designed particularly for long research papers (and groups of cognate papers) in pure and applied mathematics. Papers intended for publication in the *Memoirs* should be addressed to one of the following editors:

Ordinary differential equations, partial differential equations, and applied mathematics to JOHN MALLET-PARET, Division of Applied Mathematics, Brown University, Providence, RI 02912-9000; electronic mail: jmp@cfm.brown.edu.

Harmonic analysis, representation theory, and Lie theory to ROBERT J. STANTON, Department of Mathematics, The Ohio State University, 231 West 18th Avenue, Columbus, OH 43210-1174; electronic mail: stanton@math.ohio-state.edu.

Ergodic theory, dynamical systems, and abstract analysis to DANIEL J. RUDOLPH, Department of Mathematics, University of Maryland, College Park, MD 20742; e-mail: djr@math.umd.edu.

Real and harmonic analysis and geometric partial differential equations to WILLIAM BECKNER, Department of Mathematics, University of Texas at Austin, Austin, TX 78712; e-mail: beckner@math.utexas.edu.

Algebra and algebraic geometry to EFIM ZELMANOV, Department of Mathematics, Yale University, 10 Hillhouse Avenue, New Haven, CT 06520-8283; e-mail: zelmanov@math.yale.edu

Algebraic topology and cohomology of groups to STEWART PRIDDY, Department of Mathematics, Northwestern University, 2033 Sheridan Road, Evanston, IL 60208-2730; e-mail: s_priddy@math.nwu.edu.

Global analysis and differential geometry to CHUU-LIAN TERNG, Department of Mathematics, Northeastern University, Huntington Avenue, Boston, MA 02115-5096; e-mail: terng@neu.edu.

Probability and statistics to RODRIGO BAÑUELOS, Department of Mathematics, Purdue University, West Lafayette, IN 47907-1968; e-mail: banuelos@math.purdue.edu.

Combinatorics and Lie theory to PHILIP J. HANLON, Department of Mathematics, University of Michigan, Ann Arbor, MI 48109-1003; e-mail: hanlon@math.lsa.umich.edu.

Logic and universal algebra to THEODORE SLAMAN, Department of Mathematics, University of California at Berkeley, Berkeley, CA 94720; e-mail: slaman@math.berkeley.edu.

Number theory and arithmetic algebraic geometry to ALICE SILVERBERG, Department of Mathematics, Harvard University, 1 Oxford St.–Science Center, Cambridge, MA 02138; e-mail: silver@math.ohio-state.edu.

Complex analysis and complex geometry to DANIEL M. BURNS, Department of Mathematics, University of Michigan, Ann Arbor, MI 48109-1003; e-mail: dburns@umich.edu.

Algebraic geometry and commutative algebra to LAWRENCE EIN, Department of Mathematics, University of Illinois, 851 S. Morgan (M/C 249), Chicago, IL 60607-7045; email: lawrence.man.ein@uic.edu.

Geometric topology to JOHN LUECKE, Department of Mathematics, University of Texas at Austin, Austin, TX 78712; e-mail: luecke@math.utexas.edu.

All other communications to the editors should be addressed to the Managing Editor, PETER SHALEN, Department of Mathematics, University of Illinois, 851 S. Morgan (M/C 249), Chicago, IL 60607-7045; e-mail: shalen@math.uic.edu.

Selected Titles in This Series

(*Continued from the front of this publication*)

589 **James Damon,** Higher multiplicities and almost free divisors and complete intersections, 1996

588 **Dihua Jiang,** Degree 16 Standard L-function of $GSp(2) \times GSp(2)$, 1996

587 **Stéphane Jaffard and Yves Meyer,** Wavelet methods for pointwise regularity and local oscillations of functions, 1996

586 **Siegfried Echterhoff,** Crossed products with continuous trace, 1996

585 **Gilles Pisier,** The operator Hilbert space OH, complex interpolation and tensor norms, 1996

584 **Wayne W. Barrett, Charles R. Johnson, and Raphael Loewy,** The real positive definite completion problem: Cycle completability, 1996

583 **Jin Nakagawa,** Orders of a quartic field, 1996

582 **Darryl McCollough and Andy Miller,** Symmetric automorphisms of free products, 1996

581 **Martin U. Schmidt,** Integrable systems and Riemann surfaces of infinite genus, 1996

580 **Martin W. Liebeck and Gary M. Seitz,** Reductive subgroups of exceptional algebraic groups, 1996

579 **Samuel Kaplan,** Lebesgue theory in the bidual of $C(X)$, 1996

578 **Ale Jan Homburg,** Global aspects of homoclinic bifurcations of vector fields, 1996

577 **Freddy Dumortier and Robert Roussarie,** Canard cycles and center manifolds, 1996

576 **Grahame Bennett,** Factorizing the classical inequalities, 1996

575 **Dieter Heppel, Idun Reiten, and Sverre O. Smalø,** Tilting in Abelian categories and quasitilted algebras, 1996

574 **Michael Field,** Symmetry breaking for compact Lie groups, 1996

573 **Wayne Aitken,** An arithmetic Riemann-Roch theorem for singular arithmetic surfaces, 1996

572 **Ole H. Hald and Joyce R. McLaughlin,** Inverse nodal problems: Finding the potential from nodal lines, 1996

571 **Henry L. Kurland,** Intersection pairings on Conley indices, 1996

570 **Bernold Fiedler and Jürgen Scheurle,** Discretization of homoclinic orbits, rapid forcing and "invisible" chaos, 1996

569 **Eldar Straume,** Compact connected Lie transformation groups on spheres with low cohomogeneity, I, 1996

568 **Raúl E. Curto and Lawrence A. Fialkow,** Solution of the truncated complex moment problem for flat data, 1996

567 **Ran Levi,** On finite groups and homotopy theory, 1995

566 **Neil Robertson, Paul Seymour, and Robin Thomas,** Excluding infinite clique minors, 1995

565 **Huaxin Lin and N. Christopher Phillips,** Classification of direct limits of even Cuntz-circle algebras, 1995

564 **Wensheng Liu and Héctor J. Sussmann,** Shortest paths for sub-Riemannian metrics on rank-two distributions, 1995

563 **Fritz Gesztesy and Roman Svirsky,** (m)KdV solitons on the background of quasi-periodic finite-gap solutions, 1995

562 **John Lindsay Orr,** Triangular algebras and ideals of nest algebras, 1995

561 **Jane Gilman,** Two-generator discrete subgroups of $PSL(2, R)$, 1995

560 **F. Tomi and A. J. Tromba,** The index theorem for minimal surfaces of higher genus, 1995

559 **Paul S. Muhly and Baruch Solel,** Hilbert modules over operator algebras, 1995

558 **R. Gordon, A. J. Power, and Ross Street,** Coherence for tricategories, 1995

557 **Kenji Matsuki,** Weyl groups and birational transformations among minimal models, 1995

556 **G. Nebe and W. Plesken,** Finite rational matrix groups, 1995

(See the AMS catalog for earlier titles)